System Reliability: Methods, Models and Applications

System Reliability: Methods, Models and Applications

Editor: Anne Woolf

MURPHY & MOORE
www.murphy-moorepublishing.com

Published by Murphy & Moore Publishing
1 Rockefeller Plaza,
New York City, NY 10020, USA
www.murphy-moorepublishing.com

System Reliability: Methods, Models and Applications
Edited by Anne Woolf

International Standard Book Number: 978-1-63987-516-0 (Hardback)

Cataloging-in-Publication Data

System reliability : methods, models and applications / edited by Anne Woolf.
p. cm.
Includes bibliographical references and index.
ISBN 978-1-63987-516-0
1. Reliability (Engineering). 2. System failures (Engineering). 3. Systems engineering. I. Woolf, Anne.
TA169 .S97 2022
620.004 52--dc23

Contents

Preface

This book was inspired by the evolution of our times; to answer the curiosity of inquisitive minds. Many developments have occurred across the globe in the recent past which has transformed the progress in the field.

Reliability is defined as the ability of a system or component to function under specified settings for a stated period of time. It often plays a crucial role in the cost-effectiveness of a system. The sub-discipline of systems engineering that focuses on the capacity of equipment to function without failure is known as system reliability engineering. It deals with the prediction, prevention and management of high levels of engineering uncertainty and failure risks. Availability, testability, maintainability and maintenance fall under the umbrella of reliability engineering. It focuses on costs of failure resulting from system downtime, cost of spares, repair equipment, personnel and cost of warranty claims. Different approaches, evaluations, methodologies and advanced studies on system reliability have been included in this book. It elucidates the concepts and innovative models around prospective developments with respect to this field. As this field is emerging at a rapid pace, the contents of this book will help the readers understand the modern concepts and applications of the subject.

This book was developed from a mere concept to drafts to chapters and finally compiled together as a complete text to benefit the readers across all nations. To ensure the quality of the content we instilled two significant steps in our procedure. The first was to appoint an editorial team that would verify the data and statistics provided in the book and also select the most appropriate and valuable contributions from the plentiful contributions we received from authors worldwide. The next step was to appoint an expert of the topic as the Editor-in-Chief, who would head the project and finally make the necessary amendments and modifications to make the text reader-friendly. I was then commissioned to examine all the material to present the topics in the most comprehensible and productive format.

I would like to take this opportunity to thank all the contributing authors who were supportive enough to contribute their time and knowledge to this project. I also wish to convey my regards to my family who have been extremely supportive during the entire project.

Editor

1

Model Development for Reliability Cannibalization

Bernard Tonderayi Mangara

Abstract

This chapter looks at cannibalization as a method (procedure) of improving reliability of engineering systems. Cannibalization gives one the opportunity to use resources in the most efficient way. In this chapter, we have explored strategies to reduce the adverse effects of cannibalization on maintenance costs and personnel morale. The strategies developed in this chapter, at least, can be used to determine (1) which types of cannibalizations are appropriate, (2) cannibalization reduction goals and (3) the actions to be taken to meet the cannibalization reduction goals. In this chapter, we also presented a combined analytical and simulation model of a two-line, three-line and *k*-line system when cannibalization is not allowed and when cannibalization is allowed (with and without short interruptions to the system). It is clear from the analytical and simulation results that cannibalization can substantially increase the reliability of the systems where it is allowed. The improvement factor of unreliability obviously exists in systems where cannibalization is allowed as compared to those in which cannibalization is not allowed. Moreover, the improvement factor is larger when we have two-stage cannibalization (short interruptions) than without them.

Keywords: cannibalization, system, component, reliability, model

1. Introduction

Usually, policy dictates that systems of equipment are dispensed together with spare parts in case of components which are prone to failure. With failed components being replaced and/or repaired, the systems of equipment remain functional subservient to the said policy. Nevertheless, the two exceptions to the preceding policy are [1, 2] as follows: (1) as a result of the high acquisition and holding costs, high-technology manufacturing environments and organizations, which make use of expensive equipment, will not be able to always stock spare parts and (2) when equipment has reached its last life-cycle stage, failure rates have increased, spare

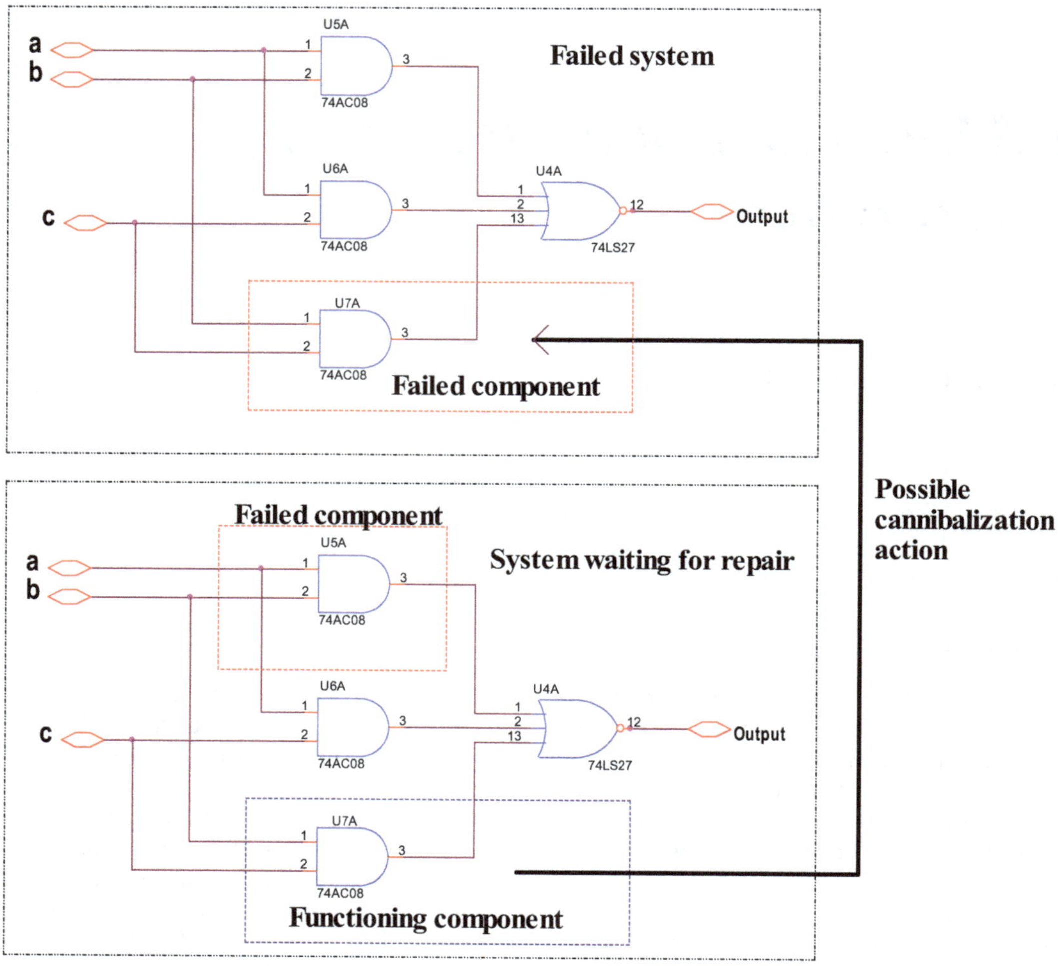

Figure 1. The cannibalization concept.

parts become increasingly difficult to acquire and there is a slump in the usage of the said equipment. Therefore, the inclination is not to acquire numerous spare parts as the said equipment will be phased out soon.

One conceivable way of maintaining systems whose spare parts have been depleted or are not available is through cannibalization. Readiness requirements and short maintenance turn-arounds in high-technology environments can be achieved through cannibalization. In general, 'cannibalization' ascribes to the process of removing a failed component from a system and replacing the said component by an operating component (of the same type) extracted from another part of the system. The concept of cannibalization is illustrated in **Figure 1**. The practice of cannibalization has a long history, especially in the military [3].

2. Strategies to conduct informed cannibalizations

Cannibalizations, when carried out imprudently, can have unintended negative effects. Cannibalization actions often bear negative connotations because of the following [4, 5]: (1) cannibalizations indicate that there are problems with the spare part supply chain, (2) there is a potential of mechanically damaging systems during cannibalizations and (3) cannibalizations increase the workload of maintenance personnel and if practiced frequently will dampen their morale. Nonetheless, the practice of cannibalization is unlikely to be eliminated completely. Therefore, one has to consider the idea of (1) designated cannibalizations and (2) methodology-informed cannibalizations. These will then cushion against the negative effects of cannibalizations.

The idea of designated cannibalizations means the designation of components in the requirements database as cannibalizable (i.e. easy to cannibalize) or non-cannibalizable (i.e. difficult to cannibalize). It can be noted that the cannibalizable or non-cannibalizable of components is mainly dependent on their type. For example, the cannibalization of a fuse is a trivial task which maintenance personnel can conduct almost invariably rather than wait for long periods of time for the spare part to be delivered. On the other hand, the cannibalization of an aircraft wing spar, for example, is very costly (i.e. in terms of energy, time and money), very dangerous and unfathomed in the world of aircraft maintenance. Cannibalization of components designated as cannibalizable will provide serviceable components when the spare stock is depleted. Cannibalizations of components designated as non-cannibalizable will not be permitted as it may be too costly in labour or time or money, or too risky in that the component (system) may incur mechanical damage during removal.

The methodology-informed cannibalization is an alternative route to estimate optimum rates of cannibalization required to meet the set readiness and operational demand goals, given the system: (1) mean uptime (MUT) (also known as MTTF), (2) MTTR, (3) mean supply response time (MSRT), (4) mean maintenance and supply time (MMST) and (5) mean downtime (MDT). This methodology addresses designated cannibalizations only deemed necessary in the industry. The theoretical model is presented in Section 2.1. This model is an extension of the model developed in Ref. [6]. In the model developed in Ref. [6], it is assumed that the sum of the supply time and maintenance time uniquely classifies the total downtime. Nonetheless, in real-life situations, three categories of downtime, that are exclusive, exist. These are (1) maintenance time, (2) supply time and (3) overlapping MMST. Hence, we extend the model of Ref. [6] to accommodate these different categories of downtime because cannibalizations affect MTTR and MSRT, as well as MMST. In Section 2.2, we provide cannibalizations policy implications based on the results of Section 2.1.

2.1. Theoretical model to show the effects of cannibalization on mission time availability of systems

We begin with the following definition of system mission time average availability. The system mission time average availability ($MTAA_{system}$) denotes the mean proportion of mission time the system is functioning. It is assumed that every time a component (system) fails, it is restored to an 'as good as new' condition through repair and the said system mission time average availability is

$$MTAA_{\text{system}} = \frac{MUT}{MUT + MDT} \tag{1}$$

where *MUT* is the mean uptime (also referred to as *MTTF*) and *MDT* is the mean downtime. The *MUT* denotes the mean functioning of the system. The *MDT* is decomposed into three mutually exclusive activities: (1) the *MTTR*;, (2) the *MSRT* and (3) the *MMST*. If we substitute these three variables into Eq. (1), we get

$$MTAA_{\text{system}} = \frac{MUT}{MUT + MTTR + MSRT + MMST} \tag{2}$$

To simplify the calculations, one assumes a zero time for the decision to cannibalize and similarly a zero time for accomplishing the cannibalizations. Therefore, *MSRT* is calculated as follows [6]:

$$MSRT = (1 - GE)(1 - c)\mu \tag{3}$$

where *GE* denotes gross effectiveness—that is, the ratio of the parts required which can be obtained in the supply chain, c denotes the ratio of the parts requests which cannot be obtained in the supply chain and have to be cannibalized and μ is the mean customer wait time (*CWT*) for spare parts.

The U.S. Navy describes the cannibalization action to be the number of cannibalizations performed per 100 flight hours. This is referred to as the cannibalization rate and is denoted $CANN_{AF}$. In this research, it will suffice to use this definition. The ratio of all the parts requests which are cannibalized, c, is determined as follows [6]:

$$c = \frac{CANN_{AF}}{100(1 - GE)\theta} \tag{4}$$

where θ is the component mean failure rate.

If we substitute Eq. (4) into Eq. (3), we get

$$MSRT = \frac{-\left(\mu\left(CANN_{AF} - (100\theta) + (100GE\theta)\right)\right)}{100\theta} \tag{5}$$

It can be noted that Eq. (5) shows a negative linear relationship between *MSRT* and $CANN_{AF}$. If we hold θ, μ and *GE* constant, it can be deduced that when the cannibalization rate is higher the MSRT becomes lower.

If we substitute Eq. (5) into Eq. (2) and solve for $CANN_{AF}$, we get

$$CANN_{AF} = \frac{\left(\theta(100GE - 100)\left(\frac{MMST + MTTR + MUT - MUT}{MTAA_{\text{system}} - \mu(GE - 1)}\right)\right)}{\mu(GE - 1)} \tag{6}$$

We impose the following mathematical constraints when calculating with Eq. (6):

$$CANN_{AF} = 0 \ if \ (1 - GE)\mu \leq \frac{MUT}{MTAA_{\text{SYSTEM}} - MUT - MTTR - MMST} \tag{7}$$

$$MTAA_{\text{system}} \le \frac{MUT}{MUT + MTTR + MMST} \tag{8}$$

It is impossible to achieve an $MTAA_{\text{system}}$ value higher than this. It can only mean that the values set for the parameters in the model are not showing rational thought.

Using Eq. (6), we plot $CANN_{AF}$ as a function of μ for different values of $MTAA_{\text{system}}$ (with all the other parameters fixed at values given and shown in **Figure 2**. It can be deduced from **Figure 2** that μ approaches infinity as $CANN_{AF}$ reaches a maximum of 12.31 for an $MTAA_{\text{system}}$ of 0.65; 12.24 for an $MTAA_{\text{system}}$ of 0.60 and 12.17 for an $MTAA_{\text{system}}$ of 0.55.

Figure 3 shows that a policy to limit cannibalization activities is required. In the example of **Figure 3** (with all the other parameters fixed), it can be seen that a cannibalization rate above 14 does not add any value as the $MTAA_{\text{system}}$ is now 1.

2.2. Policy implications of the theoretical model

In this section, we state a number of conclusions based on our simulations using Eqs. (1)–(6). These conclusions have implications on cannibalization policies. The conclusions are as follows:

a. The gradient $\frac{\partial CANN_{AF}}{\partial \mu}$ of the graph in **Figure 2** is positive. This shows that the longer μ is the higher $CANN_{AF}$ is (for different values of $MTAA_{\text{system}}$ all other parameters being held constant). Thus, cannibalization activities may be reduced by decreasing μ;

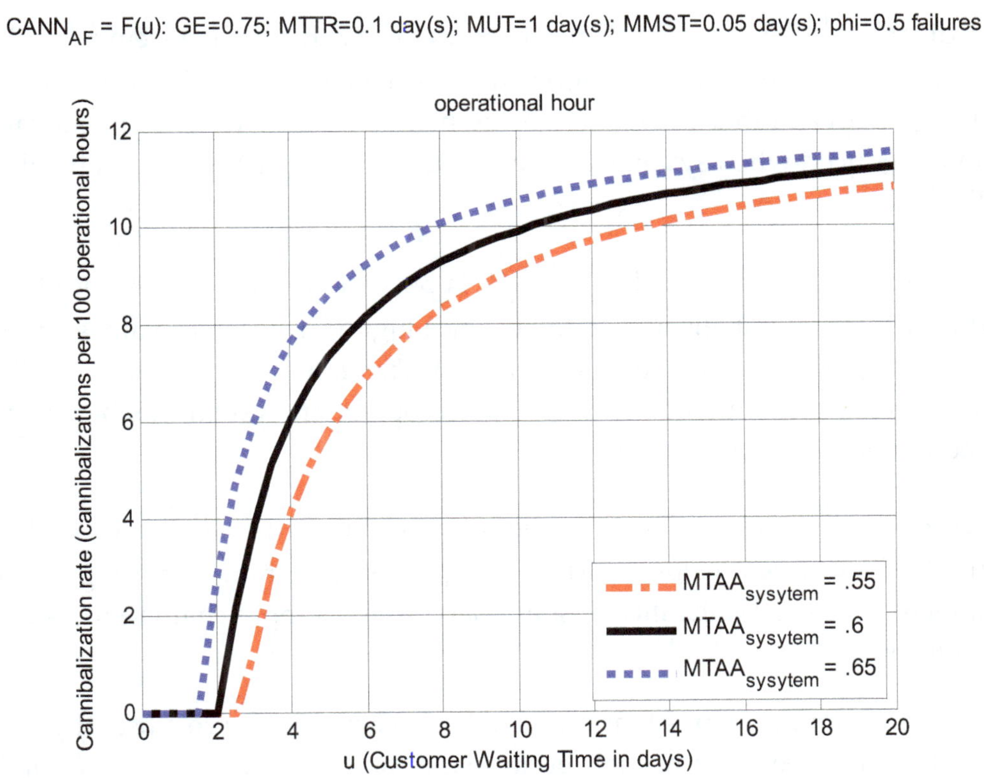

Figure 2. $CANN_{AF}$ versus μ for different $MTAA_{\text{system}}$ values.

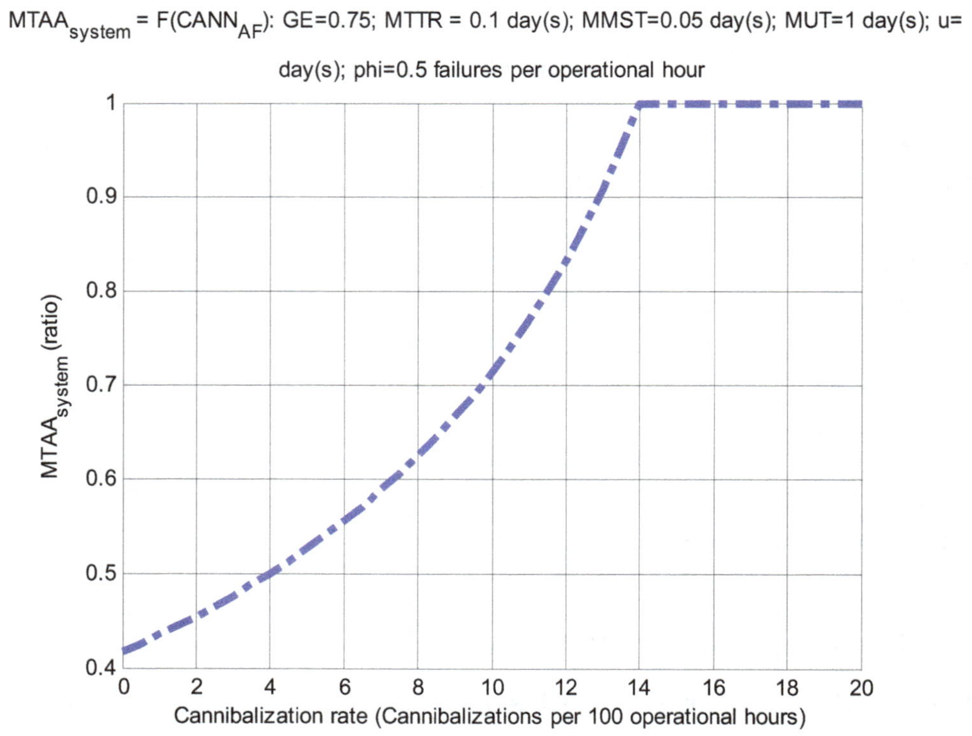

Figure 3. $MTAA_{\text{system}}$ versus $CANN_{AF}$.

b. The function $MTTR = F(CANN_{AF})$ is positive. Hence, more cannibalization activities imply longer maintenance time;

c. The gradient $\frac{\partial CANN_{AF}}{\partial MUT}$ is negative. Thus, it shows that with higher system reliability or with longer mean uptime, cannibalization rates become lower (for different values of $MTAA_{\text{system}}$ with all other parameters being held constant). Thus, cannibalization activities can be reduced when systems are designed taking cognisance of probabilistic design for reliability;

d. It can be deduced that the gradient $\frac{\partial CANN_{AF}}{\partial MTTR}$ is positive. It implies that with longer repair time (*MTTR*), the cannibalization rate becomes high (for different values of $MTAA_{\text{system}}$ with all other parameters being held constant). Therefore, cannibalization activities may be reduced with a more efficient maintenance operation system (i.e. better trained and qualified maintenance personnel);

e. The gradient $\frac{\partial CANN_{AF}}{\partial GE}$ is negative. It implies that the higher the *GE*, the lower the cannibalization rate (for different values of $MTAA_{\text{system}}$ all other parameters being held constant). Hence, increasing the availability of spare parts in the supply chain reduces cannibalization activities;

f. It can be seen from the results that cannibalization activities serve a useful purpose in the maintenance and operation of high performance and complex systems. Cannibalization activities are necessary, viable and cost-effective, only if the optimum cannibalization rate is sought for specific operating parameters; and

g. Lastly, when the actual data on cannibalization rates and other parameters (i.e. the ones related to $MTAA_{\text{system}}$) is available, the models presented in Sections 2 and 3 can be empirically tested.

3. Cannibalization revisited: theoretical model and example

This section considers a situation where repair facilities or spare components are not immediately available so that the probability of survival of a system can only be enhanced by extracting needed replacement components from another part of the system. We develop a model of cannibalization for the probability of survival (at time t) of a system with k -lines in parallel of n series-connected components when short interruptions to the system are allowed and when short interruptions to the system are not allowed. It is assumed for practical reasons that the lines are identical. Let all components be also identical, with exponentially distributed lifetimes with parameter λ. We can generalize the approach to the case of non-identical components and lines but the resulting expressions will be extremely cumbersome. We start with two lines as follows:

i. Assume that when only one line is left, the time for replacing the failed component of this remaining line (if one has a spare, e.g. from the failed line) is not allowed. Then, there is no possibility of cannibalization in this system and the survival function can be easily obtained.

ii. The time for replacing the failed component is allowed. Then, when one line fails, all $n-1$ non-failed components of the failed line can be used as spares for the operable line. The corresponding formulas (survival function) are then derived and this is cannibalization.

Then, we consider the case of three lines:

i. No time is allowed for replacing the failed component. However, cannibalization is still performed here. Indeed, when one line fails, we can use $n-1$ spares for the system of two lines (i.e. cannibalization) and when they will be exhausted, then no cannibalization, as in the case with two lines. The formula for probability of survival (at time t) is then written for the case with cannibalization and without and compared.

ii. Time for replacing the failed component is allowed. Then, the two-stage cannibalization goes. When the first line fails, $n-1$ non-failed components can be used to maintain the two lines. When this is exhausted and one of the two lines fails, the same process as in the previous case is followed and then the corresponding relationships are obtained.

We generalize the approach to the case when we have more than three lines and obtain the corresponding recurrent equations for survival probabilities in this case that can be solved numerically.

It can be noted that short interruptions to the system give us the possibility to use some components of a system as spares.

3.1. Notation

X: lifetime of a system.

$S_{kn}^{-}(t)$: probability of survival (at time t) of a system with k -lines of n -series-connected components with cannibalization (when short interruptions of the system are not allowed).

$S_{kn}^{+}(t)$: probability of survival (at time t) of a system with k -lines of n -series-connected components with cannibalization (when short interruptions of the system are allowed).

$S_{kn}^{nc}(t)$: probability of survival (at time t) of a system with k -lines of n -series-connected components when no cannibalization at all is allowed.

$q_{kn}^{-+}(t) = \frac{1 - S_{kn}^{-}(t)}{1 - S_{kn}^{+}(t)}$: improvement factor of unreliability (at time t) for k -lines of n -series-connected components with cannibalization due to allowing short interruptions.

$q_{kn}^{nc}(t) = \frac{1 - S_{kn}^{nc}(t)}{1 - S_{kn}^{-(+)}(t)}$: improvement factor due to cannibalization of unreliability (at time t) for k -lines of n -series-connected components when initially no cannibalization at all is allowed.

$$k = 1,\ 2,\ \ldots.$$
$$n = 1,\ 2,\ \ldots$$

Note: The improvement factor of unreliability shows how much the unreliability has decreased due to cannibalization being allowed.

3.2. One-line system

This section is concerned with the probability of survival (at time t) evaluation of standard series network occurring in engineering systems. The series network is the basic building block of the work in this section. In this incident, n number of components form a series network, as illustrated in **Figure 4**. The system fails if any one of the components fails. All components constituting the system must not fail in order to ensure a successful system operation.

The four wheels of a car illustrate a typical example of a series system. The car cannot be driven for practical purposes with any one of the tyres punctured. It therefore follows that these four car tyres form a series system. When one assumes independent and identical components (each component $i(i = 1,2,\ldots..,n)$ with a lifetime that is exponentially distributed with failure rate λ), it then follows that the probability of survival for the series system as shown in **Figure 4** is

$$S_{1n}^{nc}(t) = e^{-n\lambda t} \tag{9}$$

3.3. Two-line system

Now consider two identical lines of series-connected components as shown in **Figure 5**. Now, we compute the probability of survival for two cases [7, 8]: (Section 3.3.1) when no short interruptions to the system are allowed (i.e. no possibility of cannibalization: this is indeed

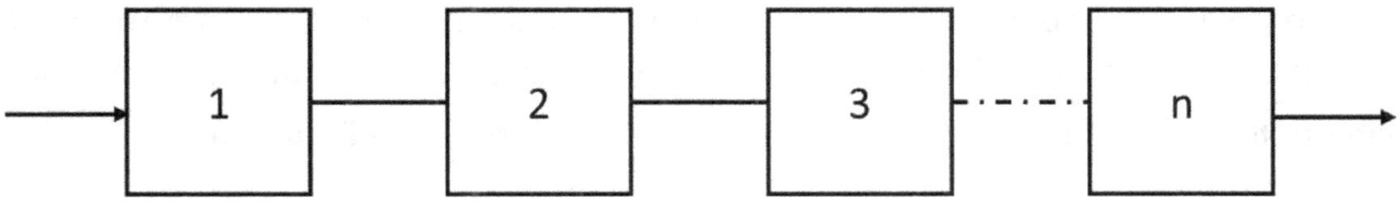

Figure 4. One line of series-connected components.

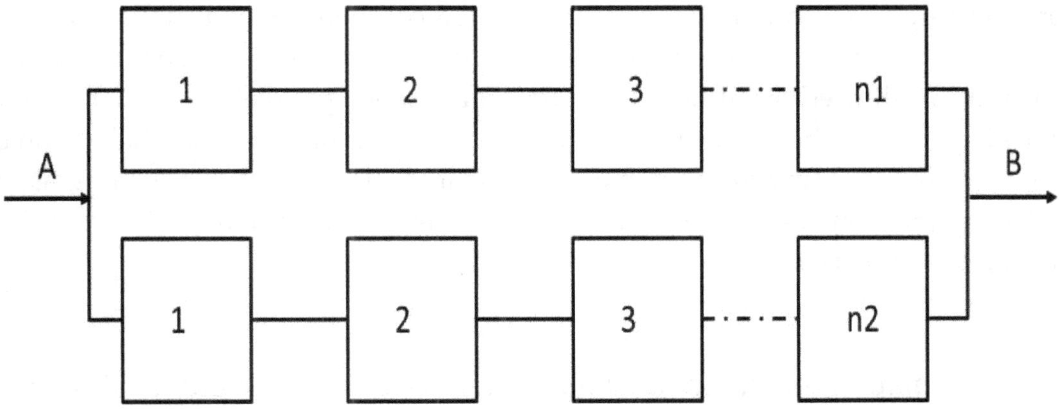

Figure 5. Two lines of series-connected components.

the only case without cannibalization, but if we have three or more lines and no interruptions, we already have cannibalization) and (Section 3.3.2) when short interruptions to the system are allowed (i.e. when cannibalization can be executed).

3.3.1. No short interruptions to the system are allowed (i.e. no possibility of cannibalization)

The formula for $Pr(X \geq t)$ (i.e. the probability of survival) is written obviously as follows:

$$S_{2n}^{-}(t) = S_{2n}^{nc}(t) = Pr(X \geq t) = 1 - (1 - e^{-n\lambda t})^2 \tag{10}$$

3.3.2. Short interruptions to the system are allowed (i.e. cannibalization is allowed)

When one line fails, the time for replacing the failed component of the remaining line is allowed and all $n - 1$ non-failed components of the failed line can be used as spares for the operable line. The cannibalization formula for $Pr\ (X \geq t)$ (i.e. the probability of survival) is written as follows:

$$S_{2n}^{+}(t) = Pr(X \geq t) = e^{-2n\lambda t} + \int_0^t \left(2n\lambda e^{-2n\lambda x} \sum_{i=0}^{n-1} e^{-n\lambda(t-x)} \frac{\left(n\lambda(t-x)\right)^i}{i!} \right) dx \tag{11}$$

where $e^{-2n\lambda t}$ in the first term in Eq. (11) means that both lines have survived (i.e. by the law of total probability); the integral corresponds to the probability that one line failed and then the remaining line has survived with $n - 1$ spares and $2n\lambda e^{-2n\lambda x}\, dx$ means the density of the first failure of $2n$ components. Then with one line left, there will be no further failures.

One can compare probabilities with and without cannibalization. More appropriately, we compare probabilities of failures. Therefore, we compute the improvement factor of unreliability for the two-line system as $q_{kn}^{nc}(t) = \frac{1 - S_{kn}^{nc}(t)}{1 - S_{kn}^{+}(t)}$ as shown in **Figure 6**.

3.4. Three-line system

Now consider three identical lines of series-connected components in a similar manner to the two-line system. Now, we compute $Pr(X \geq t)$ for three cases [7, 8] (Section 3.4.1) when no cannibalization is allowed at all (just three lines of n series parallel-connected components) (Section 3.4.2), when no short interruptions to the system are allowed (cannibalization is made possible here as we are using the operable components of the failed line as spares, as reflected in Eq. (13)) and (Section 3.4.3) when short interruptions to the system are allowed (i.e. when cannibalization is allowed).

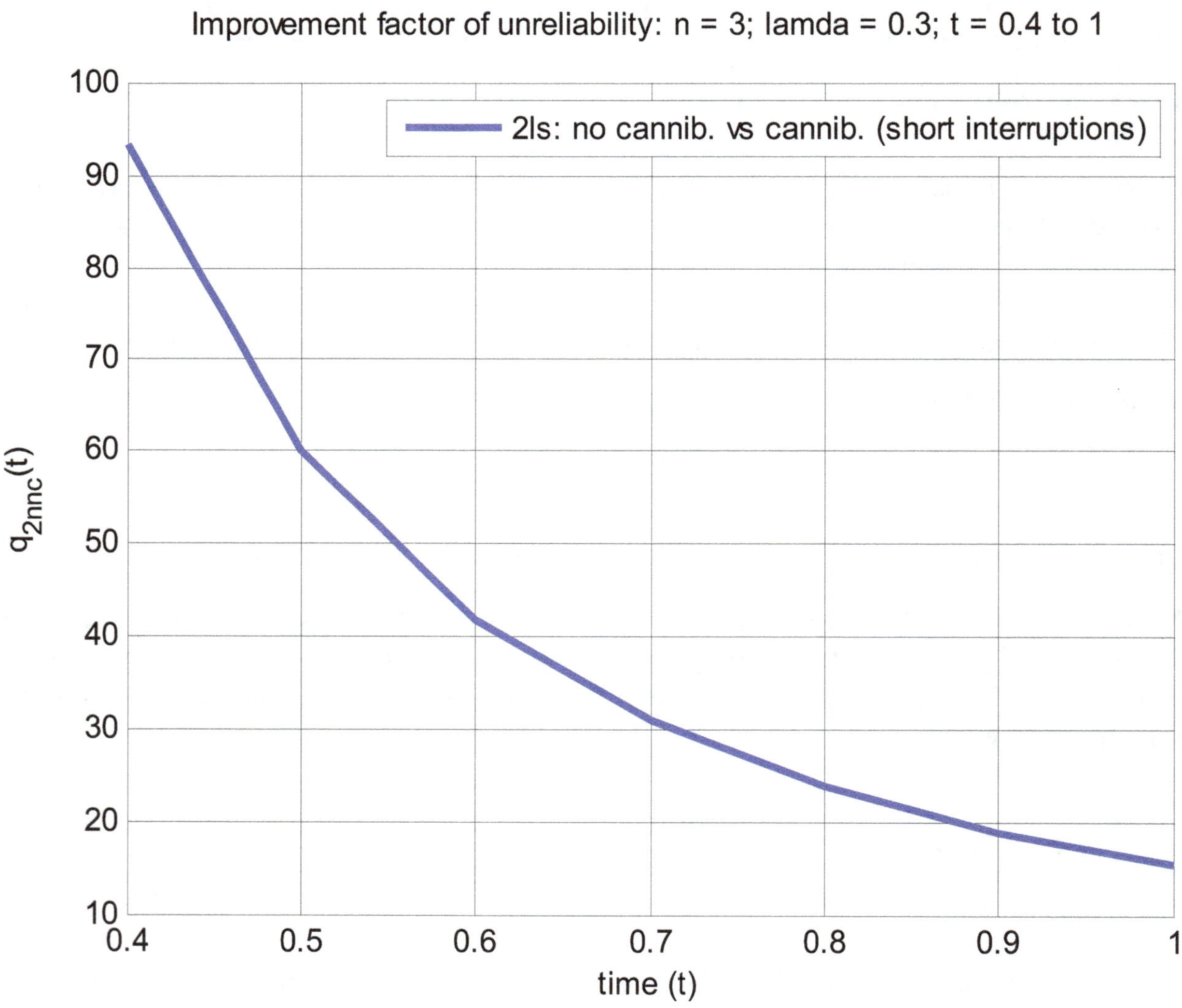

Figure 6. Improvement factor of unreliability for a two-line system (comparison of a system with no cannibalization and that with cannibalization when short interruptions to the system are allowed).

3.4.1. No cannibalization is allowed at all (just three lines of n series parallel-connected components)

The formula for $Pr(X \geq t)$ (i.e. the probability of survival) is written, obviously, as follows:

$$S_{3n}^{nc}(t) = Pr(X \geq t) = 1 - (1 - e^{-n\lambda t})^3 \tag{12}$$

3.4.2. No short interruptions to the system are allowed (i.e. cannibalization is made possible here by operable components of the failed line which are used as spares)

Cannibalization can still be done here. Indeed, when one line fails, we can use $n - 1$ spares for the system of two lines. When the $n - 1$ non-failed components are exhausted, then no cannibalization can be done. The formula is written as follows:

$$\begin{aligned} S_{3n}^{-}(t) &= Pr(X \geq t) \\ &= e^{-3n\lambda t} + \int_0^t 3n\lambda e^{-3n\lambda x} \sum_{i=0}^{n-1} e^{-2n\lambda(t-x)} \frac{\left(2n\lambda(t-x)\right)^i}{i!} \\ &\quad + \int_0^{t-x} \frac{(2n\lambda)^n}{(n-1)!} y^{n-1} e^{-2n\lambda y} e^{-n\lambda(t-x-y)} dy \Bigg) \Bigg) dx \end{aligned} \tag{13}$$

where $e^{-3n\lambda t}$ in the first term in Eq. (13) means that all three lines have survived (i.e. by the law of total probability); the integral corresponds to the probability that one line failed and then the remaining two lines survived with $n - 1$ spares, and $3n\lambda e^{-3n\lambda x}\, dx$ means the density of the first failure of $3n$ components. Then with two lines with $n - 1$ spares and the nth failure with intensity 2λ will 'ruin' it and one line will be left and no further failures.

3.4.3. Short interruptions to the system are allowed (i.e. cannibalization is allowed)

The two-stage cannibalization goes. When the first line fails, the time for replacing the failed component of the two remaining lines is allowed. $n - 1$ non-failed components of the failed line can be used to maintain the two remaining lines. When these components are exhausted and one of the two lines fails, the $n - 1$ non-failed components of the failed line can be used as spares for the remaining operable line. The corresponding cannibalization formula for $Pr(X \geq 1)$ is written as follows:

$$\begin{aligned} S_{3n}^{+}(t) &= Pr(X \geq t) \\ &= e^{-3n\lambda t} + \int_0^t 3n\lambda e^{-3n\lambda x} \sum_{i=0}^{n-1} e^{-2n\lambda(t-x)} \frac{\left(2n\lambda(t-x)\right)^i}{i!} \\ &\quad + \int_0^{t-x} \frac{(2n\lambda)^n}{(n-1)!} y^{n-1} e^{-2n\lambda y} \sum_{i=0}^{n-1} e^{-n\lambda(t-x-y)} \frac{\left(n\lambda(t-x-y)\right)^i}{i!} dy \Bigg) \Bigg) dx \end{aligned} \tag{14}$$

where $e^{-3n\lambda t}$ in the first term in Eq. (14) means that all three lines have survived; the integral corresponds to the probability that one line failed and then the remaining two lines survived with $n - 1$ spares; $3n\lambda e^{-3n\lambda x}dx$ means the density of the first failure of $3n$ components, and $\frac{(2n\lambda)^n}{(n-1)!}y^{n-1}e^{-2n\lambda y}$ is the density of the nth event from the Poisson process with rate $2n$. Then with two lines with $n - 1$ spares and the nth failure with intensity 2λ will 'ruin' it and one line will be left and no further failures.

Here, we can compare probabilities with no cannibalization at all (just three lines of n -series parallel-connected components) and that with cannibalization (with and without short interruptions). More suitably, we compare probabilities of failures. Hence, we compute the improvement factor of unreliability for the three-line system (i.e. that for no cannibalization and cannibalization with and without short interruptions) as $q_{3n}^{nc}(t) = \frac{1 - S_{3n}^{nc}(t)}{1 - S_{3n}^{(+)}(t)}$ and $q_{3n}^{-+}(t) = \frac{1 - S_{3n}^{-}(t)}{1 - S_{3n}^{+}(t)}$ as depicted in **Figures 7** and **8**, respectively.

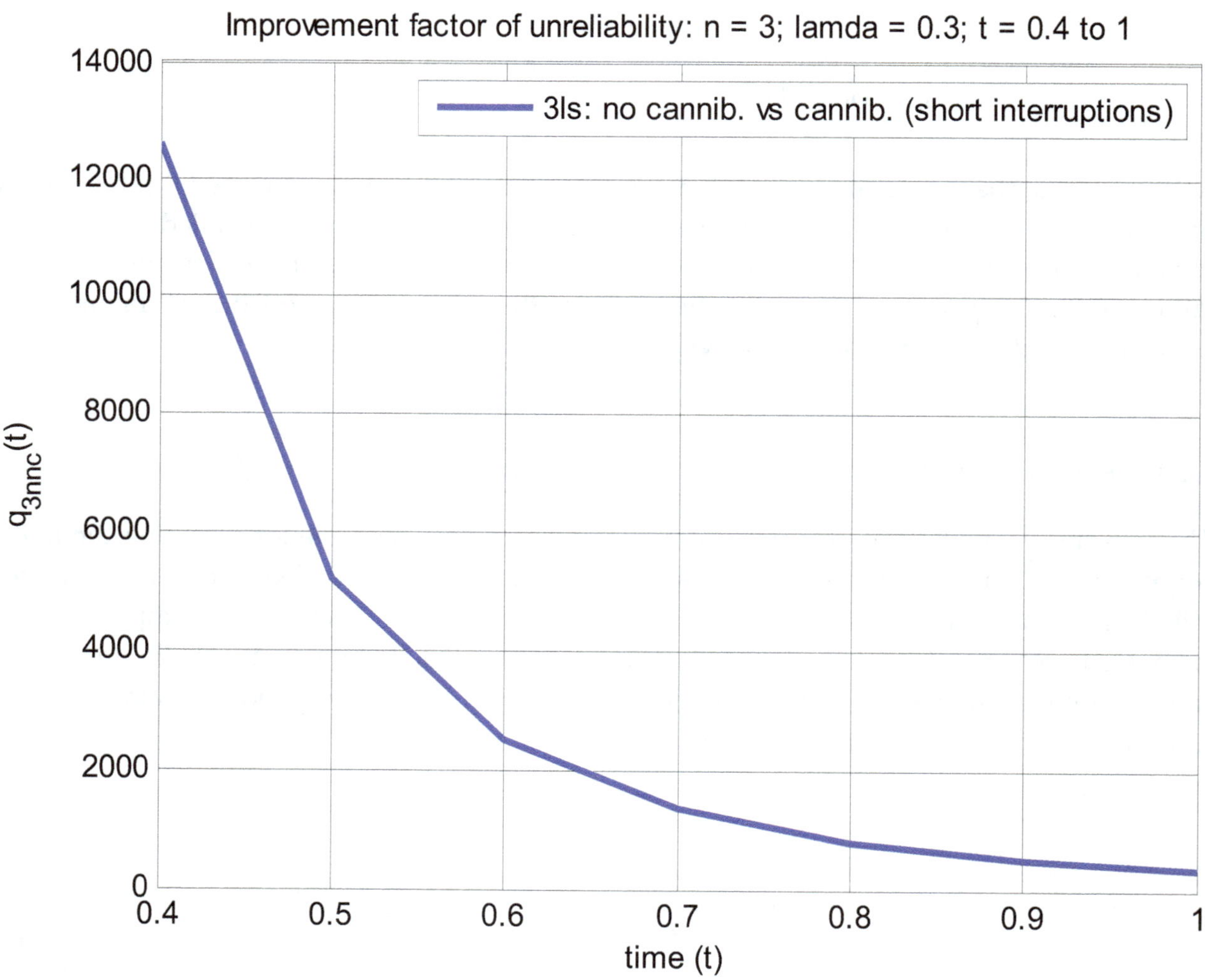

Figure 7. Improvement factor of unreliability for a three-line system (comparison of a system with no cannibalization and that with cannibalization when short interruptions to the system are allowed).

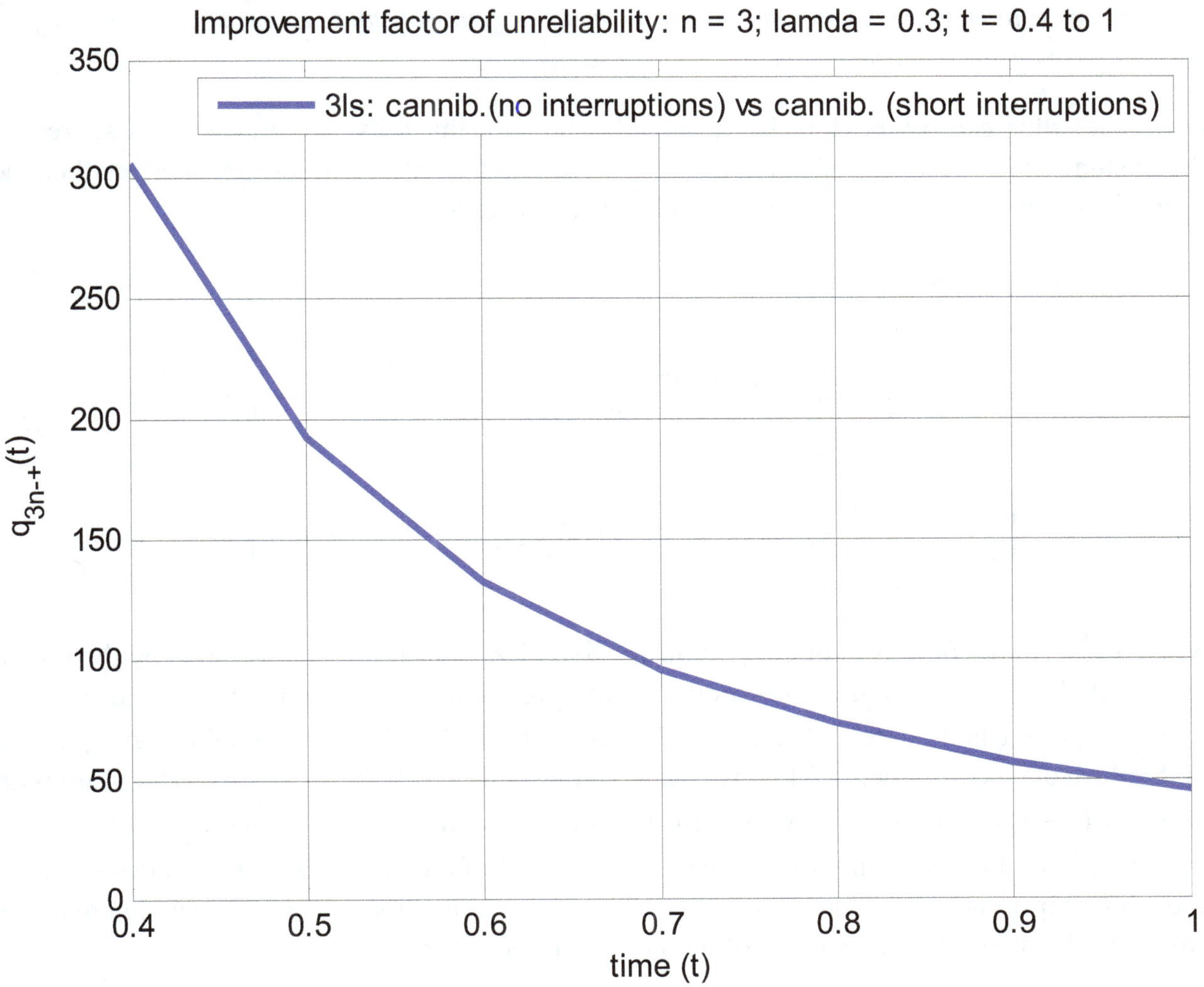

Figure 8. Improvement factor of unreliability for a three-line system (comparison of a system with cannibalization when no short interruptions to the system are allowed and that with cannibalization when short interruptions to the system are allowed).

3.5. *k*-Line system

Now consider k -identical lines of series-connected components in a similar manner to the two-line system. Now, we compute $Pr(X \geq t)$ for three cases [7, 8]: (Section 3.5.1) when no cannibalization is allowed at all (just k -lines of n -series parallel-connected components) (Section 3.5.2), when no short interruptions to the system are allowed (cannibalization is made possible here as we are using the operable components of the failed line as spares, as reflected in Eq. (16)) and (Section 3.5.3) when short interruptions to the system are allowed (i.e. when cannibalization is allowed).

3.5.1. No cannibalization is allowed at all (just k-lines of n-series parallel-connected components)

The formula for $Pr(X \geq t)$ (i.e. the probability of survival) is written as follows:

$$S_{kn}^{nc}(t) = Pr(X \geq t) = 1 - (1 - e^{-n\lambda t})^k \tag{15}$$

3.5.2. No short interruptions to the system are allowed (i.e. cannibalization is made possible here by operable components of the failed lines which are used as spares)

Cannibalization can still be done here. Indeed, when one line fails, we can use $n - 1$ spares for the system of $k - 1$ lines. When the $n - 1$ non-failed components are exhausted, then no cannibalization can be done. The formula is written as follows:

$$S_{kn}^{-}(t) = Pr(X \geq t)$$

$$= e^{-kn\lambda t} + \int_0^t \left\{ kn\lambda e^{-kn\lambda x} \sum_{i=0}^{n-1} e^{-(k-1)n\lambda(t-x)} \frac{\left((k-1)n\lambda(t-x)\right)^i}{i!} \right. \tag{16}$$

$$\left. + \int_0^{t-x} \frac{\left((k-1)n\lambda\right)^n}{(n-1)!} y^{n-1} e^{-(k-1)n\lambda y} S_{(k-1)n}^{-}(t-x-y)dy \right) \right\} dx$$

where $e^{-kn\lambda t}$ in the first term in Eq. (16) means that all k -lines have survived (i.e. by the law of total probability); the integral corresponds to the probability that one line failed and then the remaining $k - 1$ lines survived with $n - 1$ spares, and $kn\lambda e^{-kn\lambda x}dx$ means the density of the first failure of kn components. Then with $k - 1$ lines with $n - 1$ spares and the nth failure with intensity $(k - 1)\,\lambda$ will 'ruin' it and one line will be left and no further failures. $S_{(k-1)n}^{-}(t - x)$ is recurrent from the probability of survival (at time $t - x$) of the system with $k - 1$ lines when no short interruptions to the system are allowed (i.e. cannibalization is made possible here as we are using the operable components of the failed line as spares).

3.5.3. Short interruptions to the system are allowed (i.e. cannibalization is allowed but it was made possible in b) by operable components of the failed lines which are used as spares)

The $k - 1$ stage cannibalization goes. When the first line fails, the time for replacing the failed component of the $k - 1$ remaining lines is allowed. $n - 1$ non-failed components of the failed line can be used to maintain the $k - 1$ remaining lines. When these components are exhausted and one of the $k - 1$ lines fails, the $n - 1$ non-failed components of the failed line can be used as spares for the remaining operable lines. The process is repeated until one (line) remains, and the time for replacing the failed component of this remaining line (if one has a spare) is not allowed. The corresponding cannibalization formula for $Pr(X \geq t)$ is written as follows:

$$S_{kn}^{+}(t) = Pr(X \geq t)$$

$$= e^{-kn\lambda t} + \int_0^t \left\{ kn\lambda e^{-kn\lambda x} \sum_{i=0}^{n-1} e^{-(k-1)n\lambda(t-x)} \frac{\left((k-1)n\lambda(t-x)\right)^i}{i!} \right. \tag{17}$$

$$\left. + \int_0^{t-x} \frac{\left((k-1)n\lambda\right)^n}{(n-1)!} y^{n-1} e^{-(k-1)n\lambda y} S_{(k-1)n}^{+}(t-x-y)dy \right) \right\} dx$$

where $e^{-kn\lambda t}$ in the first term in Eq. (17) means that all k -lines have survived; the integral corresponds to the probability that one line failed and then the remaining $k-1$ lines survived with $n-1$ spares; $kn\lambda e^{-kn\lambda x}dx$ means the density of the first failure of kn components, and $S^{+}_{(k-1)n}(t-x)$ is the probability of survival (at time $t-x$ of the system with $k-1$ lines when short interruptions to the system are allowed. Then with $k-1$ lines with $n-1$ spares and the nth failure with intensity $(k-1)\lambda$ will 'ruin' it and one line will be left and no further failures. Thus, Eq. (17) is a recurrent relationship.

Again, we can compare probabilities with and without cannibalization. More usefully, we compare probabilities of failures. Hence, we compute the improvement factor of unreliability for the k-line system (i.e. that for no cannibalization and cannibalization with and without short interruptions) as $q^{nc}_{kn}(t)=\frac{1-S^{nc}_{kn}(t)}{1-S^{(+)}_{kn}(t)}$ and $q_{kn}^{-+}(t)=\frac{1-S^{-}_{kn}(t)}{1-S^{+}_{kn}(t)}$ as depicted in **Figures 9** and **10**, respectively.

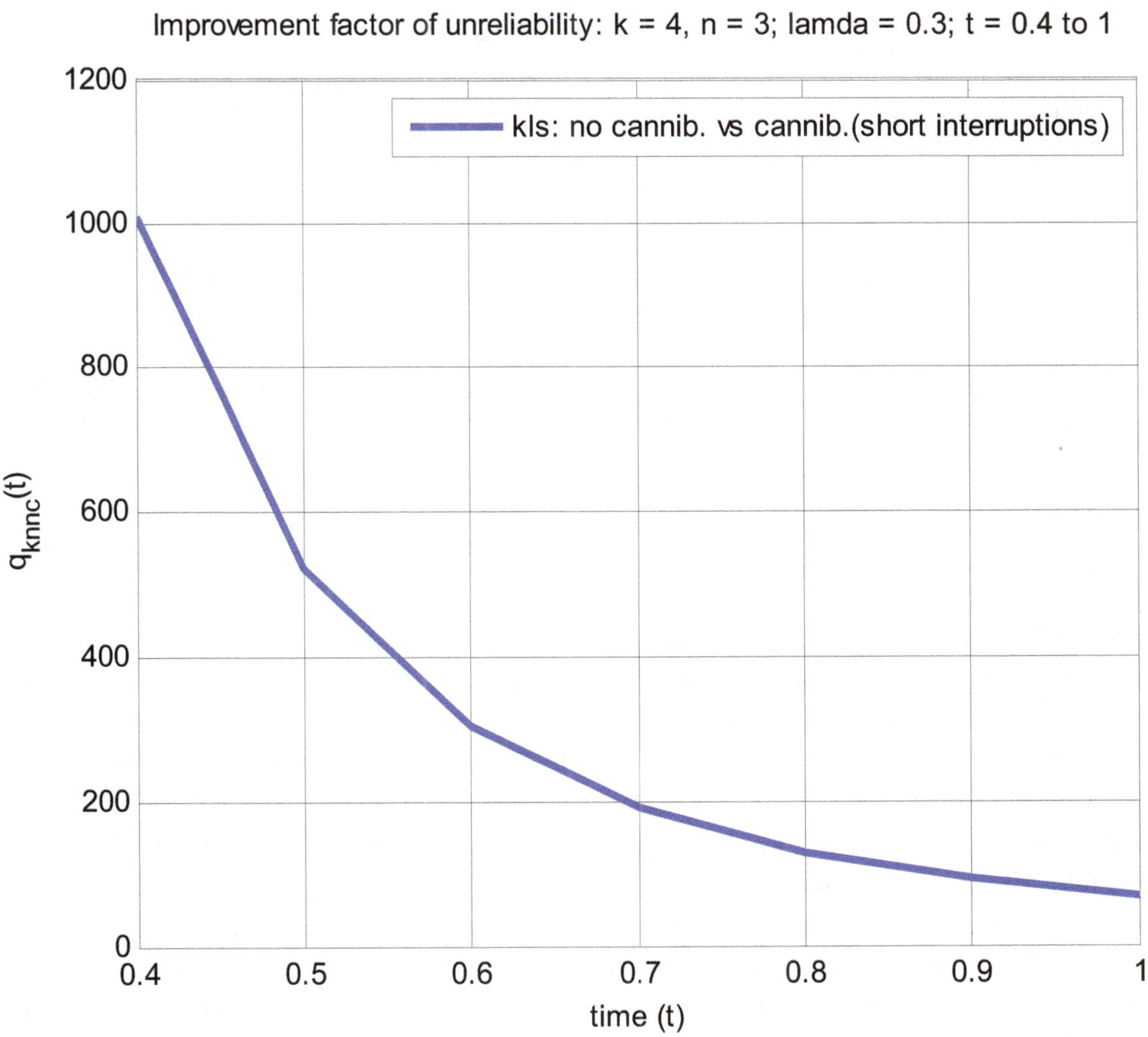

Figure 9. Improvement factor of unreliability for a k-line system (comparison of a system with no cannibalization and that with cannibalization when short interruptions to the system are allowed).

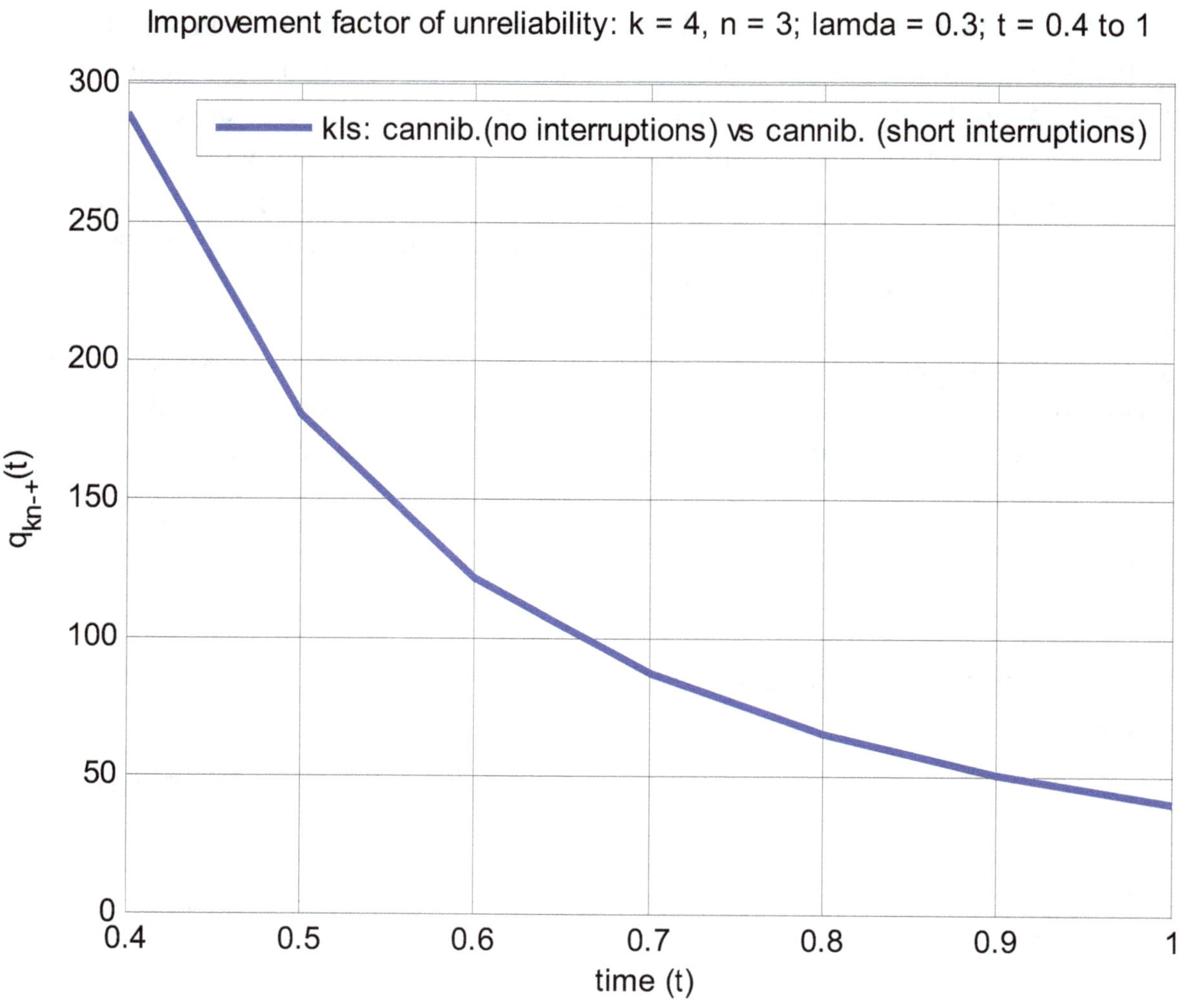

Figure 10. Improvement factor of unreliability for a *k*-line system (comparison of a system with cannibalization when no short interruptions to the system are allowed and that with cannibalization when short interruptions to the system are allowed).

3.6. Computation results

Figures 6–10 show the improvement factors of unreliability for the two-line system, three-line system and *k*-line system, respectively. We are looking at reliable systems, where the survival functions should be close to 1. For illustrative purposes, we choose the values of the failure rate and time accordingly. Therefore, it is more effective to compare unreliability of the systems with no cannibalization (i.e. $1 - S_{kn}^{nc}(t)$) and with cannibalization (i.e. with and without short interruptions) (i.e. $1 - S_{kn}^{+}(t)$ and $1 - S_{kn}^{-}(t)$). The improvement factor of unreliability is obtained by dividing the unreliability of the said system without cannibalization (worse quantity) by that of the said system with cannibalization (without short interruptions) (better quantity) and thus is larger than one. This is then compared to that obtained by dividing the unreliability of the said system with cannibalization with short interruptions (worse quantity) by that of the said system with cannibalization (without short interruptions) (better quantity). It can be seen from **Figures 6–10** that the improvement factors of unreliability of the systems in which cannibalization is allowed are better than in those in which it is prohibited.

From simulation results of **Figures 6–10**, it can also be shown that very larger values lead to asymptotically equivalent system performance levels.

4. Conclusions

In this chapter, we have explored strategies to mitigate the unfavourable effects that cannibalizations have on the costs of maintaining systems of equipment and also on the morale of maintenance personnel. The methodologies developed in this chapter, at least, can be used to (1) ascertain those cannibalizations that are appropriate, (2) start implementation of goals to reduce cannibalization and (3) enlist actions to be taken in order for cannibalization reduction goals to be met.

We also presented a combined analytic and simulation model of a two-line, three-line and *k*-line system when cannibalization is not allowed and when cannibalization is allowed (with and without short interruptions to the system). It is clear from the analytic and simulation results that cannibalization can substantially increase the reliability of the systems where it is allowed. The improvement factor of unreliability obviously exists in systems where cannibalization is not allowed as compared to those in which cannibalization is allowed. Moreover, the improvement factor is larger when we have two-stage cannibalization (short interruptions) than without them.

Author details

Bernard Tonderayi Mangara

Address all correspondence to: aragnam@gmail.com

Central University of Technology, Free State (CUT), Bloemfontein, South Africa

References

[1] Byrkett DL. Units of equipment available using cannibalization for repair-part support. IEEE Transactions on Reliability. 1985;**R34**(1):25–28

[2] Salman S, Cassady CR, Pohl EA, Ormon SW. Evaluating the impact of cannibalization on fleet performance. Quality and Reliability Engineering International. 2007;**23**:445–457

[3] Hirsch WM, Meisner M, Boll C. Cannibalization in multicomponent systems and the theory of reliability. Naval Research Logistics Quarterly. 1968;**15**:331–359

[4] United States General Accounting Office. Report to the Chairman, Subcommittee on National Security, Veterans Affairs, and International Relations, Committee on Government Reform,

House of Representatives [Internet]. November 2001. Available from: https://www.gao.gov/assets/240/233055.pdf [Accessed: March 8, 2017]

[5] de Oliveira N, Ghobbar A A. Cannibalization: How to measure and its effect in the inventory cost. In: Grant I, editor. 27th International Congress of the International Council of the Aeronautical Sciences (ICAS 2010); 19-24 September 2010; Nice, France. Nice, France: International Council of the Aeronautical Sciences; 2010. p. 1-12

[6] Hoover J, Jondrow JM, Trost RS, Ye M. A Model to Study: Cannibalization, FMC, and Customer Waiting Time [Internet]. February 2002. Available from: https://www.cna.org/CNA_files/PDF/D0005957.A2.pdf [Accessed: March 9, 2017]

[7] Finkelstein M. Failure Rate Modelling for Reliability and Risk. UK: Springer-Verlag London; 2008. p. 290. DOI: 10.1007/978-1-84800-986-8

[8] Finkelstein M, Cha JH. Stochastic Modeling for Reliability: Shocks, Burn-in and Heterogeneous Populations. UK: Springer London; 2013. p. 388. DOI: 10.1007/978-1-4471-5028-2

2

Reliability Prediction Considering Multiple Failure Mechanisms

Joseph B. Bernstein

Abstract

The multiple temperature operational life (MTOL) testing method is used to calculate the failure in time (FIT) by a linear combination of constant-rate failure mechanisms. This chapter demonstrates that, unlike other conventional qualification procedures, the MTOL testing procedure gives a broad description of the reliability from sub-zero to high temperatures. This procedure can replace the more standard single-condition high-temperature operational life (HTOL) by predicting the system failure rate by testing a small number of components over more extreme accelerated conditions for much shorter times than is conventionally used. The result is a much more accurate result for the failure rate, calculating the mean time to failure (MTTF) based on much shorter time-scale testing only a fraction of the number of components. Rather than testing 77 parts for 1000h, a failure rate prediction can be obtained from as few as 15 parts tested for only 200h with reliable results.

Keywords: MTTF, MTOL, HTOL, FIT, failure rate, multiple mechanisms

1. Introduction to MTOL

Traditional high-temperature operational life (HTOL) test strategy is based on the outdated JEDEC standard that has not been supported or updated for many years. The major drawback of this method is that it is not based on a model that predicts failures in the field. Nonetheless, the electronics industry continues to provide data from tests of fewer than 100 parts, subjected to their maximum allowed voltages and temperatures for as many as 1000h. The result based on zero, or a maximum of one, failure out of the number of parts tested does not actually predict. This null result is then fit into an average acceleration factor (AF), which is the product of a thermal factor and a voltage factor. The result is a reported failure rate as described by the standard failure in time (FIT) model, which is the number of expected failures per billion part hours of operation. FIT is still an important metric for failure rate in today's technology; however, it does not account for the fact that multiple failure mechanisms simply cannot be averaged for either thermal or voltage acceleration factors.

One of the major limitations of advanced electronic systems qualification, including advanced microchips and components,is providing reliability specifications that match the variety of user applications. The standard HTOL qualification that is based on a single high-voltage and high-temperature burn-in does not reflect actual failure mechanisms that would lead to a failure in the field. Rather, the manufacturer is expected to meet the system's reliability criteria without any real knowledge of the possible failure causes or the relative importance of any individual mechanism. More than this, as a consequence of the non-linear nature of individual mechanisms, it is impossible for the dominant mechanism at HTOL test reflect the expected dominant mechanism at operating conditions, essentially sweeping the potential cause of failure under the rug while generating an overly optimistic picture for the actual reliability.

Two problems exist with the current HTOL approach, as recognized by JEDEC in publication JEP122G:

1. Multiple failure mechanisms actually compete for dominance in our modern electronic devices and
2. Each mechanism has a vastly different voltage and temperature acceleration factors depending on the device operation.

This more recent JEDEC publication recommends explicitly that multiple mechanisms should be addressed in a sum-of-failure-rates approach. We agree that a single point HTOL test with zero failures can, by no means, account for a multiplicity of competing mechanisms.

In order to address this fundamental limitation, we developed a special multiple-mechanism qualification approach that allows companies to tailor specifications to a variety of customer's needs. We call this approach the multiple temperature operational life (MTOL) test at multiple conditions and match the results with the foundry's reliability models to make accurate FIT calculations based on specific customer's environments including voltage, temperature, and speed. The basic strategy is outlined in **Figure 1**. Time to fail models are put into the matrix as failure rates (λ_i) for each given set of conditions of temperature, voltage, frequency, etc. Then, the left-hand side of the matrix takes measured failure rates as extrapolated from measurements tested on the actual system under investigation. Hence, the relative acceleration factor for each mechanism is calculated based on measurement data rather than from zero-failures, as is traditional done through HTOL.

This new MTOL system allows the FIT value to be calculated with the assumption of not just one but multiple degradation mechanisms that are characterized by multiple acceleration factors. This chapter will describe the advantages of considering multiple failure mechanisms and how they can be linearly combined with a simple matrix solution that accounts for each mechanism proportionally based on data rather than based on a zero-failure result.

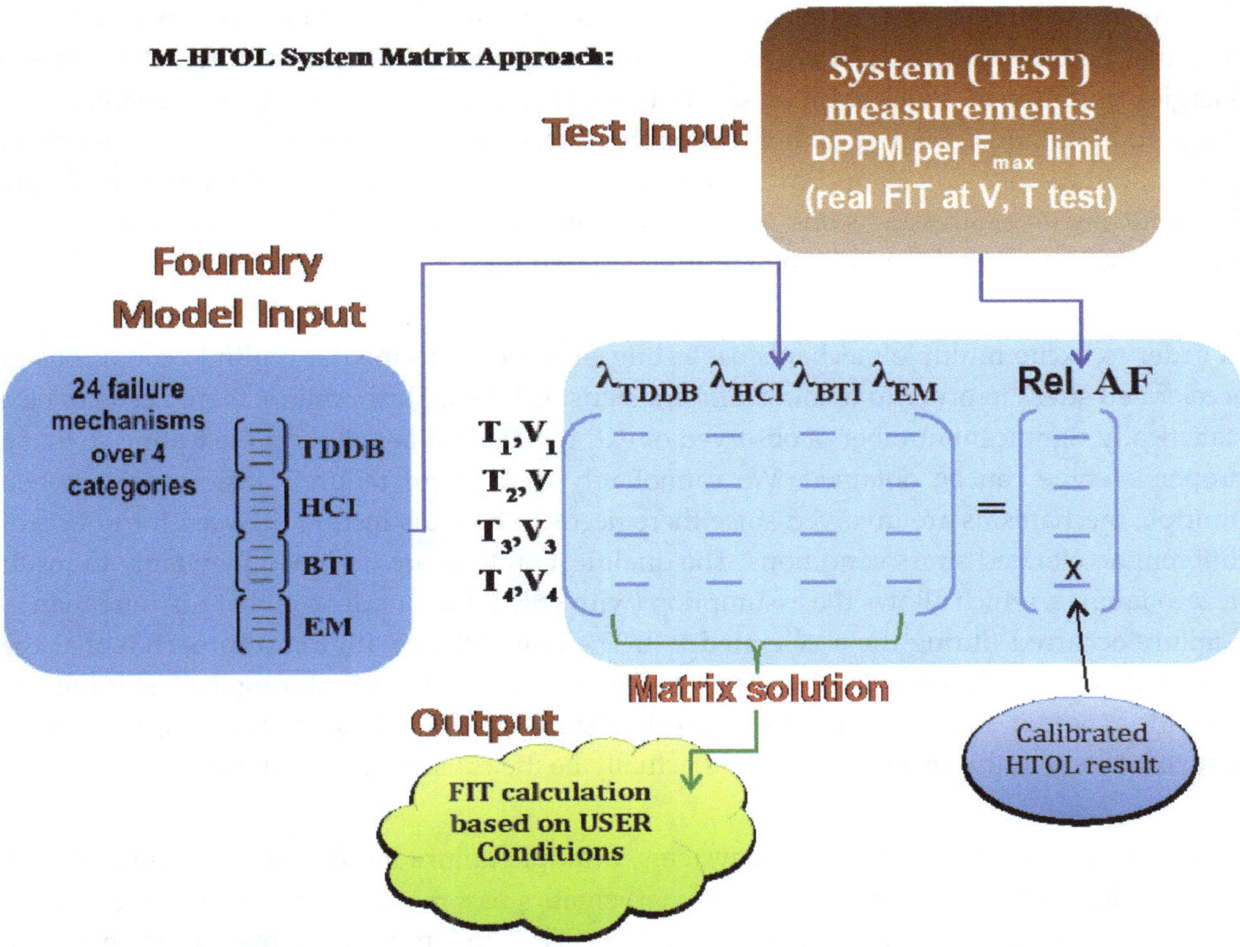

Figure 1. Matrix methodology for reliability prediction.

1.1. Limitation of traditional HTOL

The semiconductor industry provides an expected FIT for every product that is sold based on operation within the specified conditions of voltage, frequency, heat dissipation, etc. Hence, a system reliability model is a prediction of the expected MTBF's, or as we will use here, mean time to fail (MTTF), for a system that is not replaced as the sum of the FIT rates for every component.

A FIT is defined in terms of an acceleration factor (AF) and MTTF as:

$$FIT = \frac{\#failures}{\#tested * time(hours) * AF} 10^9 = {10^9}/{MTTF(hours) * AF} \quad (1)$$

where #failures and #tested are the numbers of actual failures that occurred as a fraction of the total number of units subjected to an accelerated test per total test time in hours. From a statistical perspective, this calculation would be correct if there is only a single known mechanism that is completely characterized by a single acceleration factor, AF. However, if multiple mechanisms are present, there is no way to average the acceleration factor, and thus, the denominator cannot be characterized as one AF for any set of operating conditions. The true AF must be based on the physics of the actual mechanisms, including different activation energies for different physical processes. Without testing at multiple accelerated conditions, a standard HTOL qualification cannot distinguish effects of more than one thermally activated process, rather only give an approximation for the dominant mechanism at the test condition. The test consists of stressing some number of parts, usually around 77, for an extended time, usually 1000 h, at an accelerated voltage and temperature.

In order to excite multiple mechanisms, testing must be performed at multiple conditions of accelerated stress in order to obtain sufficient statistical data. Furthermore, there needs to be a statistically significant number of observed or extrapolated failures during the testing so that a proper average can be obtained. We cannot rely on a "zero failure" pass criterion when multiple mechanisms are involved since there needs to be a distinction between the effects of different accelerated stress conditions. The qualification tests are designed inevitably to result in zero failures, which allows the assumption (with only 60% confidence!) that no more than ½ a failure occurred during the accelerated test. The only fallacy with this approach is that the assumption is that the only dominant mechanism that would be seen during the test is the one with the reported AF. However, if that mechanism is not modelled or observed, there is no way to prove that this mechanism would actually be the cause of a field failure.

We don't need to prove that in most systems, multiple failure mechanisms contribute to the overall reliability of a system. Reliability mathematics assumes that the influences are time-independent, occurring at a constant rate, while each is independent of the others. In reality, most systems experience failures at approximately a constant rate, at least for the first few "random" occurrences. When we consider that the defects responsible for earlier failures are generally distributed in time, the assumption of multiple failure mechanisms makes valid sense as to why the random failures occurring during the useful life of a product will be, in fact, caused by not a single mechanism, but rather by a proportional combination of all the likely failure and wear-out mechanisms. However, due to the physics involved with each cause of failure, each will be accelerated differently depending on the thermal, electrical, or environmental stresses that are responsible for each mechanism. Hence, when an accelerated test is performed at an arbitrary voltage and temperature for acceleration based only on a single failure mechanism, then, only that mechanism will be accelerated. When the failure rate (FIT) is calculated based on the non-occurrence of a failure (i.e., zero failure assumption), then it is naturally over-estimating the reliability by whatever factor was not introduced by the second or third mechanism that was not accounted for in the model.

Unfortunately for the test and qualification industry, the final test procedure and failure rate calculation have not kept pace with the depth of understanding that we have today about the actual failure mechanisms. Also, manufacturing processes are so tightly controlled that each known mechanism is designed to be theoretically non-existent in the field. Thus, naturally, since there is no single mechanism that will cause a known end-of-life, so it is logical that multiple mechanisms will affect the final failure rate. Furthermore, HTOL tests are known to reveal multiple failure mechanisms during final qualification, which would suggest also that no single failure mechanism would dominate FIT in the field. Thus, finally, in order to make a more accurate model for FIT, a preferable approximation should be that all mechanisms contribute and the resulting overall failure distribution resembles combination of *constant failure rate processes* that is consistent with the mil-handbook and JEDEC standards.

1.2. MTOL methodology

The key innovation of the multiple temperature operational life (MTOL) testing method is its ability to separate different failure mechanisms so that predictions can be made for any user defined operating conditions. This is opposed to the common approach for assessing device reliability today, using high-temperature operating life (HTOL) testing [1], which is based on the assumption that just one dominant failure mechanism is responsible for a failure of the device in the field [2]. However, it is known that, in reality, multiple failure mechanisms act simultaneously on any system that causes failure based on more than a single mechanism at any time [3].

Our new approach, MTOL, deals with this issue [4]. This method predicts the reliability of electronic components by combining separately measured FITS of multiple failure mechanisms [5]. Our data reveal that different failure mechanisms act on a component in different regimes of operation causing different mechanisms to dominate, depending on the stress and the particular technology. When multiple mechanisms are known to affect the failure a product, then JEDEC standard publication JEP-122G states that "**When multiple failure mechanisms and thus multiple acceleration factors are involved, then a proper summation technique, for example, sum-of-the-failure rates method, is required.**" The only question that is not answered by the JEDEC standard is how to "sum" the failure rates. As a practical solution to reaching the desired goal of a calculated failure rate that combines multiple mechanisms, we have proposed the following approach.

Because failure rates are linear and sum linearly only if they are all considered as constant rate processes, they can be combined linearly to calculate the actual reliability as measured in FIT of the system based on the physics of degradation at specific operating conditions. In a more recent publication [6], we present experimental results of the MTOL method tested on both 45 and 28nm FPGA devices from Xilinx that were processed at TSMC (according to the Xilinx data sheets). The FPGAs were tested over a range of voltages, temperature and frequencies. We measured ring frequencies of multiple asynchronous ring oscillators simultaneously during stress in a single FPGA. Hundreds of oscillators and the corresponding frequency counters were burned into a single FPGA to monitor of statistical information in real time. Since the frequency of a ring oscillator itself monitors the device speed and performance, there is no recovery effect, giving a true measure for the effects of all the failure mechanisms in real time. Our results produced an acceleration factor (AF) for each failure mechanism as a function of core voltage, temperature and frequency.

The failure rates of all of the mechanisms were then combined using a matrix to normalize the AF of the mechanisms to find the overall failure in time or FIT of the device. In other words, we found an accurate estimate of the device's mean lifetime and thus the reliability that can be conveniently transposed to other technologies and ASICs and not necessarily only FPGAs, as was the basis of our previous work. In this chapter, we show that the MTOL methodology is general and can apply to any system that is characterized by multiple failure mechanisms, which can individually be treated as approximately occurring at a constant rate, having its own FIT per mechanism.

2. Multiple mechanism considerations

The acceleration of the rate of occurrence of a single failure mechanism is a highly non-linear function of temperature and/or voltage as is well known through studies of the physics of failure [3–5]. The temperature acceleration factor (AF_T) and voltage acceleration factor (AF_V) can be calculated separately for each known mechanism in a combined model. The total acceleration factor of the different stress combinations for each mechanism will be the product of the acceleration factors of temperature and voltage or any other stress-related factor that could include current, frequency, humidity, etc.

$$AF = AF_V * AF_T \tag{2}$$

Calculated acceleration factors (AF) are universally used as the industry standard for device qualification. However, it only approximates a single dielectric breakdown type of failure mechanism and does not correctly predict the acceleration of other mechanisms. Similarly, an acceleration factor can be determined using any other type of stress applied, for example, vibration, radiation, number of cycles, etc. However, when only a single AF is assumed to contribute to the expected time to fail based on the high temperature, high voltage acceleration, there is no way to account for the effect of multiple mechanisms.

The goal here is to improve the approach from standard HTOL to a one where a true "sum of failure rates" model is considered based on a proportional contribution of each mechanism based on its relative influence. Each one mechanism acts on the system in combination with others to cause an eventual failure. When more than one mechanism affects the reliability of a system or component, then the relative acceleration of each one must be defined and calculated at the applied condition. Every potential failure mechanism should be identified, and its unique AF should then be relatively known at a given temperature and voltage so the FIT rate can be approximated separately for each mechanism. Thus, the actual FIT will be the sum of the failure rates per mechanism, as is described by:

$$\mathrm{FIT}_{\mathrm{total}} = \mathrm{FIT}_1 + \mathrm{FIT}_2 + \ldots + \mathrm{FIT}_i \tag{3}$$

whereby each mechanism is described by its own FIT, which leads to its own expected failure unit per mechanism, FIT_i. Therefore, it is impossible to accelerate more than one mechanism with a single set of accelerated stress conditions. Thus, requiring that more than a single test is necessary to determine what would be the actual FIT that would be found in any given expected operating conditions.

The qualification of device reliability, as reported by a FIT rate, must be based on an acceleration factor, which represents the failure model for the tested device. Since multiple mechanisms are known to lead to degradation and thus failure in any complex system, it is obvious that a single mechanism model with a single AF assumption will never produce a useful result for reliability prediction. This will be explained by way of example. Suppose there are two identifiable, constant rate competing failure modes (assume an exponential distribution). One failure mode is accelerated only by temperature. We denote its failure rate as AF_{T1} other failure mode is only accelerated by voltage, AF_{V2}, and the corresponding failure rate is denoted as

$$FIT = {10^9}/{MTTF_1 * AF_{T1}} + {10^9}/{MTTF_2 * AF_{V2}} \quad (4)$$

where the measured Mean Time To Fail (MTTF) (measured in hours) would be different for each mechanism. However, since only one condition of Voltage and Temperature is applied, yet the calculated FIT is based on a combination of two mechanisms, each with its own acceleration factor, then there is now way to determine which mechanism dominates. Because the effective acceleration factor for any given set of test conditions is related to the inverse of the acceleration factor, without separately testing each mechanism, the resulting FIT will have no relation to the actual tested results.

Due to the exponential nature of the acceleration factor as a function of V or T, or any other stress-inducing parameter, including vibration, humidity, radiation, etc, if only a single parameter is changed, then it is not likely for more than one mechanism to be accelerated significantly compared to the others. As we will see in the next section, at least three mechanisms should be considered, many more perhaps depending on the system and use environment. Also, each voltage and temperature dependencies must be considered separately for each mechanism in order to make a reasonable reliability model for the whole device.

3. MTOL test system example

A test system was built in off-the-shelf Xilinx FPGA evaluation boards. The system ran hundreds of internal oscillators at several different frequencies asynchronously, allowing independent measurements across the chip and the separation of current versus voltage induced degradation effects. In order to create a measurable accelerating system, ring oscillators (ROs) consisting of inverter chains were used. The last inverter in the chain is connected to the first, forming a cycle/ring (**Figure 2**). When the number of stages is odd, every sampled cell in the chain will invert its logic level. Additionally, as no clock is fed into the RO, the frequency of the alternating logical states depends just on the internal delay of the cells and the latency of the connections between them, where the frequency of each RO is given by $½NT_p$, where N is the number of inverters and T_p is the propagation delay of each inverter. Each inverter chain was implemented as a complete logical cell using predefined Xilinx primitives, and thus, each ring oscillator was made up of the basic components of the FPGA. When degradation occurred in the FPGA, a decrease in performance and frequency of the RO could be observed and attributed to either increase in resistance or change in threshold voltage for the transistors.

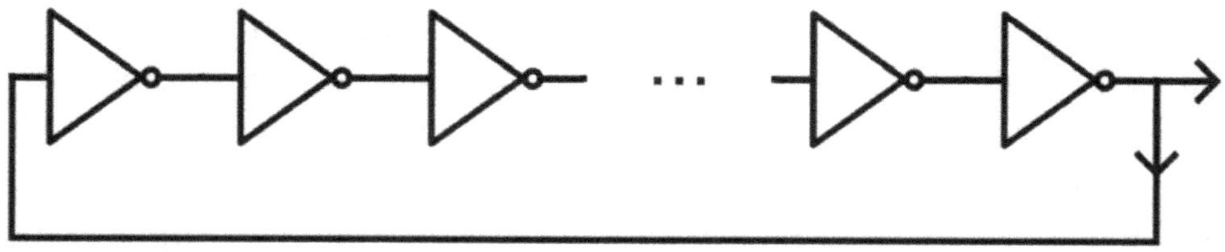

Figure 2. Ring oscillator is made of $2N+1$ Inverters connected in a chain.

For optimal testing and chip coverage, different sized ROs were selected, ranging from three inverters, giving the maximum frequency possible in accordance with the intrinsic delays of the FPGA employed (400–700 MHz), and up to 1001-inverter oscillators, giving a much lower frequency (around 800 KHz). The system implemented on the chip starts operating immediately when the FPGA core voltage is connected. Using a wide range of ROs enabled us to measure the frequency and the internal delay of a real, de-facto system on a chip. This allows seeing the frequency dependence of each failure mechanism without any recovery effect. The set of ROs consisted of:

- 150 oscillators of 3 stages
- 50 oscillators of 5 stages
- 20 oscillators of 33 stages
- 3 oscillators of 333 stages
- 1 oscillator of 1001 stages

It is important to note, here, that the size of the ring determines the interdependence of any degradation. The shortest oscillators containing only three stages will have the greatest variability as well as the highest frequency. This is because a shorter critical electrical path will be much more sensitive to minor variations that lead to greater or smaller degradation over time. This means that the lower frequency oscillators containing as many as 1001 stages will average out the effects of individual degradations. Furthermore, the random statistical variability of individual devices will be exaggerated by the statistical distribution in wear-out slopes seen at high frequencies. Thus, we made 150 of the smallest ring size devices, which would need to be averaged to find the average degradation at those frequencies exhibiting more random times to fail. Interestingly, we see that the variability of three-ring oscillators is quite diverse, nearly randomly distributed about an average, whereas the lower frequency rings are much more narrowly distributed, indicating a more predictable time to fail, as compared to circuits having a much shorter critical path.

3.1. Testing methods

The testing system was synthesized and downloaded to the FPGA card. The test conditions were predefined for allowing separation and characterization of the relative contributions of the various failure mechanisms by controlling voltage, temperature, and frequency. Extreme core voltages and environmental temperatures, beyond the specifications, were imposed to cause failure acceleration of individual mechanisms to dominate others at each condition, for example, sub-zero temperatures, at very high operating voltages, to exaggerate HCI.

For each test, the FPGA board was placed in a temperature-controlled oven, dedicated to the MTOL testing, with an appropriate voltage set at the FPGA core. The board was connected to a computer via USB and the external clock signal was fed into the chip. The tests performed for 200–500h, while the device was working in the accelerated conditions. Frequencies of every ring oscillator, of different sizes, were measured. Initial sampling started after one working-hour in the accelerated environment, and then, samples were taken automatically at 5-min intervals. The frequency measurement data were stored in a database from which one could draw statistical information about the degradation in the device performance.

The acceleration conditions for each failure mechanism allowed us to examine the specific effect of voltage and temperature versus frequency on that particular mechanism at the system level and thus define its unique physical characteristics even from a finished product. A close inspection of test results yielded more precise parameters for the acceleration factors (AF) equations and allowed adjusting them to the device under test. Finally, after completing the tests, some of the experiments with different frequency, voltage and temperature conditions were chosen to construct the MTOL Matrix.

3.2. Separation of mechanisms

Our tests for various mechanisms included exposing the core of the FPGA to accelerating voltages above nominal. About 45nm defines the nominal voltage at 1.2V and for 28 and 20nm, 1.0V. Our method of separating mechanisms allowed the evaluation of actual activation energies for the three failure mechanisms, which are hot carrier injection (HCI), bias temperature instability (BTI) and electromigration (EM). We plotted the degradation in frequency and attributed it to one of the three failure mechanisms.

We need to justify our approach for accounting for current in the devices. Both-and HCI have J^{γ} factors; however, in a completely packaged CMOS digital circuit, there is no way to directly measure current, I, or as current density, J. We assume, in our experiments, that the stored gate charge strictly determines current transferred for each switching transition, that is, from a 0 to a 1 and vise verse. Whatever the current is for any state-transition will be same for each transition and the current will be, therefore, directly proportional to the frequency. Hence, the degradation for each transition will be directly proportional to the measured frequency, f. The voltage exponent will depend on the frequency, but the exponent, γ, measured will be the effective voltage acceleration parameter and comes into the equations for EM and HCI at $f^{*}V^{\gamma}$.

The results of our experiments give both E_A and γ for the three mechanisms we studied at temperatures ranging from −50 to 150°C. The Eyring model [1] is utilized here to describe the failure in time (FIT) for all of the failure mechanisms. The specific FIT of each failure mechanisms follows these formulae:

$$FIT_{HCI} = fV^{\gamma_{HCI}}e^{-\frac{Ea_{HCI}}{kT}} \tag{5}$$

$$FIT_{BTI} = e^{\gamma_{BTI}V}e^{-\frac{Ea_{BTI}}{kT}} \tag{6}$$

$$FIT_{EM} = fV^{\gamma_{EM}}e^{-\frac{Ea_{EM}}{kT}} \tag{7}$$

The degradation slope, α, is measured as the degradation from initial frequency as an exponential decay, approximated by taking the difference in frequency, and divided by initial frequency over the time. In our experiments, we found that when the decay was dominated by BTI, the decay was proportional to the fourth root of time, while HCI and EM, being diffusion-related mechanisms, have decay that is proportional to the square root of time [2], as seen in **Figure 3**.

In the 45nm boards each oscillator, the ring frequency was measured and plotted against the square root of time in 45nm devices. The slope, α, was then converted to a FIT for each test as determined by extrapolating the degradation slope to 10% degradation from its initial value. Each set is plotted as an exponential decay dependent on the square root of time as shown by example in **Figure 3**. This slope is then used to find the time to fail as seen in the development of FIT below (Eqs. (8)–(11)). We defined the exponent as $1/n$ so that we can apply the degradation based on the square-root law, as it does the degradation that is dominated by HCI or EM. For 28 and 20nm technology and below, we found that $n=4$ fits much more closely, as seen later with our 1000h evaluation.

$$\alpha_{(slope)} = \frac{\Delta f}{f_0 \times \Delta\sqrt{t}} \tag{8}$$

$$TTF = \left({}^{0.1}/_{\alpha}\right)^{1/n} \tag{9}$$

$$FIT = {}^{10^9}/_{TTF} \tag{10}$$

$$FIT = 10^9 (10 * \alpha)^n \tag{11}$$

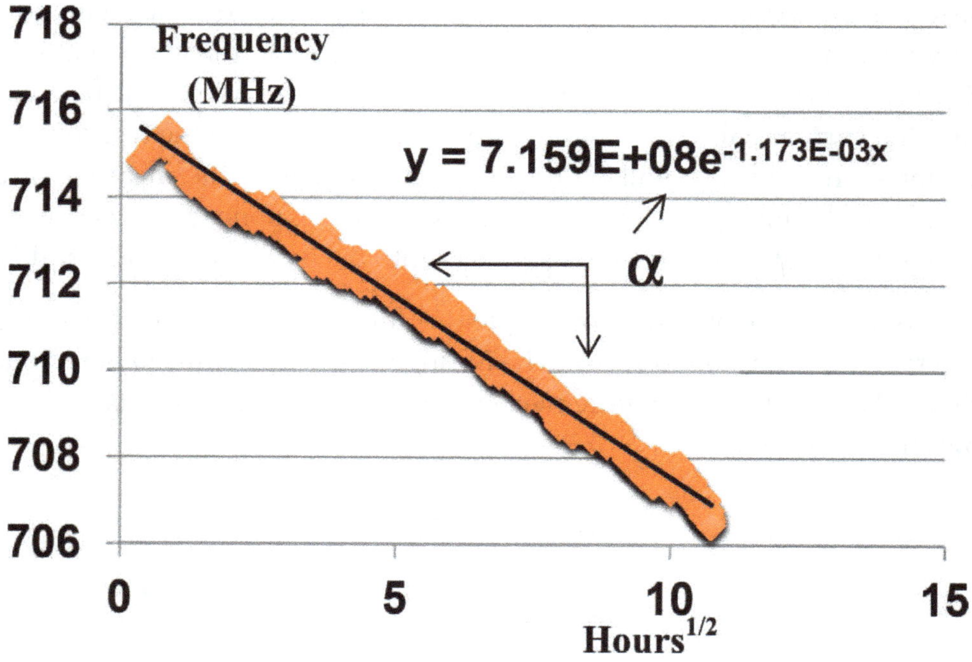

Figure 3. Typical frequency versus square root of time showing degradation slope α.

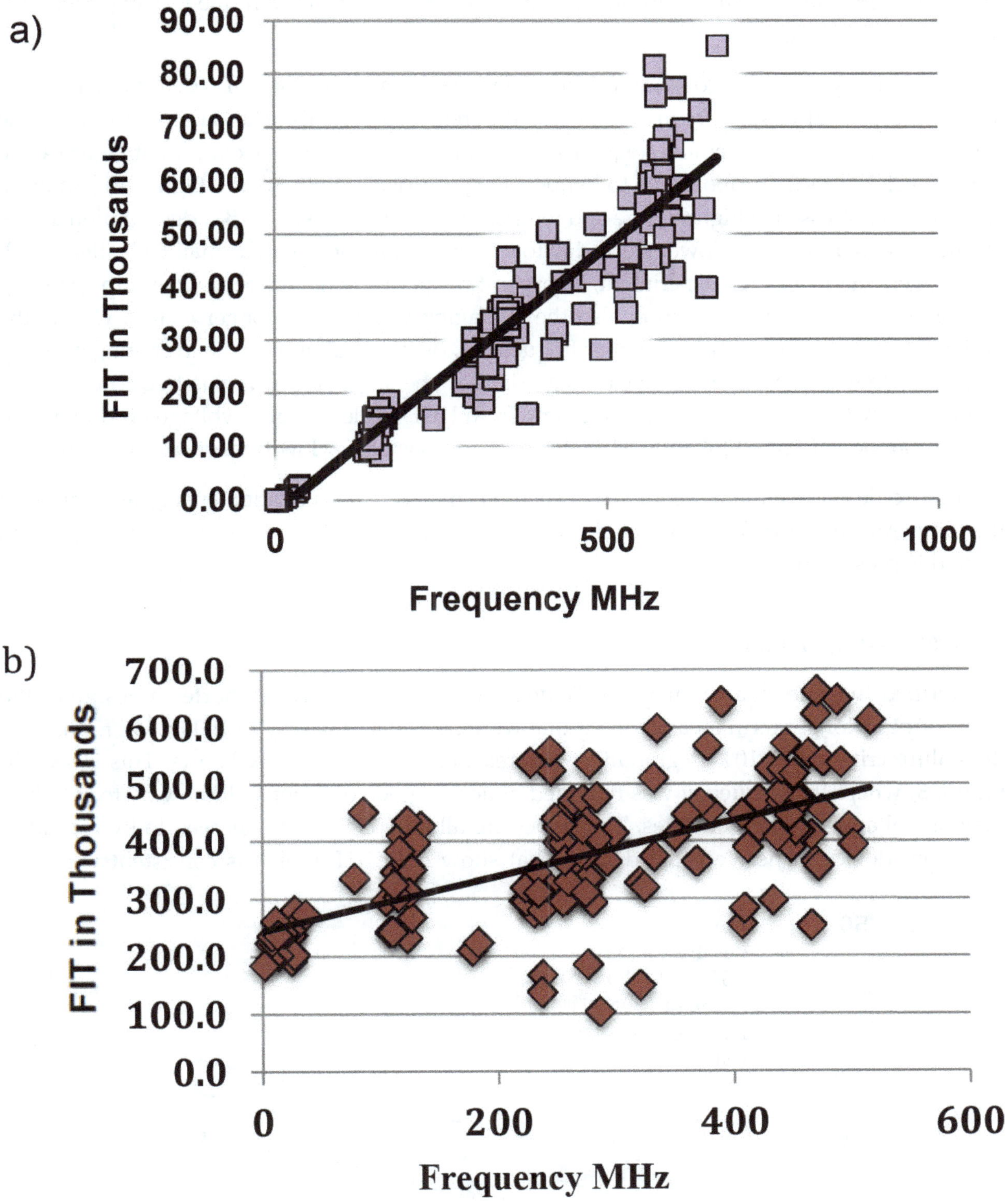

Figure 4. Failure rate, FIT/1000, versus frequency in MHz for (a) HCI, stressed at −35°C with 2.0V core voltage and (b) BTI, stressed at 145°C with 2.4V at the core.

The time to fail (TTF) was then calculated for as the square of the inverse slope times the failure criterion, which is 10% degradation in the 45nm technology [1]. Hence, the FIT for each slope is simply determined as the $(10^*\alpha)^2$, where $n = 2$. The average FIT is the metric to determine the reliability since that corresponds to the MTTF in Eq. (9). This FIT value is plotted as a function of the frequency in order to determine the failure mechanisms and to fit the model parameters for each mechanism.

Two typical degradation plots are shown in **Figure 4**(**a** and **b**), the FITs, determined by the slopes, are plotted against frequency in two different experiments. The data demonstrate the clear advantage of RO generated frequencies in a single chip [4]. In the examples of **Figure 4**, we see that FIT is directly proportional to frequency [6], consistent with Eq. (5). **Figure 4(b)** shows a chip that was stressed at high voltage and temperature showing a strong BTI degradation at low frequency and a much shallower slope due to EM in combination with a small HCI effect. Such curves were made for each experiment, incorporating all the oscillators across the chip spanning the range of frequencies, reflecting also the averaging effect of the longer chains. Hence, the variability is much lower than at higher frequencies, demonstrating that the averaging of many variations results in a consistent mean degradation. The slope of FIT versus frequency is then related at low temperatures as occurring only from HCI, while at higher voltages and temperatures, it can be due to BTI [6] and EM. BTI is only responsible for low frequency degradation.

In order to determine the dependence of each mechanism, the activation energy as relating to the temperature factor (T_F) and voltage acceleration factors (V_F) is determined from Eqs. (2) to (4) and is presented in Ref. [6].

3.3. 1000h extrapolation

We verified that the measurement to 1% degradation over relatively shorter times gives the same slope as longer term measurement that were carried all the way to 1000h. We found that the failure criterion of 10% degradation was reached in these ring oscillators. This is seen in **Figure 5**, where the frequency was recorded at accelerated conditions all the way to 1000h at various voltage and temperatures. The slopes are all very close to $t^{1/4}$, as seen to by typical of the 28nm devices. These are the devices that show only BTI, which is consistent with a ¼

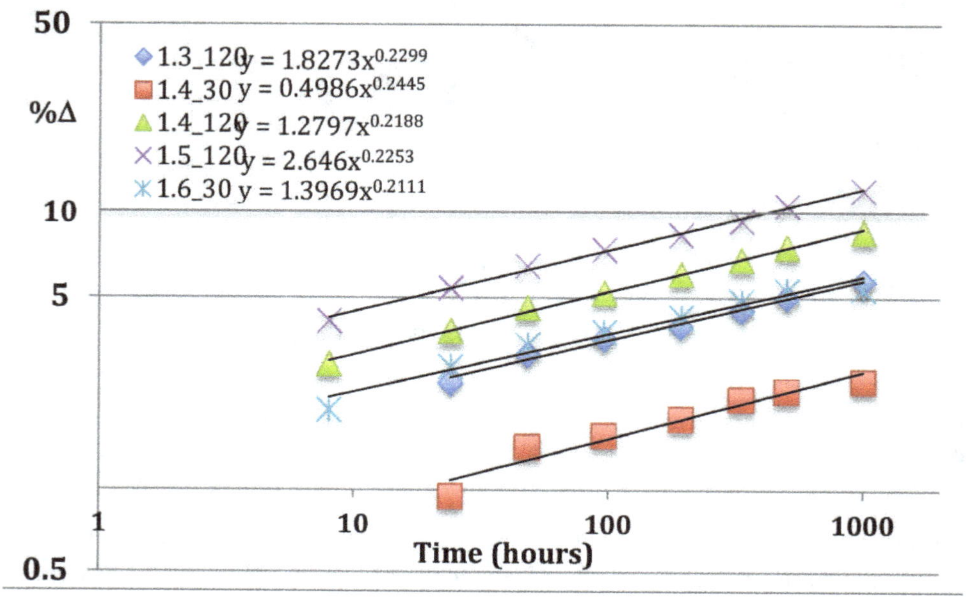

Figure 5. 1000h degradation data for 28nm devices over a range of core voltages from 1.3 to 1.6V at 30 and 120°C as indicated.

power law signature. Furthermore, we see that the initial degradation extrapolates to the 10% failure criterion, verifying the approach to measure for only a few hundred hours instead of requiring a complete 1000-h test.

4. Linear matrix solution

We assume here that the linear, Poisson, model for constant rate is associated with the probability of failure for each separable mechanism. As we showed in Eq. (3) above, each FIT adds linearly to the other FITs in order to obtain an average system failure rate. By observation of the procedure in **Figure 1**, it is clear that each FIT will have its own value that is uniquely determined by the acceleration factor for each mechanism depending on the voltage (*V*), temperature (*T*), and operational fequency (*F*). For this example, we found that there was no evidence of a time-dependent dielectric breakdown (TDDB), and therefore, we included only HCI, BTI, and EM.

This approach is exactly what JEDEC describes as a sum of failure rates methodology as it sums the expected failure rate of each mechanism distinctly from the other mechanisms. The combination results from actual accelerated life tests where there is an extrapolated mean time to fail based on the known operating conditions of *V*, *T* and *F*. Hence, we are sure to test at a large range of Temperatures, including very high and very low temperatures as well as core voltage as high as will be practical to achieve reliable operation.

Of course, we assume that each component is composed of multiple sub-components, for example, a certain percentage is effectively ring-oscillator, static SRAM, DRAM, etc. Each type of circuit, based on its operation, can be seen to affect the potential steady-state (defect related) failure mechanisms differently based on the accelerated test conditions. However, unlike traditional reliability systems, rather than treat each sub-system or component as an individual source with a known failure rate, we separate the system into distinct mechanisms that is known to have its own acceleration factor with voltage, temperature, frequency, cycles, etc. Hence, the standard system reliability FIT can be modeled using traditional MIL-handbook-217 type of algorithms and adapted to known system reliability tools; however, instead of treating each component as individuals, we propose treating each complex component as a series system of various mechanisms, each with its own reliability.

The matrix is arranged as in **Table 1**. The three most left-hand columns show the temperature, *T*, voltage, *V*, and frequency, *F*, used for the accelerated test. The measured value for FIT is then put in the third column from the right, after the relative (un-normalized) calculations for each mechanism are placed below the column describing the mechanism. Here, they are labeled HCI, BTI, and EM. Any three rows can be used to solve the matrix, and the product of the solution parameters is then put in the FIT column on the right-hand side of the matrix. The three rows that are used to solve the matrix will then have exactly the same solution as the measured FIT values used to calibrate the matrix.

The second from the right-hand column shows the ratio of the extrapolated failure rate and the calculated FIT. These values serve to show the closeness of fit to the model parameters by comparing the other measured FIT values with the calculations. This matrix will have a unique solution that will fit the percentages of each mechanism (P_i) with the measured failure rate, FIT.

T (°C)	V	F (GHz)	HCI	BTI	EM	Measured	Ratio	FIT
−62.5	1.2	1	99.99%	0.01%	0.00%	30	94%	2.83E+01
125	1.2	1	0.00%	86.86%	13.14%	997.4	102%	1.01E+03
153	1.2	1	0.00%	63.79%	36.21%	3672	100%	3.67E+03
−35	2.5	0.5	100.00%	0.00%	0.00%	23,750,000	100%	2.38E+07
154	1.2	0	0.00%	100.00%	0.00%	2420	100%	2.42E+03
140	2.2	0	0.00%	100.00%	0.00%	66,200	102%	6.76E+04
−22.5	2.8	1	100.00%	0.00%	0.00%	240,000,000	101%	2.43E+08
7.3	3	1	100.00%	0.00%	0.00%	156,000,000	106%	1.66E+08

Table 1. Test results showing proportions of failure mechanisms for given V, T, and F compared with the calculated as well as the measured failure rate (FIT).

	Inverse matrix		P_i
−4.36972E−29	4.76285E−18	−1.10403E−20	1.13118E−10
0	0	10,946.04333	26,489,424.87
1.19767E+14	−2040.515932	−1.15932E+14	1.59226E+17

Table 2. Inverse matrix (left three columns) and respective P values (right-hand column).

Once the parameters for the three mechanisms have been calculated and verified against the other test data, a full set of extrapolated values for FIT can then be calculated using the equations for each mechanism times the same P values used to fit the three exemplary rows. **Table 2** shows the inverse matrix of the values under the three mechanisms with the corresponding P values for HCI, BTI, and EM, respectively.

Since the matrix is linear, as are the calculations for FIT at any given T, V and F, then the full matrix of actual FIT calculations is simply the sum of each P value times the calculated relative significance of each mechanism. A calculated reliability curve is shown in **Figure 6** across the full range of expected FIT versus temperature for any set of operational conditions shown in **Table 3**.

The unique solution that solves all three equations with the three extrapolated acceleration factors gives a percentage contribution for each of the failure mechanisms. We report the reliability as FIT, which is 10^9/MTTF for each condition. The percentages for each mechanism are shown, based on the relative contributions that were extrapolated from the physics of failure equations normalized to the measured FIT of each test. Seeing the dispersion of FIT values per test proves that the approximation of a constant rate, meaning a random distribution in time, is the proper statistical model for these results. **Figure 6** shows the resulting FIT as plotted versus temperature (°C) for the measured 45nm technology FPGA.

One advantage of plotting our data as failure rate versus temperature allows one to determine effective activation energy as a function of temperature and stressor parameters, V and F. The principle follows the assumption that the failure rate, λ, is exponentially dependent on the activation energy divided by the absolute temperature, T:

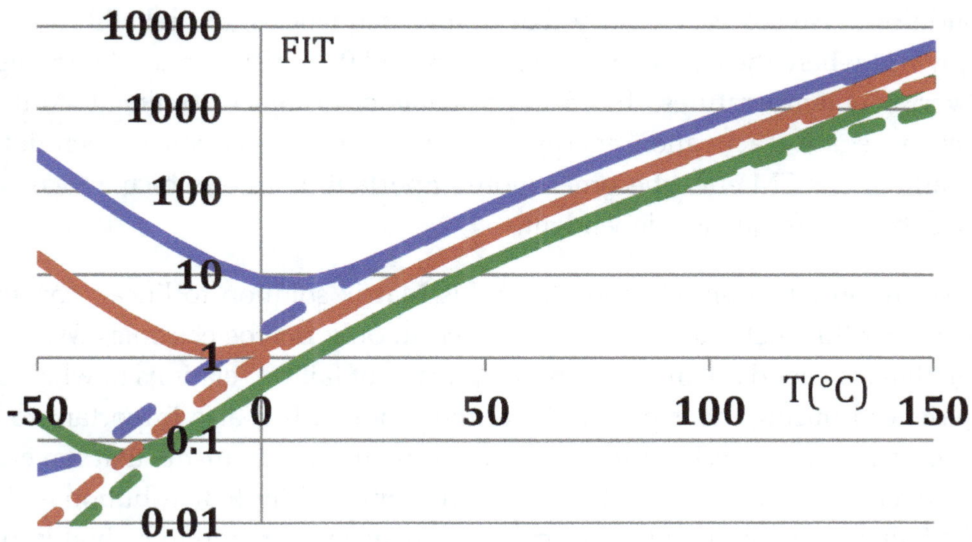

Figure 6. Reliability curves for 45nm technology showing FIT versus temperature for voltages. These curves are for 1.0, 1.2, and 1.4V core voltage at 10MHz (dashed) and at 1GHz (solid).

T (°C)	*V*	*F* (GHz)	HCI	BTI	EM	FIT
−50	1.2	2	1.45382E+11	2.84438E−10	1.99008E−27	16.5
−10	1.2	2	6,131,362,305	1.61083E−08	2.65294E−23	1.1
20	1.2	2	1,006,254,891	1.61337E−07	6.00169E−21	4.4
30	1.2	2	596,524,778.1	3.14239E−07	2.8808E−20	8.4
40	1.2	2	365,644,331.1	5.86524E−07	1.2509E−19	15.6
50	1.2	2	231,020,972.9	1.05325E−06	4.95957E−19	28.0
80	1.2	2	68,110,854.71	4.99845E−06	1.9353E−17	135.5
100	1.2	2	33,650,811.61	1.22819E−05	1.60476E−16	350.9

Table 3. Calculated FIT based on the solved matrix for typical use conditions.

$$FIT = \lambda = \exp\left(-\frac{E_{A,eff}}{kT}\right) \tag{12}$$

If we assume that $E_{A,eff}$ is a function of *V* and *F*, we can calculate this from our solutions shown above,

$$-\frac{d\lambda}{d\left(\frac{1}{kT}\right)} = E_{A,eff} * \lambda \tag{13}$$

Hence, if we plot the change in λ by (1/kT) divided by λ at any temperature, *T*, we get

$$E_{A,eff} = \left(\frac{1}{\lambda}\right) * \frac{-d\lambda}{d\left(\frac{1}{kT}\right)} \tag{14}$$

The advantage of this representation allows a designer to consider the temperature rage as a function of stressor factors that would affect the reliability of a product, especially under extreme conditions. We see very clearly that at low frequencies, the reliability is completely dominated by BTI where the activation energy is around 0.53 eV, whereas at very high temperatures and very low temperatures, the effect of frequency becomes dominant. At the very low temperatures, a negative activation energy is seen for higher frequency operation, while at high temperatures, the EM effect becomes more important, both of which are current-related effects; hence, they are frequency dependent.

What is most important to understand about this Matrix solution to linear, constant, failure rate models is that this methodology is not limited to only microelectronics. We must understand that all that is needed are the appropriate physics of failure relations to whatever stresses will be experienced during the expected life of the product. It is also important to know that this method of combining mechanisms is limited to failure mechanisms that have a generally constant rate over time. That is to say that the slope from a Weibull distribution is close to 1. If, however, the failure distribution of a particular mechanism is known to be highly predictable, that is with a wear-out characteristic, having a Weibull slope of 2 or more, then this methodology will not properly work to combine mechanisms. On the other hand, if one mechanism is known to dominate or be the limitation, that one mechanism can be separated from the other more random mechanisms, as shown in **Figure 7** and based on our extrapolation from **Table 3**.

One clear proof from this graph is that it is not possible to choose simply one accelerating temperature and voltage, or any one condition for any accelerated test, expecting that a simple extrapolation can be made based on a single failure mechanism. The mechanisms interact such that any accelerated test will surely give incorrect results, and, thus, the traditional HTOL test is obviously not sufficient for reliability prediction. Furthermore, the MTOL, multiple stressor qualification will give accurate prediction for the failure rate under any given operating conditions from a fraction of the number of samples tested over a much shorter period of time. Hence, this methodology will save a large proportion of the standard qualification procedure and give much more accurate and meaningful results.

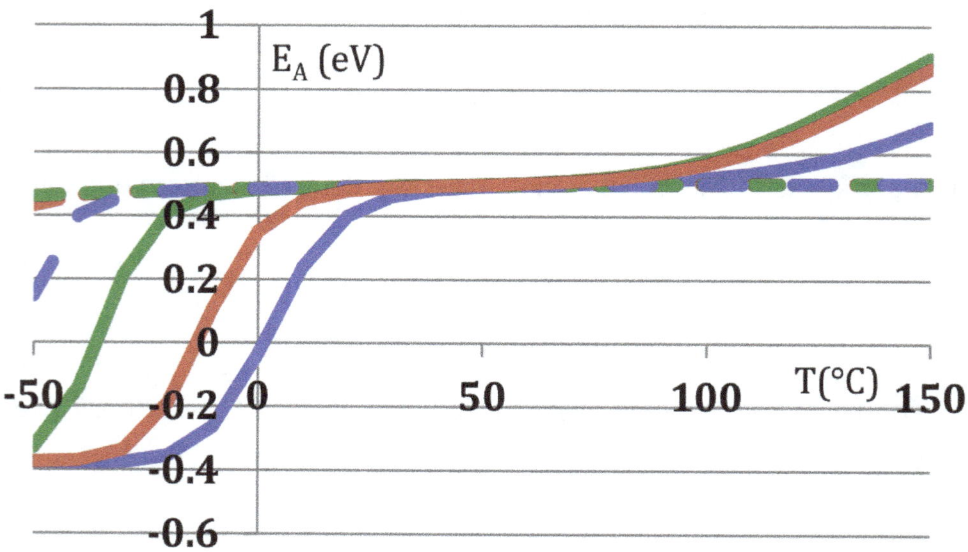

Figure 7. Activation energy versus temperature based on the data above in **Figure 6** for the same voltages, 1.0, 1.2, and 1.4 V on the 45 nm FPGA technology.

Author details

Joseph B. Bernstein

Address all correspondence to: joeybern@gmail.com

Ariel University, Ariel, Israel

References

[1] Xilinx, Device Reliability Report, UG116 (v10.3.1), 8 September 2015 (As an example)

[2] Bernstein JB. Reliability prediction from Burn-in Data fit to reliability models. 2014 (ISBN–10: 0–128007–47–8) Academic Press

[3] Bernstein JB. et al. Physics-of-failure based handbook of microelectronic systems, Reliability Information Analysis Center, Utica, NY; 2008 (ISBN-10: 1 933904–29-1)

[4] Bernstein JB, Gabbay M, Delly O. Reliability matrix solution to multiple mechanism prediction. Microelectronics Reliability. 2014;**54**:2951–2955

[5] Bernstein JB, Reliability Prediction for Aerospace Electronics Descriptive Note : Final rept. 15 Jul 014-14 Apr 2015 Accession Number : ADA621707

[6] Joseph B, Bernstein A, Bender BE. Reliability prediction with MTOL. Microelectronics Reliability. 2017;**68**:91–97

3

Coordination and Selectivity of Protection Devices with Reliability Assessment in Distribution Systems

Marco Antônio Ferreira Boaski, Caio dos Santos,
Mauricio Sperandio, Daniel Pinheiro Bernardon,
Maicon Jaderson Ramos and Daniel Sperb Porto

Abstract

This chapter provides an overview of the reliability of electricity distribution networks, and its evaluation that is linked with the protection system. In this way, the characteristics of network protection are presented, along with the peculiarities in coordination and device selectivity adjustments. For the assessment of the reliability, we have the methodology of logic-structural matrix (LSM) that integrates the constitution of the network with historical data of faults, so that with this, a model can be elaborated that can evaluate the impact of changes in the system directly on the reliability indicators.

Keywords: distribution network reliability, coordination and selectivity, protection system

1. Introduction

With the latest technologies and new concepts that have emerged, the electricity distribution system has become more flexible, but all this is reflected in general modifications of operation, planning, study, and analysis [1]. One of the great influencers in this environment is the smart grids, which changed the vision of a strongly static network and inserting versatility in the structure of the networks, using technologies, automation, and methodologies in a coupled manner [2, 3]. The uses of smart grid are wide ranging from load control, self-healing, voltage control, among others [4].

With the possibilities of this development, it inserts new complexities in the network and this can have both positive and negative impacts, due to which there are several factors or functionalities that must be reevaluated. Considering the main function of the power utilities being the transportation of energy to the final customer, we seek to use these new technologies and methods to increase the quality and safety of the service provided. In this way, one of the criteria most considered by the companies is the reliability of the network, both for strategic issues of network operation, and for the existence of financial penalties for not meeting targets [5–7].

The reliability of distribution networks is influenced by several factors like operation, maintenance, and planning [8], but as the reliability of the networks is directly linked to the interruptions in the power supply, one of the systems which has more impact is the protection [9]. Along with new automation, control, and communication technologies, some protection instruments have also advanced [10, 11], enabling reclosing, and different performance curves. In this way, the methodologies for adjusting the coordination and selectivity parameters of the devices must also consider a greater variety of factors, both those that can impact and those that can be impacted by this system.

Thus, it is proposed to evaluate the reliability of networks through their indicators considering the protection system. This will explore the main protection devices and how to obtain the main quality indicators, as well as verify the functionalities and main steps for coordination and selectivity of the device, in addition to modeling the network together with the factors that will directly impact the reliability of the distribution system.

2. Distribution network protection devices

The distribution systems of electric energy are composed of long networks, and are mostly aerial and made up of bare cables, due to the extension and exposure can cause failures or faults of various natures. Therefore, means for the protection of these systems become essential, for that, the protection devices are used. Moreover, the most used in distribution networks are the fuses and reclosers [12], which are discussed in the following subsections in detail.

2.1. Fuses

The protective devices that are most used in power distribution networks are the fuses. This is due to its low cost when compared to other protection devices and its satisfactory operation for one of the major problems of the distribution networks that refer to the overcurrent from short circuits by contact of the cables between each other, the vegetation or to the ground, and among others.

As a fuse-type protective device, it bases its operation on a metal link with specific characteristics of time versus current; when it reaches the maximum tolerable current, the heat melts the active element and releases the opening of circuit [13].

The construction of a fuse is divided into three elements: *Base*—Consists of an insulating material and serves as an interconnection between the moving parts and the support structure of the device. *Fuse tube*—Consists of an insulating material, serves as a support for the fuse link and is the moving part that promotes the opening between the terminals when the link fuses. *Fuse link*—Consisting of metal alloys with specific characteristics of melting temperature, it is the active element of protection of the equipment, having different types of links, with faster or slower curves depending on the application, the most common are types H and K. One example of fuse is represented in **Figure 1**.

2.2. Reclosers

Compared to fuses, reclosers have a relatively high cost, but these devices are more sophisticated and offer wider capabilities in protection, due to the measurement, automation, and the control possibilities. The reclosers have been increasingly used by electric power distributors; this happens due to the possibility of control and telecommunication of this device [14]. Through these characteristics, one can have a real-time control of the network allowing maneuvers for various purposes, thus contributing to the evolution of the smart grid.

The reclosers allow timed automatic reclosers; though this feature associated with fast curves, it is possible to minimize power interruptions caused by transient faults [15] such as contact of tree branches. In most devices, the timing system can be adjusted with fast or slow reclosing operations, or combinations thereof, depending on the need and philosophy of the company. This protection against transient faults also includes the networks protected by the fuses downstream of the reclosers. The construction of reclosers is more complex, and has a chamber for interrupt of electric arc, and there is current sensors and in some cases voltage sensors. One example of recloser is shown in **Figure 2**.

Regardless of the control and communication factors, the operating principle of the recloser is also related to the lack of current and its performance has a behavior described by current curves versus time. However, there are commercial models that are also made up of potential transformers; with this and the ability to communicate, these can protect the distribution system from other possible types of failures.

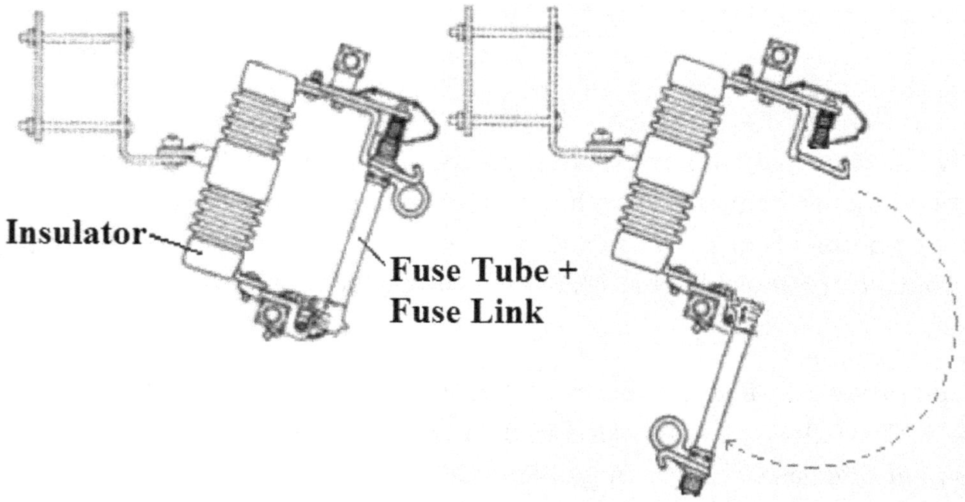

Figure 1. Medium voltage fuse.

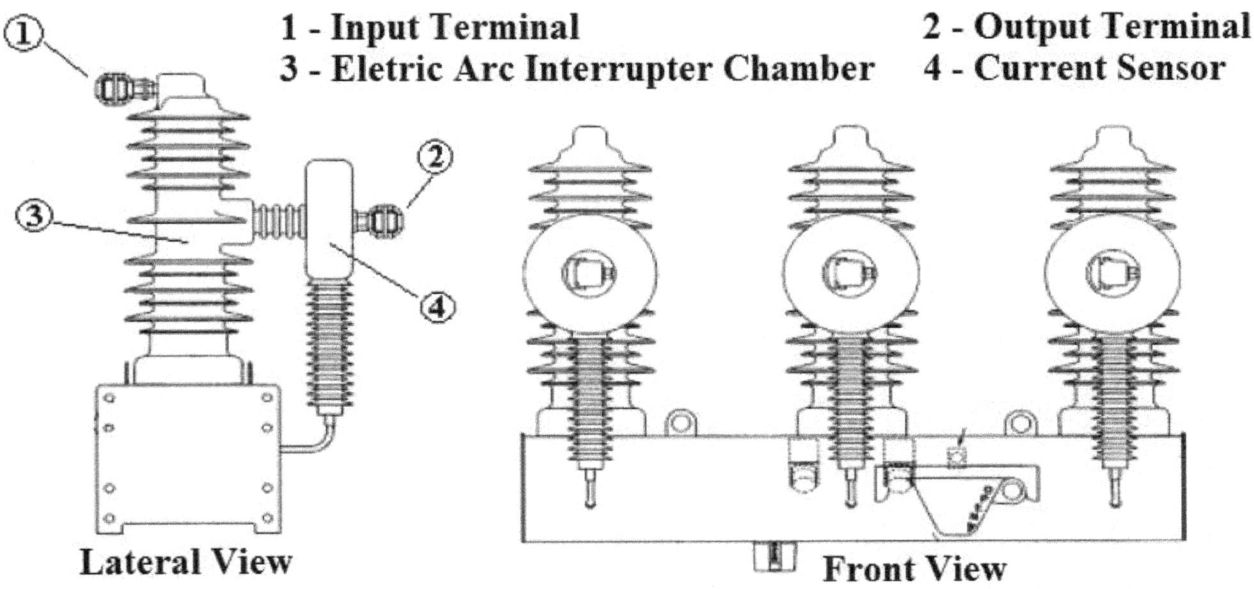

Figure 2. Recloser.

2.3. Coordination and selectivity

For proper protection of the electrical system, in addition to the use of appropriate devices, must be taken care with the sequence of operation of such equipment. This is necessary due to the fact that along the same network there are several devices and if this sequence of operation does not agree, there will be disconnection of undue loads, thus compromising the reliability of the system and may negatively influence the continuity indicators of the system from the power utility [16, 17].

The selectivity that refers to the sensitivity of the protection devices is evaluated first. This need arises since, that in addition to the interruption capacity of the devices, they need to be sensitized to operate in the minimum short circuit and at the same time allow the normal load current.

The criteria for selectivity of fuses can be summarized by Eq. (1) below:

$$K \cdot I_n \leq I_e \leq \frac{1}{4} \cdot I_{ccmin} \tag{1}$$

where

I_n–nominal load current of the circuit(A);

I_e–nominal current of the fuse link;

I_{ccmin}–minimum value of short-circuit current (A); and

K–demand growth rate.

For reclosers, since they have both phase and neutral settings, the settings can be expressed with the following Eqs. (2) and (3)

$$1.5 \cdot I_L \leq I_{pf} \leq \frac{I_{2\Phi F}}{2} \quad (2)$$

$$(0.1 \approx 0.3) \cdot I_c \leq I_p \leq \frac{I_{1\Phi m}}{2} \quad (3)$$

where

I_L–nominal load current of the circuit (A);

I_{pf}–pick-up current of phase, from recloser;

$I_{2\Phi F}$–biphasic current of short circuit;

I_p–pick-up current of neutral, from recloser; and

$I_{1\Phi m}$–single-phase current of short circuit.

In the other step, concerning coordination refers to the evaluation of the sequence of operation of the devices besides evaluating which devices downstream will be sensitized for upstream faults. This process seeks the best sequence of equipment performance, seeking to minimize the area affected by the fault or defect. In order to have the correct sequence, some criteria are used, depending on the type of equipment involved.

For coordination between fuses, we can cite three main criteria:

1. The nominal current of the protected link must always be higher than the rated current of the protective link.
2. Ideally, the protected fuse link (source side) must be coordinated with the protective link (load side) so as not to open first for the maximum short-circuit current at the point of installation of the protective link.
3. The coordination between two serial fuse links is guaranteed if the interruption time of the protective link is at most 75% of the minimum time of fusion of the protected link.

For coordination between fuses and reclosers, we can cite three main criteria that are explained as follows:

1. For all possible fault values within the circuit section protected by the fuse link, the minimum link fusion time must be greater than the recloser opening time multiplied by a factor K characteristic of the recloser, which varies depending on the number of fast operations adjusted and the reclosing time of the circuit.
2. For all possible fault values within the circuit section protected by the fuse link, the total time of the link interruption must be less than the minimum opening time of the recloser in its timed curve (slow curve), by adjusting the recloser for two or more timed operations.
3. In the case of not being able to coordinate between recloser and link for the whole range of short-circuit currents, it will be guaranteed at least the coordination for the condition of

single-phase faults involving contact impedance and maintained the selectivity to 80% maximum circuit.

3. Reliability in distribution networks

The term reliability has broad meaning and can refer to different applications in the same system. The authors of Ref. [18] define that the reliability of a system refers to the correct operation with full performance and no failure. Thus, the reliability is influenced by several factors, and may be manageable, such as planning, maintenance, and operation or unforeseen events such as storms or accidents.

For the main current regulations, the concept of reliability in the distribution of electricity is linked to interruptions in energy supply, which may be temporary (momentary) or sustained (permanent) [19]. This can be considered a subgroup of the perturbations that affect the quality of energy, as shown in **Figure 3**. The main disturbances that affect the quality can be frequency variations, noise, transient, harmonic distortions, temporary variation of voltage amplitude and properly the interruptions [20].

For the measurement and evaluation of the reliability of the distribution networks, the reliability indicators are used, these will be explained next. The evaluation of the power utilities is through these indicators, failure to comply with the stipulated values causes a penalty for the company and a discount to the final consumer. Thus in power systems, it is important to keep the reliability indicators at good levels [21].

3.1. Reliability indicators

The calculation of the reliability indicators are values that synthesize statistical aggregates, which can be calculated from the historical interruptions that occur in certain regions or groups of consumers of the distribution system [22, 23].

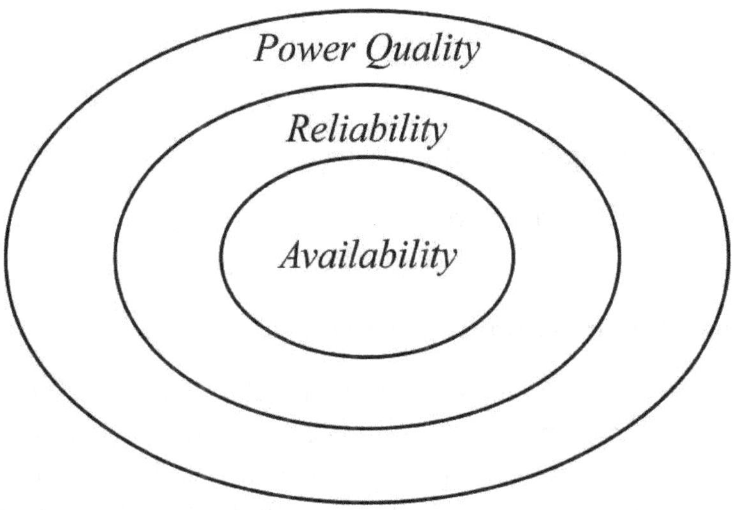

Figure 3. Power quality and subgroups.

The main reliability indicators are defined in the guide [24], among which the most used are:

- SAIFI—System average interruption frequency index: This index considers the number of interruptions on a media that consumer or a group or consumers suffer during a period.

$$\text{SAIFI} = \frac{\sum_i N_i}{N_T} \tag{4}$$

where

N_i: number of consumer units affected at each interruption; and

N_T: total number of consumer units of the group.

- SAIDI—System average interruption duration index: This index considers the duration of the interruptions, thus having an average of hours that a given consumer has his energy supply interrupted during the period.

$$\text{SAIDI} = \frac{\sum_i (r_i \cdot N_i)}{N_T} \tag{5}$$

where

N_i: number of consumer units affected at each interruption;

r_i: duration of each interruption; and

N_T: total number of consumer units of the group.

- ENS—Energy not supplied index: This index relates the total of energy not supplied in all group due interruptions in all the events in a given period.

$$\text{ENS} = \sum_i r_i \cdot L_i \tag{6}$$

where

r_i: duration of each interruption; and

L_i: energy not supplied in each interruption.

3.2. Modeling of reliability in distribution networks

To model the distribution network to calculate its reliability, it is important to be aware that there may be faults that impact different amounts of customers, such as an external fault that affects all consumers or a one fault on a transformer or in a point of the network that only affects a small group [25]. Then there will be faults with different levels of comprehensiveness.

For this modeling, there are different methods such as block classification and analytical simulation [26] and the logic-structural matrix (LSM). The method that best fits with the protection equipment is the logic-structural matrix, due to the presence of the maneuverable equipment installed in the network, on the different groups of consumers. By switching, the equipment can isolate faults with consumers, or reestablish them, based on the maneuvers allowed for fault isolation and network reconfiguration.

The logic-structural matrix is composed of the main data of the following:

- *Annual failure rate* (λ): This media is obtained through the history of failures of that group.
- *Mean time to restore power supply* (*TR*): This measure is the average time of the restoration of the energy supply, this time is composed of several phases such as the time of displacement of the maintenance team, time of repair, and even the time of waiting between the event of the failure and the authorization of the displacement of the maintenance team
- *Number of customers* (*N*): It refers properly to the number of consumers fed in that region, being able to be fed by transformers or directly connected in the primary network.
- *Load active power* (*L*): Active load, transformers or consumers connected directly to the primary network.

The LSM is composed as follows: the columns are equivalent to the protection or switching equipment of the system and each row is equivalent to the points of the system (these points can be divided according to the needs of the company; they can be transformers, primary consumers or networks extensions). In the cells of the logical-structural matrix, there are initial values of the mean time to power restoration. In order to define these values, it is required to analyze how long it takes to restore the power supply for the corresponding consumers (matrix line), when they are faced with a failure in the distribution network assuming the protective and switching equipments installed on the network (matrix column) [27, 28].

In the presence of switching equipment, one must evaluate the possibilities for switching, isolating defects or transferring loads through these devices. The first possibility is sectionalizing, what corresponds to the isolation of the segment under failure and other associated nodes downstream of a normally closed (NC) from nodes upstream. The mean time to isolate (TI) is computed for consumers on all these upstream nodes. The second option is the transfer of the nodes downstream from the NC switch; when an upstream fault occurs, then the mean time to transfer (TT) is considered for the consumers downstream. The last possibility depends on the existence of a normally open (NO) switch downstream from the NC, and the adjacent feeder must have available technical capacity to receive the loads that will be transferred. For manual switches, the TI and TT also include: mean time of wait (TW) and mean time to travel (TTr). For automatic switches, the TI and TT are much shorter, because there are not TW and TTr. Normally TR > TT > TI.

The protection devices prevent upstream faults or defects from affecting the nodes in downstream of the device. This way, downstream nodes do not have their power supply interrupted, then in the cells of LSM, in these nodes can be placed the number 0.

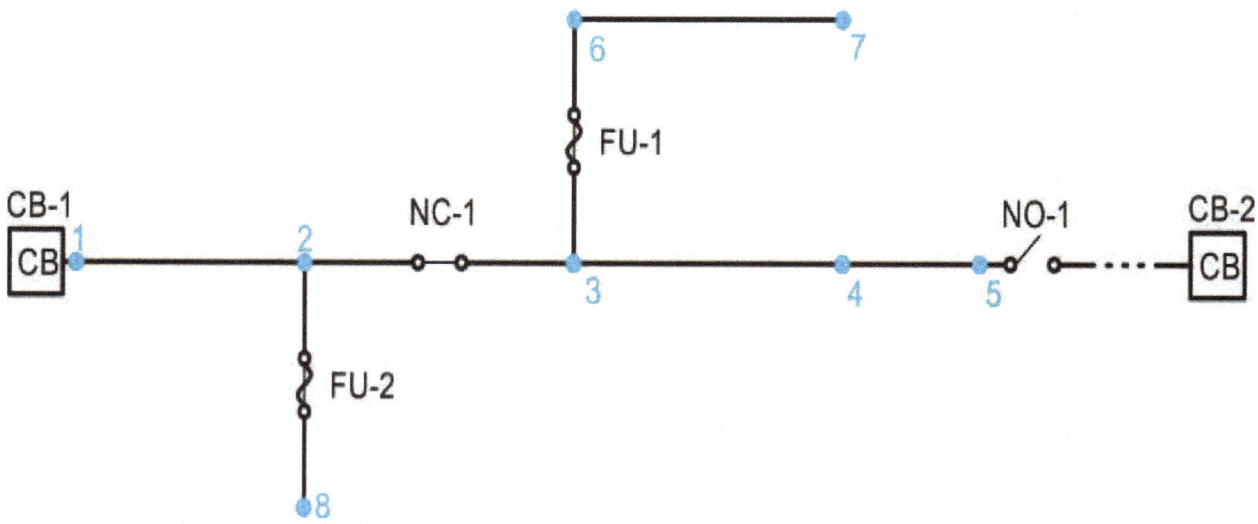

Figure 4. Example distribution network.

To illustrate, the logical-structural matrix for the simplified distribution network is shown in **Figure 4**. It is assumed that the NO switch at node 5 is connected to another feeder with the technical capability to receive loads downstream of the NC switch.

Table 1 shows the construction of logical-structural matrix for the example in **Figure 4**, considering the mean time to power restoration of node *i* (TR*i*), and constants time to isolation (TI), and time to transfer (TT) for each device.

One can note that, for the outage of the circuit breaker (CB-1), the total time to restore power for all consumers is computed, except to those downstream of the NC switch, for which the transfer time to another feeder is considered. For failures downstream of the NC switch, the time to isolate the fault for upstream consumers of the switch and the total time to restore

Nodes (distribution transformers or primary customers)	**Protective and switching equipment**			
	Circuit breaker CB-1	**Switch NC**	**Fuse FU-1**	**Fuse FU-2**
1	TR1	TI	0	0
2	TR1	TI	0	0
3	TT	TR2	0	0
4	TT	TR2	0	0
5	TT	TR2	0	0
6	TT	TR2	TR3	0
7	TT	TR2	TR3	0
8	TR1	TI	0	TR4

Table 1. Logical-structural matrix to the distribution network of **Figure 4**.

power to its downstream customers is computed. Regarding the outage of fuses (FU-1 and FU-2), it only affects its downstream consumers, so the total time to restore power is computed. The upstream nodes are not affected by the fault, not suffering interruption, since the fuse is coordinated to blow before the circuit breaker trips (trip saving scheme).

Then, the matrix values are multiplied by the failure rate of the respective equipment (λi), as shown in **Table 2**.

The reliability index is then calculated from the LSM. To calculate the expected value of SAIDI, the terms of each row of **Table 2** are added and then multiplied by the respective amount of consumers in that row, and then the results of all lines are added together and divided by the total number of customers served [28], as follows:

$$\mathrm{ESAIDI} = \frac{\sum_{i=1}^{n}\left(\sum_{j=1}^{m} M_{i,j}\right) \cdot N_i}{N_C} \tag{7}$$

where

ESAIDI = expected value of system average interruption duration index (h/year);

$M_{i,j}$ = element in row i and column j of LSM;

Ni = number of consumers for the row i;

N_C = total number of customers served;

n = number of rows; and

m = number of columns.

Nodes (distribution transformers or primary customers)	**Protective and switching equipment**			
	Circuit breaker CB-1	**Switch NC**	**Fuse FU-1**	**Fuse FU-2**
1	TR1 $\lambda 1$	TI $\lambda 2$	0	0
2	TR1 $\lambda 1$	TI $\lambda 2$	0	0
3	TT $\lambda 1$	TR2 $\lambda 2$	0	0
4	TT $\lambda 1$	TR2 $\lambda 2$	0	0
5	TT $\lambda 1$	TR2 $\lambda 2$	0	0
6	TT $\lambda 1$	TR2 $\lambda 2$	TR3 $\lambda 3$	0
7	TT $\lambda 1$	TR2 $\lambda 2$	TR3 $\lambda 3$	0
8	TR1 $\lambda 1$	TI $\lambda 2$	0	TR4 $\lambda 4$

Table 2. Logical-structural matrix with times versus failure rate.

The expected value of ENS is straight forward obtained by replacing the number of consumers in Eq. (8) by its respective load, active power of the distribution transformers, ignoring the total number of customers served

$$\text{EENS} = \sum_{i=1}^{n}\left(\sum_{j=1}^{m} M_{i,j}\right) \cdot L_i \tag{8}$$

where

EENS = expected value of energy not supplied (kWh/year);

$M_{i,j}$ = element in row i and column j of LSM;

L_i = average load, maximum demand of active power multiplied by the respective load factor, associated to row i (kW);

n = number of rows; and

m = number of columns.

To obtain the expected value of SAIFI, the process is similar to the SAIDI, requiring only replacement of the logical-structural matrix average times (TR, TI, and TT) by 1, so are considered only the failure rates.

$$\text{ESAIFI} = \frac{\sum_{i=1}^{n}\left(\sum_{j=1}^{m} M^{*}_{i,j}\right) \cdot N_i}{N_C} \tag{9}$$

where

ESAIFI = expected value of system average interruption frequency (failures/year);

$M^{*}_{i,j}$ = element in row i and column j of LSM, without considering the mean times;

N_i = number of customers for the row i;

N_C = total number of customers served;

n = number of rows; and

m = number of columns.

4. Assessment of protection considering the reliability

In order to evaluate the influence of protection on the reliability of the distribution system, the methodology is shown in the flowchart of **Figure 5** can be used. This generic model seeks that the protection devices can be adjusted for both coordination and selectivity quality. For a better understanding of how the impact of selectivity and coordination on reliability will be, we have an example below, the circuit is shown in **Figure 6**.

Considering a theoretical system, we have the base network shown in **Figure 5** and having the following considerations in the protection:

- *Circuit breaker* (CB-1): It is selective for short circuits in all nodes of the network.
- *Recloser* (R-1): It is selective for short circuits between nodes 4 and 5.
- *Fuse* (FU-1): It is selective for short circuits between nodes 8 and 9.
- *Fuse* (FU-2): It is selective for short circuits between nodes 6 and 7.

Faults may occur with transient (momentary) character as being sustained (permanent) [29]. With the coordination of protection devices, in addition to avoiding a complete shutdown of the system, the effect of transient faults on the network can be reduced through the reclosers. The transient faults are most of the faults that occur in the distribution network, reaching values between 80 and 90% of total faults [30]. To better illustrate this, two LSMs can be established, one for sustained faults (**Table 3**) and the other for momentary faults (**Table 4**).

With the idea of visualizing the impact of device coordination and selectivity on reliability, a simple change that could be implemented would be to selectively the R-1 device for nodes 6 and 7 together with the coordination between the R-1 device and FU-2. With these changes, we would have the following result in the sustained LSM (**Table 5**) and momentary LSM (**Table 6**).

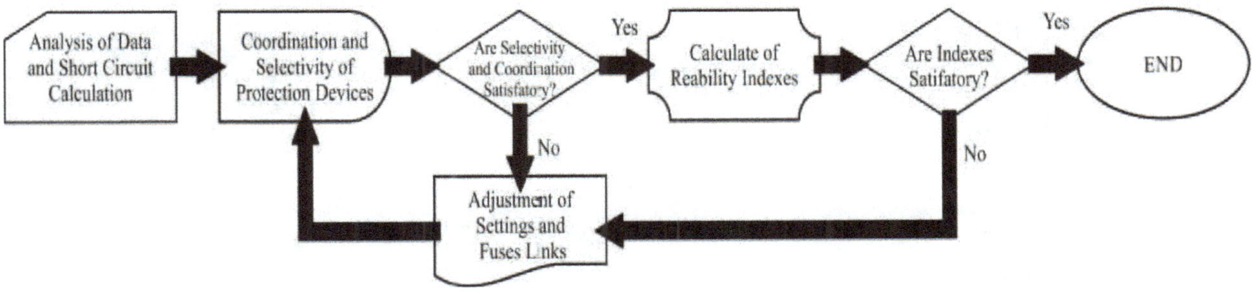

Figure 5. Methodology of assessment of protection and reliability.

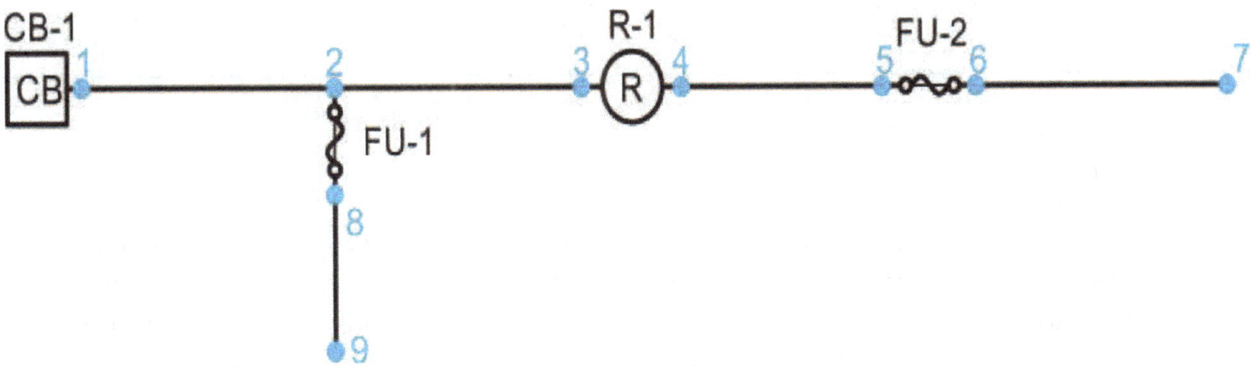

Figure 6. Example distribution network.

Nodes	Protective equipment			
	Circuit breaker CB-1	Recloser R-1	Fuse FU-1	Fuse FU-2
1	TR1 λ_{P1}	0	0	0
2	TR1 λ_{P1}	0	0	0
3	TR1 λ_{P1}	0	0	0
4	TR1 λ_{P1}	TR2 λ_{P2}	0	0
5	TR1 λ_{P1}	TR2 λ_{P2}	0	0
6	TR1 λ_{P1}	TR2 λ_{P2}	0	TR4 λ_{P4}
7	TR1 λ_{P1}	TR2 λ_{P2}	0	TR4 λ_{P4}
8	TR1 λ_{P1}	0	TR3 λ_{P3}	0
9	TR1 λ_{P1}	0	TR3 λ_{P3}	0

Table 3. Initial LSM considering sustained faults.

Nodes	Protective equipment			
	Circuit breaker CB-1	Recloser R-1	Fuse FU-1	Fuse FU-2
1	TR1 λ_{M1}	0	0	0
2	TR1 λ_{M1}	0	0	0
3	TR1 λ_{M1}	0	0	0
4	TR1 λ_{M1}	TA2 λ_{M2}	0	0
5	TR1 λ_{M1}	TA2 λ_{M2}	0	0
6	TR1 λ_{M1}	TA2 λ_{M2}	0	TR4 λ_{M4}
7	TR1 λ_{M1}	TA2 λ_{M2}	0	TR4 λ_{M4}
8	TR1 λ_{M1}	0	TR3 λ_{M3}	0
9	TR1 λ_{M1}	0	TR3 λ_{M3}	0

TA2: Reclosing action time, considering the ability to disconnect and reconnect the circuit with the reclosing device, thus decreasing the shutdown time (TR > TA).

Table 4. Initial LSM considering momentary faults.

In short, under the conditions of permanent failure, the system would keep practically the same levels in the reliability indicators. The main impact would be directly on the temporary faults, where for faults in nodes 6 and 7 the re-establishment time would now be only the recloser R-1, no longer the replacement time of the FU-2 fuse. This impact would be strongly noticed in the SAIDI indicator, which considers the duration of faults.

Nodes	Protective equipment			
	Circuit breaker CB-1	Recloser R-1	Fuse FU-1	Fuse FU-2
1	TR1 λ_{P1}	0	0	0
2	TR1 λ_{P1}	0	0	0
3	TR1 λ_{P1}	0	0	0
4	TR1 λ_{P1}	TR2 λ_{P2}	0	0
5	TR1 λ_{P1}	TR2 λ_{P2}	0	0
6	TR1 λ_{P1}	TR2 λ_{P2}	0	TR4 λ_{P4}
7	TR1 λ_{P1}	TR2 λ_{P2}	0	TR4 λ_{P4}
8	TR1 λ_{P1}	0	TR3 λ_{P3}	0
9	TR1 λ_{P1}	0	TR3 λ_{P3}	0

Table 5. Final LSM considering sustained faults.

Nodes	Protective equipment			
	Circuit breaker CB-1	Recloser R-1	Fuse FU-1	Fuse FU-2
1	TR1 λ_{M1}	0	0	0
2	TR1 λ_{M1}	0	0	0
3	TR1 λ_{M1}	0	0	0
4	TR1 λ_{M1}	TA2 λ_{M2}	0	0
5	TR1 λ_{M1}	TA2 λ_{M2}	0	0
6	TR1 λ_{M1}	TA2 λ_{M2}	0	TA4 λ_{P4}
7	TR1 λ_{M1}	TA2 λ_{M2}	0	TA4 λ_{P4}
8	TR1 λ_{M1}	0	TR3 λ_{M3}	0
9	TR1 λ_{M1}	0	TR3 λ_{M3}	0

Table 6. Final LSM considering momentary faults.

5. Conclusions

In this chapter, the main issues of reliability in the electric power supply as well as the characteristics and adjustments in coordination and selectivity of protection devices were presented, as well as a brief evaluation of their direct impact on the reliability indicators. In addition, it became clear how important it is to keep reliability levels in the pattern for both operational and financial issues. Finally, it is possible to see a theoretical example where the influence of the protection system on reliability was exposed, where, with the coordination and selectivity of a recloser, a great part of the impact of the temporary faults of a circuit can be reduced. The result, although theoretical, contributed to the validation of the importance of coordination and selectivity of the protection devices of an energy distribution network. This study can contribute directly in the two target areas of the power utilities, since it brings a broad vision of the system, besides the description of a way to model the network and to calculate the main indicators of reliability.

Acknowledgements

The authors would like to thank the technical and financial support of RGE Sul Power Utility by project "Solução Inovadora para Gerenciamento Ativo de Sistemas de Distribuição" (P&D/ANEEL), Coordination for the Improvement of High Level Personnel (CAPES) and the National Center of Scientific and Technological Development (CNPq).

Author details

Marco Antônio Ferreira Boaski[1]*, Caio dos Santos[1], Mauricio Sperandio[1], Daniel Pinheiro Bernardon[1], Maicon Jaderson Ramos[2] and Daniel Sperb Porto[2]

*Address all correspondence to: ferreirab.marco@gmail.com

1 Federal University of Santa Maria, Santa Maria, Brazil

2 RGE Sul Power Utility, Brazil

References

[1] Mello APC, Bernardon DP, Pfitscher LL, Sperandio M, Toller BB, Ramos MJS. Intelligent system for multivariables reconfiguration of distribution networks. In: Presented at the IEEE PES. Conference on Innovative Smart Grid Technologies Latin America (ISGT LA); IEEE; São Paulo, Brazil; 2013

[2] Brown RE. Impact of smart grid on distribution system design. In: Presented at the Power and Energy Society General Meeting—Conversion and Delivery of Electrical Energy in the 21st Century; Pittsburgh, USA 2008 IEEE

[3] Siqueira IP. Estimating the Impact of Wide-Area Protection Systems on Power System Performance and Reliability. In: Presented at the 13th International Conference on Development in Power System Protection 2016 (DPSP); IET 2016

[4] Bernardon DP, Mello APC, Pfitscher LL, Canha LN, Abaide AR, Ferreira AAB. Real-time reconfiguration of distribution network with distributed generation. Electric Power Systems Research. 2014;**107**:59-67

[5] Siirto OK, Safdarian A, Lehtonen M, Fotuhi-Firuzabad M. Optimal distribution network

automation considering earth fault events. IEEE Transactions on Smart Grid. 2015;**6**(2): 1010-1018

[6] Elkadeem MR, Alaam MA, Azmy AM. Optimal automation level for reliability improvement and self-healing MV distribution networks. Presented at the 2016 Eighteenth International Middle East Power Systems Conference (MEPCON); 2016; Cairo, Egypt: IEEE; Changsha, China 2017

[7] Xu X, Mitra J, Wang T, Mu L. An evaluation strategy for microgrid reliability considering the effects of protection system. IEEE Transactions on Power Delivery. 2016;**31**(5):1989-1997

[8] Silva LGW, Pereira RAF, Abbad JR, Mantovani JRS. Optimised placement of control and protective devices in electric distribution systems through reactive tabu search algorithm. Electric Power Systems Research. 2008;**78**(3):372-381

[9] Hamman ST, Hopkinson KM, Fadul JE. A model checking approach to testing the reliability of smart grid protection systems. IEEE; Power & Energy Society. 2016

[10] Velásquez MA, Quijano N, Cadena AI. Multi-objective planning of recloser-based protection systems on DG enhanced feeders, In: Presented at the 2015 IEEE PES Innovative Smart Grid Technologies Latin America (ISGT LATAM); Montevideo, Uruguay 2015

[11] Bo Z, Zhao Y, Wang L, Ma X, Ding S, Wei F. The protection cooperation of distribution network based on protection intelligent center and the validation of its reliability. In: Presented at the 2015 5th International Conference on Electric Utility Deregulation and Restructuring and Power Technologies (DRPT); IEEE; Changsha, China 2015

[12] Supannon A, Jirapong P. Recloser-fuse coordination tool for distributed generation installed capacity enhancement. Presented at the Innovative Smart Grid Technologies—Asia (ISGT ASIA); 2015 IEEE; Bangkok, Thailand 2015

[13] Khorshid-Ghazani B, Seyedi H, Mohammadi-ivatloo B, Zare K, Shargh S. Reconfiguration of distribution networks considering coordination of the protective devices. IET Generation, Transmission & Distribution. 2011;**11**(1):82-92

[14] Ramos MJS, Bernardon DP, Comassetto L, Resener M. Analysis of coordination and selectivity of protection systems during reconfigurations of distribution energy systems in real time. In: Presented at the 2013 IEEE PES Conference on Innovative Smart Grid Technologies (ISGT Latin America); São Paulo, BR. IEEE; 2013

[15] Jamali S, Borhani-Bahabadi H. A new recloser time-current-voltage characteristic for fuse saving in distribution networks with DG. IET Generation, Transmission & Distribution. 2017

[16] Comassetto L, Bernardon DP, Canha LN, Abaide AR. Automatic coordination of protection devices in distribution system. Electric Power Systems Research. 2008;**78**(7):1210-1216

[17] IEEE, 242-2001-IEEE Recommended Practice for Protection and Coordination of Industrial and Commercial Power Systems. IEEE; New York, USA 2001

[18] Blischke WR, Murthy DNP. Reliability: Modeling, Prediction, and Optimization. Wiley-Interscience New York, USA 2000

[19] Meneses A, Pinto LC. Quality of supply at the Portuguese Electricity Transmission Grid. In: Presented at the 11th International Conference on Electrical Power Quality and Utilisation (EPQU); IEEE; Lisbon, Portugal 2011

[20] Brown RE. Eletric Power Distribution Reliability. 2nd ed. CRC Press; New York, USA 2009

[21] Wu Y, Li M, Ni M, Zhang Q. Reliability assessment of line protection based on reliability graphs and Monte Carlo simulation. In: Presented at the 2015 5th International Conference on Electric Utility Deregulation and Restructuring and Power Technologies (DRPT); IEEE; Changsha, China 2015

[22] Sailaja CVSS, Prasad PVN. Determination of optimal distributed generation size for losses, protection co-ordination and reliability Evaluation Using ETAP. In: 2016 Biennial International Conference on Presented at the Power and Energy Systems: Towards Sustainable Energy (PESTSE); IEEE; Bangalore, India 2016

[23] Tomsovic FSK. Optimal distribution protection design: quality of solution and computational analysis, International Journal of Electrical Power & Energy Systems.1999;**21**:327-335

[24] IEEE. 1366-2003-IEEE Guide for Electric Power Distribution Reliability Indices. IEEE; 2004

[25] Rocha LF, Borges CLT, Taranto GN. Reliability evaluation of active distribution networks including islanding dynamics. IEEE Transactions on Power Systems. 2016;**32**(2):1545-1552

[26] Zhang J, Kang Q, Huang D, Xiong X, Liu Y, Ma J, et al. Reliability evaluation of the new generation smart substation considering relay protection system. Presented at the 2016 IEEE PES Asia-Pacific Power and Energy Engineering Conference (APPEEC); IEEE Xian, China 2016

[27] Bernardon DP, Sperandio M, Garcia VJ, Canha LN, Abaide AR, Daza EFB. AHP decision making algorithm to allocate remotely controlled switches in distribution networks. IEEE Transactions on Power Delivery. 2011;**26**(3):1884-1892

[28] Neto NK, Abaide AR, Bernardon DP, Canha LN, Oberto M, Pressi R. The application of the logical structural matrix for reliability analysis in a distribution system planning environment. In: Presented at the 23° International Conference on Electricity Distribution-CIRED; Lyon, France 2015

[29] Ferreira GD, Bretas AS, Cardoso G. Optimal distribution protection design considering momentary and sustained reliability indices. In: Presented at the Proceedings of the International Symposium Modern Electric Power Systems (MEPS) 2010; IEEE; Wroclaw, Poland 2010

[30] Blackburn JL, Domin TJ. Protective Relaying: Principles and Applications. CRC-Press New York, USA 2006

4

Power System Reliability: Mathematical Models and Applications

Rabah Medjoudj, Hassiba Bediaf and Djamil Aissani

Abstract

This chapter deals with power systems reliability including technical, economical, and decisional aspects. Knowing that almost 90% of failures occur in the distribution systems, great interest was dedicated to this part of the system, and the first work was oriented to reliability indices defined as objectives to attempt and as performance measures in the electricity market. Some works deal with the managers' behavior, and the customers reactions are modeled using economic criteria in uncertain future and inspired from game theory. When studying components, degradation models were introduced and combined with the effects of socks to study the reliability changing during system operation. In some works, the correlation between maintenance policies and reliability aspects was highlighted. In a recent work, considering the importance of new technologies integration and renewable energy insertion to power systems, it was revealed that reliability aspects and energy sustainability are two fundamental issues of progress in a given society.

Keywords: power systems reliability, distribution functions, degradation modeling, maintenance, decision-making

1. Introduction

In general way, power system reliability addresses the issues of service interruption and power supply loss. In several cases, it is defined as an objective to attempt in terms of indices directly related to the customer. Typical reliability index values for US utilities are SAIFI, SAIDI, and CAIDI. Over time, they become standard values for evaluating the reliability of electrical systems and used in several publications. Medjoudj et al. [1], in their recent publication, defined other indices as reliability subcriteria in their decision-making attributes, giving the best model of a smart energy grid. These indices are discussed in Section 2 with an application to a real case study. After this classical definition, some works have integrated data analysis and processing taking into account the calculation of distribution parameters such as those used for Weibull and Weibull-Markov processes. These works were vulgarized with applications in power system reliability by Van Casteren et al. [2] and Medjoudj et al. [3]. The most techniques used in power system reliability optimization and processing is the reliability centered preventive maintenance (RCM). Several publications have highlighted that in most cases of multicomponent systems, the maintenance actions arrive very early without any effects on the system or very late, that is, the need of curative maintenance with its negative consequences. This issue is also treated in the case of a differentiated service of reliability in the case of customers with different requirements of the reliability level. These two concepts are discussed in Section 3. Recent publications have highlighted the interest of combining reliability attributes for maintenance actions in the case of degrading systems and components [4]; however, a novel work developed recently at the LaMOS unit research dealing with multiple degradation processes is applied to power switchgear and is discussed in Section 4. The discussions and the conclusion highlighting the place of reliability in a power energy smart grid are given in Section 5.

2. Power system reliability indices

In the beginning, the methods used were classical to evaluate reliability indices of distribution systems such as failure frequency, mean failure times, mean time between failure and energy not supplied. These indices help decision makers to define technical and management measures to perform systems. After that was introduced the notion of loss of load probability (LOLP), which has a lot of applications in load modeling and electrical parameters dimensioning. It is significant for any power enterprise to analyze customer satisfaction. A variety of indices have been developed to measure reliability and its cost in power systems area such as loss of load probability (LOLP), loss of load expectation (LOLE), expected frequency of load curtailment (EFLC), expected duration of load curtailment (EDLC), expected duration of a curtailment (EDC), and expected energy not supplied (EENS) [1].

2.1. Loss of load probability

LOLP is an expected value, sometimes calculated on the basis of the peak hourly load of each day and sometimes on each hour's load (24 in a day). Moreover, in the beginning, LOLP is used to characterize the adequacy of generation to serve the load on the bulk power system; it does not directly model the reliability of the transmission and distribution system where the majority of outages actually occur [5]. Nourelfeth and Ait Kadi [6] have recently made that the LOLP is usually used to estimate the reliability index. Considering S and D as the supply and the load demand respectively, they compute the reliability of a multistate system (MSS) as:

$$R = \Pr(S \geq D) Or\ R = 1 - LOLP \tag{1}$$

Using the well-known formulation of the LOLP given in several publications and discussed in the following section, they generalized the MSS reliability index R

$$R = \frac{1}{\sum_{j=1}^{M} T_j} \sum_{j=1}^{M} \Pr(S \geq D_j) T_j \tag{2}$$

where the operation period T is divided into M intervals, and each interval has duration T_j and a required demand level D_j. In the same context, as advancement in reliability applications, Taboada et al. [7] have generalized the use of LOLP as a reliability index. Using a series-parallel system, they calculate the availability of each part of the power including transmission and distribution system using LOLP model.

2.1.1. *Loss of load probability formulation*

The generation system reliability calculations are based mainly on two analytical methods, which differ by the load model used. The first method is the calculation of the LOLP; there the load is given by the load duration curve. The second method is the frequency and duration approach, by which, besides the probability, the frequency and duration of load levels higher than the generation capacity can be determined. The loss of load probability method associates each value of MW outage with a certain cumulative probability, thus producing a capacity outage table. The expected load loss of the system is obtained from the capacity outage table, and a daily load peak variation curve is derived from the daily load curves. The LOLP, number of days on which capacity is insufficient, is obtained by adding the probability that the amount of capacity on forced outage, on day i is greater than or equal to the reserve on day i, for all days of the period being studied.

For a system state where the remaining generating capacity is C_j, the percentage of time t_j during which the load demand exceeds C_j can be determined from the load curve L. The overall probability that the load demand will not be met is called the loss of load probability and is given by the Eq. [4]:

$$LOLP = \sum_j P[C = C_j]\ P[L > C_j] = \sum_j \frac{p_j\ t_j}{100} \tag{3}$$

where p_j is the probability associated to the number of the failed generating units at time t_j, and it is formulated through the following development.

In addition to complete failures, generating units may experience partial failure, when they continue to operate but at reduced capacity levels. They are also taken out of service from time to time for preventive maintenance. Using a simple two-state model for the operation of a unit, its failure probability is given depending on its failure outage rate (FOR), which can be assumed as the unit steady-state unavailability denoted $\overline{A}$. If at time t_j, r units have failed from a total of n identical and independent installed units in the generating system, the probability $p_j$is given by:

$$p_j = \binom{n}{r} \overline{A}^r (1 - \overline{A})^{n-r} \tag{4}$$

A case of unequal size of the units can appear.

2.1.2. Purposes

2.1.2.1. First purpose

It is well known that availability is a measure of success used primarily for repairable systems. For nonrepairable systems, availability $A(t)$ equals reliability $R(t)$. In repairable systems, $A(t)$ will be equal to or greater than $R(t)$. In the optimistic case, the availability is greater than the reliability. Following the Levitin and Lisnianski development for a multistate generating system (MSGS), the availability expectation is the function of demand D and may be defined as [8]:

$$E_A(D_j, T_j) = \frac{1}{\sum_{j=1}^{M} T_j} \sum_{j=1}^{M} A(D_j) T_j \tag{5}$$

The index $(1 - E_A)$ is often used and treated as loss of load probability and can be written as:

$$LOLP = (1 - E_A)\ Or\ E_A = 1 - LOLP \tag{6}$$

2.1.2.2. Second purpose

This purpose highlights the correlation between the system reliability, the energy availability, and the loss of load probability. To understand this correlation, we consider a multistate generating repairable system MSGS connected to a load L, and on a given period of time, we draw two curves representing the evolutions of the system available capacity (SAC) and the hourly system load (HSL), respectively, as shown in **Figure 1**. Depending on the states of generating units (up or down) that involve partial or total failure of a simple unit or of several units, the appearance of dips in the same curve reflects units' breakdowns, and the resumption to the initial level of capacity indicates that repairs were made. One of the most reliability indices that concerns more the utility than the customer is the energy not supplied (ENS) given by the dashed lines under the curve. Their corresponding time intervals denote durations, where the consumption exceeded the production, and therefore, we have loss of load. The decreasing level of system reliability is highlighted by degrading state, corresponding to each decreasing in the SAC curve behavior.

The generating system failures can occur in two ways: either through unit failures or through load increases. There is a loss of load when the demand is greater than the supply. However, there is a loss of supply when a failure occurs in the upstream of the load point. It is important

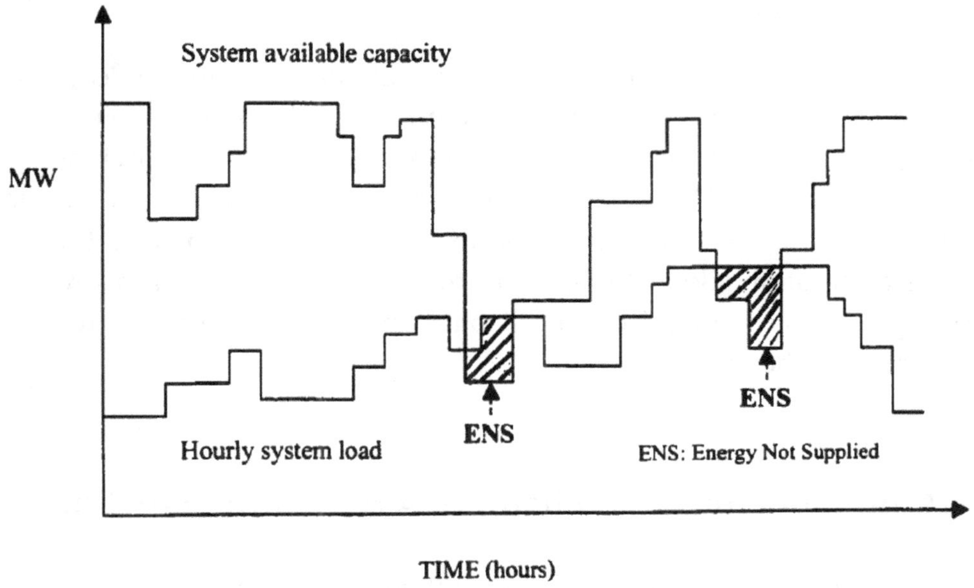

Figure 1. Superimposition of the system available capacity and the load model.

to retain this difference. In this context, two relevant questions are to be asked and the answers were given in [9, 10] with a case study application, such as: What happens when load increases? What is the consequence of generating system failure?

2.2. Frequency and duration indices

Almost every electricity utility computes reliability indices on an annual basis. The most important reliability indices involving decision-making criteria are given as follows [1]:

The Expected Frequency of Load Curtailment in (fault/yr):

$$EFLC = \sum_{k=1}^{n} \lambda_k \tag{7}$$

The Expected Duration of Load Curtailment in (hrs/yr):

$$EDLC = \sum_{k=1}^{n} \lambda_k T_k \tag{8}$$

The Expected Energy Not Supplied in (kWh/yr):

$$EENS = L.(EDLC) \tag{9}$$

where λ_k, T_k are failure rate and failure duration of an item k and L is the load curtailed at a considered load point, respectively. Application is done for a part of the distribution system of Algiers city (Algeria). Considering the electrical characteristics (network topology, section length, power value at load points and the fault search method) and reliability parameters mentioned earlier, the overall system reliability indices are computed.

2.3. Reliability indices improvement

To improve the reliability level, technical and organizational measures are considered during system planning and operation. The actions currently carried out are as follows: intensifying the operations of maintenance; networks reorganization, looping and meshing systems, and automation of networks. In [11], some options are added such as load transfer between feeders, undergrounding circuits, and replacement of aging equipment. From a practical standpoint, this application allows to highlight the goodness of each measure to the system performances by a simple comparison of reliability indices. The results of reliability indices improvement are published in [1].

3. Interruptions modeling and reliability service differentiation

In several studies dealing with electrical distribution system reliability, the objective often sought by the energy distributor is the balance between the required reliability level and its cost. In the following, we develop two important points of view in the reliability of the electrical systems: that relating to modeling and that of the differentiation of electricity prices according to the level of reliability required with a minimal guaranteed reliability level for customers without any prior requirement [12].

3.1. Interruption modeling using the Weibull-Markov process

In the last decade, a novel vision of interruption modeling in power systems was developed and consists of the Weibull-Markov process. The purpose is to model the failure and operating data according to Weibull distribution proprieties, while retaining those assigned to the Markov model where the system occupies discrete states. This process was initially developed by Van Castaren [2] and was applied successfully by Pivatolo [13] and Medjoudj et al. [3]. Applications were made to highlight maintenance policies gathered on three types of actions: namely nondestructive action which does not improve reliability level but slows the system degradation. This action, denoted $(1a)$ as a minor maintenance, is characterized by an improvement factor m_1. A second action is considered and denoted $(2b)$ and can touch some of the components of a system up to their replacement. To this action is associated an improvement factor m_2, and the maintenance is a major one. The third and final proposed maintenance action is on the renewal of equipment, and it is assumed to be perfect, and after its implementation, the system is assumed as good as new, and it is denoted $(2p)$. From a practical standpoint, this action is highlighted by taking m_1 and m_2 equal to the unity. This notion is introduced by Tsai et al. [14] for a mechatronic system and applied for power systems by Medjoudj et al. [3]. In this part of the section, we introduce the concept of the differentiated reliability with an application to the case of an electrical MV/LV substation. Starting from the expression of reliability function expressed by a desired threshold, the need of performing preventive maintenance action at time is decided regarding the behavior of this function at the coming stage of maintenance. Then, the choice of the type of action to perform is dictated by the value of the maximum benefit brought by this action. Threshold reliability is allocated to the opposite risk of system failure occurrence.

3.1.1. Reliability data analysis

Considering the formulations of mean up time and mean down time of an item is given respectively by:

$$MUT = t_m - t_b \int_0^{t_m} h(t)dt \tag{10}$$

and

$$MDT = t_a - t_b \int_0^{t_m} h(t)dt, \tag{11}$$

where, t_m, t_a and t_b are respectively, the preventive maintenance(PM) interval, the PM and corrective maintenance (CM) times on replacement; the operational availability is defined as:

$$A = \frac{t_m - t_b \int_0^{t_m} h(t)dt}{t_m + t_a} \tag{12}$$

Subsequently, the PM interval for maximizing the availability can be derived by differentiating Eq. 9 to time t_m, such as $\frac{dA}{dt} = 0$, and the differential result is:

$$(t_m + t_a)h(t_m) - \int_0^{t_m} h(t)dt = \frac{t_a}{t_b} \tag{13}$$

For data treatment and statistical processing, forced and planned outages are collected over 17 years of system operation continuously at the national company of electricity and gas center (SONELGAZ) of Bejaia city, Algeria. For an MV/LV transformer which is a critical item of an electrical substation, the estimated parameters and the adequate probability distribution functions are listed in **Table 1**.

The obtained results show that, based on the Kolmogorov Smirnov (KS) test [15], d_{ks} is lower than $d_{(n,0.05)}$, the Weibull distribution is not rejected; however, with the exponential law, d_{ks} is greater than $d_{(n,0.05)}$, the hypothesis is not accepted. In **Table 2** are gathered reliability indices, where both the feeder failure frequency (F_i) and the transformer failure rate (h) are added. The nonacceptation of the exponential distribution is comforted by the review results of reference [16], where the authors state that the exponential law, usually used to describe failures, is not always 100% suitable for electricity distribution systems.

In this study, it is assumed that the substation failures are due either to the transformer or to the internal cable connector failures. A three states diagram (working, failure and maintenance) is dressed for the life cycle modeling of the substation as shown in **Figure 2**.

Let X_{12}, X_{13}, X_1, X_2, X_3 be the random variables representing the duration of the operation until failure, the duration of the operation until maintenance, the duration of the operation (state S_1), the duration of the interruption (state S_2) and the duration of the maintenance (state S_3), respectively. The estimated parameters of the random variables following the Weibull distributions are listed in **Table 3**.

Component or Subsystem	n	Distribution	Parameters	d_{ks}	$d_{(n,0.05)}$	Decision
MT/LV Transformer	17	Weibull	$\beta = 2.459579$; $\eta = 5.909754 \times 10^6$	0.2264	0.308	Not rejected
(Tr)		Exponential	$\lambda = 0.007785467$	0.3136		Rejected

Table 1. Distribution functions parameters estimation.

Component or Subsystem	*MUT* (hours)	*MDT* (hours)	*MTBF* (hours)	*A*	F_i (1/year)	*h* (1/year)
MT/LV Transformer	87358.33	9.97	87368.32	0.9998	0.10024	0.00778

Table 2. Reliability indices of the power transformer.

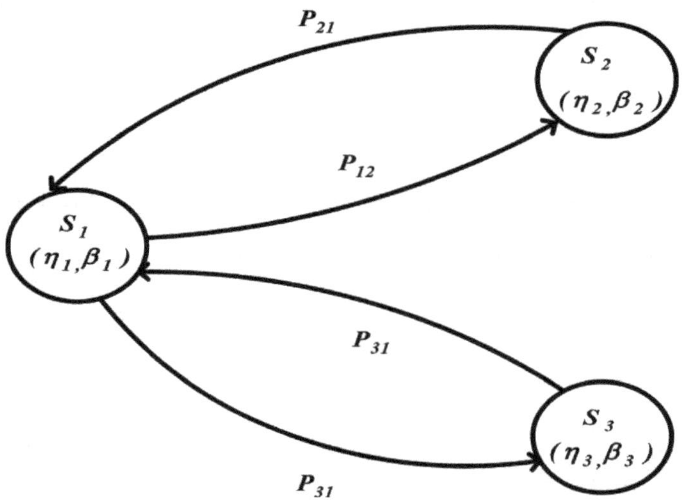

Figure 2. Three states diagram.

Variable	N	Distribution	Parameters	d_{ks}	$d_{(n,0.05)}$	Decision
X_{12}	27	Weibull	$\beta = 1.0644$; $\eta = 3.2827 \times 10^6$	0.1356	0.25438	Not rejected
X_{13}	27	Weibull	$\beta = 1.6689$; $\eta = 7.2589 \times 10^5$	0.2173	0.25438	Not rejected
X_2	27	Weibull	$\beta = 0.6764$; $\eta = 196.00159$	0.1219	0.25438	Not rejected
X_3	27	Weibull	$\beta = 1.03894$; $\eta = 123.6005$	0.1219	0.25438	Not rejected

Table 3. Parameters estimation of the random variables following Weibull distributions.

3.1.2. Reliability under preventive maintenance

The improvement of maintenance to reliability is developed using two factors, and the selection of the action to do for the components on every *PM* stage is decided by maximizing system benefit in maintenance. Depending on the percent of the survival parts of system when it is maintained, the reliability function is:

$$R_j(t) = R_{0,j}.R_{V,j}(t) \tag{14}$$

where $R_{0,j}$ is the initial reliability of the jth stage and $R_{V,j}(t)$ is the reliability degradation of surviving parts on this stage. Considering periodical PM which interval is t_m, the reliability of surviving parts is defined as:

$$R_{V,j}(t) = R\left(\frac{1}{m_1}(t-(j-1)t_m\right) \tag{15}$$

With: $(j-1)t_m \le t \le jt_m$ and m_1, $(0 < m_1 \le 1)$ is the improvement factor of action $(1a)$.

To model the reliability of systems followingPM, the effects of various actions on $R_{0,j}$ and $R_{V,j}$ must be evaluated.

$$R_{0,j} = R_{f,j-1} = R_{0,j-1}.R(t_m) \tag{16}$$

where $R_{0,j-1}$, $R_{f,j-1}$ indicate the initial and final reliability values of the system on the $(j-1)^{th}$ stage.

Action $(1b)$ can improve the surviving parts of the system and also recover the failed parts. Generally, the impact of this action on the failed parts can be measured by an improvement factor m_2, which is also set between 0 and 1 representing the restored level except the surviving parts. According to the definition, the initial reliability on the action $(1b)$ can be expressed as:

$$R_{0,j} = R_{f,j-1} + m_2\left(R_0 - R_{f,j-1}\right) \tag{17}$$

where R_0 denotes the initial reliability of the new system.

The system reliability is expressed as:

$$R_j(t) = R_{0,j}\, exp\, [(t-(j-1)t\_m)/(m\_1\ \eta)]^{\beta} \tag{18}$$

where $R_{0,j}$ is the initial reliability of the jth stage and $m_1$is the improvement factor of action $(1a)$. The benefit of component maintenance on the jth stage is defined as [14]:

$$B_{i,k} = \frac{\int_{t_j}^{\infty} R_{i,j+1}(t)dt - \int_{t_j}^{\infty} R_{i,j}(t)dt}{C_{i,k}} \tag{19}$$

where i, k denote, respectively, the ith subsystem or component and the maintenance action considered and $C_{i,k}$, the action cost. The advantageous one will correspond to the maximum of the benefit, that is, $B_i^* = \text{Max}(B_{i,k})$. Once the action of maintenance is defined and retained, the availability of the system at any stage is processed as:

$$A_{s,j} = \frac{T - t_{b,m} \sum_{i=1}^{n} \int_{t_{j-1}}^{t_j} h_{i,j}(t)dt}{T + \sum_{i}^{n} t_{i,k,a}} \quad (20)$$

where n is the number of components or subsystems and $t_{i,k,a}$ is the time of the PM actions $(1a)$, $(1b)$ and $(2p)$ and T, the cycle time. In the following are described the different types of PM actions in the case of the power transformer.

- Action $(1a)$: cleaning, lubricating, tightening and oil-level verification,
- Action $(1b)$: oil and internal cable connectors' $(I.C.C)$ replacement,
- Action $(2p)$: transformer replacement.

Parameters are needed to compute the benefits such as the distribution function parameters (β, η) PM and CM times (t_a and t_b), maintenance actions costs (C_{1a}, C_{1b}, and C_{2P}) listed in **Table 4**, and the threshold value of reliability (R_{th} = 0.8).

The obtained results are: (t_m $(I.C.C)$ = 715 days, t_m (Tr) = 2415 days); however, the maintenance interval for the system is $T_m = \min\{715, 2415\} = 715$ days. The maintenance action to retain is based on the maximum benefit value, and the results at different maintenance stages are listed in **Table 5** using the following notations:

*: no maintenance is needed,

0: nothing to do (after an inspection),

1: action $(1a)$ is carried out,

2: action $(1b)$ is carried out,

3: action $(2p)$ is carried out,

$R((j+1)T_m)$: The instantaneous reliability at $(j+1) \times T_m$.

The results listed in **Table 5** can be interpreted as follows: at every maintenance stage, verify for each component if its reliability for the coming stage is greater or equal to the R_{th}.

• If the condition is realized, the decision is doing nothing. For the example of the transformer, at $j = 1$, the reliability is $R((j+1) \times T_m) = R(2 \times T_m) = 0.9272 > R_{th} = 0.80$; however, no maintenance is needed for the stage $j = 1$.

• If no, compute the benefit for each action proposed and choose the maximum value. For example, for the threshold value $R_{th} = 0.95$, at the first stage, $j = 1$, the reliability at the coming

β	η	$t_a(days)$	$t_b(days)$	m_1	m_2	C_{1a} \$	C_{1b} \$	C_{2p} \$
2.45	4103.99	3.5	28	0.80	0.90	600	1500	8600

Table 4. The useful parameters for benefit evaluation.

Stage	Action proposed	$R((j+1)T_m)$			Benefit $			Action Retained		
		$R_{th} = 0.8$	$R_{th} = 0.9$	$R_{th} = 0.95$	$R_{th} = 0.8$	$R_{th} = 0.8$	$R_{th} = 0.95$	$R_{th} = 0.8$	$R_{th} = 0.8$	$R_{th} = 0.95$
1	$(1a)$	0.9272	0.9272	0.9272	■	■	0.4378	0	0	2
	$(1b)$				■	■	0.8285			
	$(2p)$				■	■	0.1601			
2	$(1a)$	0.8155	0.8155	0.9234	■	2.0944	0.6371	0	1	2
	$(1b)$				■	1.7618	0.9393			
	$(2p)$				■	0.3254	0.1827			
3	$(1a)$	0.6618	0.8286	0.9230	3.0284	1.2526	0.6601	1	2	2
	$(1b)$				2.5859	1.8010	0.9394			
	$(2p)$				0.4710	0.3326	0.1844			
4	$(1a)$	0.7287	0.9220	0.9222	1.7524	■	0.6830	2	0	2
	$(1b)$				2.1457	■	0.9393			
	$(2p)$				0.3945	■	0.1861			
5	$(1a)$	0.9200	0.8056	0.9222	■	1.5305	0.7058	0	1	2
	$(1b)$				■	1.0851	0.9393			
	$(2p)$				■	0.3429	0.1877			
6	$(1a)$	0.8038	0.8032	0.9219	3.1444	1.5599	0.7514	1	1	2
	$(1b)$				2.6712	0.3430	0.9392			
	$(2p)$				0.4881	0.3188	0.1910			
7	$(1a)$	0.6455	0.8721	0.9215	1.7271	1.3359	0.7741	2	1	2
	$(1b)$				2.1735	0.6047	0.9392			
	$(2p)$				0.4016	0.3687	0.1927			
8	$(1a)$	0.7181	0.8376	0.9211	1.7271	1.3359	0.7741	2	1	2
	$(1b)$				2.1735	0.6047	0.9392			
	$(2p)$				0.4016	0.3687	0.1927			
9	$(1a)$	0.9190	0.7916	0.9207	■	1.3303	0.7967	0	1	2
	$(1b)$				■	0.9271	0.9392			
	$(2p)$				■	0.4300	0.1944			
10	$(1a)$	0.8024	0.7385	0.9203	■	1.3084	0.8193	0	1	2
	$(1b)$				■	1.2177	0.9391			
	$(2p)$				■	0.4867	0.1960			

Table 5. Maintenance plan depending on reliability thresholds.

stage of maintenance $R((j+1) \times T_m) = R(2 \times T_m) = 0.9272 < R_{crit} = 0.95$ and actions $((1a), (1b))$ benefits are $(0.4378, 0.8285)$, respectively. The action $(1b)$ is retained looking at the maximum value of the benefit.

The risk management is highlighted by thresholds of reliability. Depending on the reliability level reached, or fixed a priori, maintenance operations can be decided. The objectives are the determination of maintenance frequencies on an item and consequently their costs. It will be remarked that a high level of reliability is required (i.e., the risk of failure is minimized), the maintenance frequency increases, and subsequently, the cost increases.

4. Competing failure processes of oil circuit breaker

The components constituting a high voltage oil circuit breaker (HVOCB) are subject to various degradations, namely the aging of the insulating oil in the arc's extinguishing chamber, the contacts wear out and the sharp breakdown of bus bars supports. In this section, we have modeled the behavior of this item as subject to a competing three degradation processes by using the Markov state diagram as given in **Figure 3**. The states were defined using thresholds of degradation parameters. To the degradation processes was associated a shock process highlighting the effects of short circuit arrivals on the HVOCB when defaults occur at the downstream feeder. The novelty in this work is outlined by the use of three-dimensional matrix to show the possible states, where the HVOCB can sojourn.

4.1. Case of three degradation processes modeling

We consider that the processes of degradation are modeled using continuous probability functions, and the operating condition of the system is characterized by a number of states which space is noted by Ω_μ.

Following the Li and Pham theory [17], we consider the three state spaces Ω_1, Ω_2,and Ω_3 corresponding to the degradation processes $Y_1(t), Y_2(t),$ and $Y_3(t)$, respectively. After obtaining the state spaces Ω_1, Ω_2, and Ω_3, we develop a methodology to establish a relationship between the states of the system Ω_μ, the set of degradation states and catastrophic state due to shocks arrivals $\{\Omega_1, \Omega_2, \Omega_3, F\}$.

The study deals with the three processes of degradation $Y_1(t), Y_2(t),$ and $Y_3(t)$ combined with the shock process denoted $D(t)$ as given in **Figure 3**. The sets of states are represented by $\Omega_1 = M_1, \ldots, 1_1, 0_1$, which corresponds to the degradation 1 with$(M_1 + 1)$ states, $\Omega_2 = M_2, \ldots,$

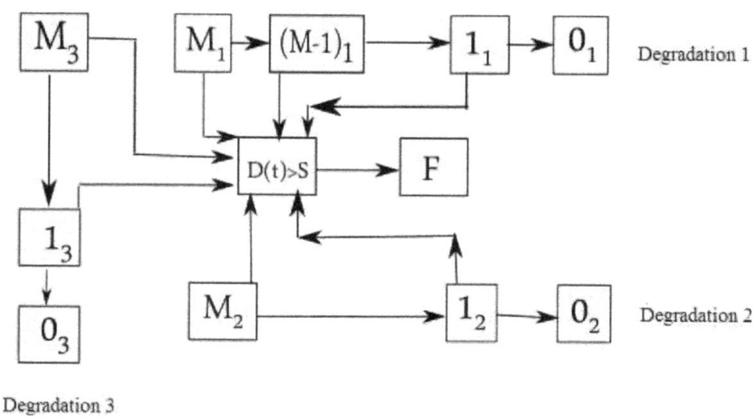

Figure 3. Diagram of transition states of a system subject to four failure processes.

$1_2, 0_2$, corresponds to the degradation 2 with $(M_2 + 1)$, and $\Omega_3 = M_3, \ldots, 1_3, 0_3$ corresponds to the degradation 3 with $(M_3 + 1)$ states. **Figure 3** shows the transition between states of a system submitted to four failure processes.

The equivalence relations between degradation states $\Omega_1 = M_1, \ldots, 1_1, 0_1$; $\Omega_2 = M_2, \ldots, 1_2, 0_2$ and $\Omega_3 = M_3, \ldots, 1_3, 0_3$, and their corresponding intervals are given as follows:

Degradation Process 1:

$0 < Y_1(t) \leq W_M \Longrightarrow$State M_1.

$W_M < Y_1(t) \leq W_{M-1} \Longrightarrow$State $(M-1)_1$.

.

.

.

$W_2 < Y_1(t) \leq W_1 \Longrightarrow$State 1_1.

$G_1 = W_1 < Y_1(t) \Longrightarrow$State 0_1.

Degradation Process 2:

$0 < Y_2(t) \leq A_M \Longrightarrow$State M_2.

$A_M < Y_2(t) \leq A_{M-1} \Longrightarrow$State $(M-1)_2$.

.

.

.

$A_2 < Y_2(t) \leq A_1 \Longrightarrow$State 1_2.

$G_2 = A_1 < Y_2(t) \Longrightarrow$State 0_2.

Degradation Process 3:

$0 < Y_3(t) \leq Z_M \Longrightarrow$State M_3.

$Z_M < Y_3(t) \leq Z_{M-1} \Longrightarrow$State $(M-1)_3$.

.

.

.

$Z_2 < Y_3(t) \leq Z_1 \Longrightarrow$State 1_3.

$G_3 = Z_1 < Y_3(t) \Longrightarrow$State 0_3.

The states' space of the system is defined by $\Omega_\mu = \{M, \ldots, 1, 0, F\}$ with $(M+2)$ states. In this part, we develop a function that generates the relation between the states' space of the system

Ω_μ and degradation states' space $\{\Omega_1, \Omega_2, \Omega_3, F\}$. For example, at a given time t, it is assumed that the degradation process 1 is at the state $i_1 \in \Omega_1$ and the degradation process 2 is at state $j_2 \in \Omega_2$ and the process of degradation 3 is in the state $k_3 \in \Omega_3$. It is assumed that at the present time, the system is not at fault condition (catastrophic state F). Thus, state F can be ignored for the moment, we must therefore seek a function relationship between Ω and$\{\Omega_1, \Omega_2, \Omega_3\}$. Instead of Ω_μ and $\{\Omega_1, \Omega_2, \Omega_3, F\}$. The operation can be described by the mathematical function formulated as follows:

$$f : R = \Omega_1 \times \Omega_2 \times \Omega_3 \rightarrow \Omega = \{M, \ldots., 1, 0\} \tag{21}$$

where $R = \Omega_1 \times \Omega_2 \times \Omega_3 = \{(i_1, j_2, k_3) \mid i_1 \in \Omega_1, j_2 \in \Omega_2, k_3 \in \Omega_3\}$

The function f is defined by: $\Omega_1 \times \Omega_2 \times \Omega_3 \ \Rightarrow f \Rightarrow H.$

The matrix H represented in **Figure 4** gives information about the resulting states space and component of $(M+1)$ elements corresponding to each space leaving by the function f. The line at the top of the matrix H represents the states of the degradation process 1, the right column of the matrix represents the states of the degradation process 2, and the top page of the matrix H represents the degradation process 3. The elements of the matrix H represent the states $f(i_1, j_2,\ k_3) = L$.

We note that in the matrix H, some elements are zeros and it can be assumed that when degradation 1 is in a certain advanced state $I_1(0_1 < I_1 < M_1)$, the degradation 2 is also in a certain weak state $I_2(0_2 < I_2 < M_2)$, and the degradation 3 is also in a certain weak state $I_3(0_3 < I_3 < M_3)$, it is considered as a failure condition. We also notice that $f(M_1; M_2; M_3) = M$, and initially, the system is in a perfect state.

We define time until failure by: $T = \inf\{\ t : Y_1(t) > G_1, Y_2(t) > G_2, Y_3(t) > G_3$ where D > S$\}$.

It is important to know that the life of the system depends on a single process among the three degradations and of that of the shock. However, the system failure is caused by the process

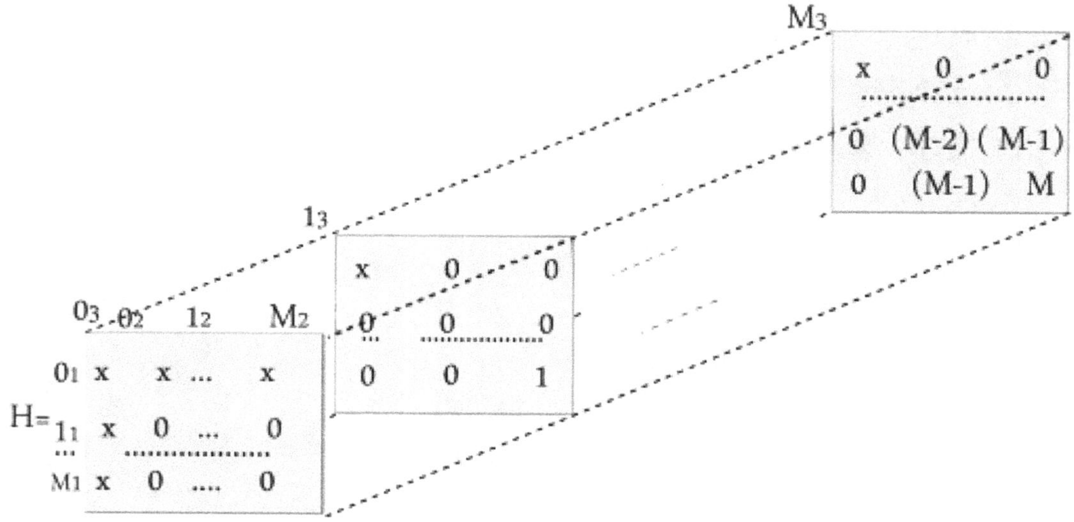

Figure 4. Three-dimensional matrix of system' states, where $I_1 \in \Omega_1$, $J_2 \in \Omega_2$, $K_3 \in \Omega_3$, and $L \in \Omega$.

that occurs first, exceeding its critical value corresponding to the level which can bring the system back to failure.

4.2. System case study modeling and application

Initially, the system is considered in good states of operation$(M_1, M_2, \text{and } M_3)$. It can pass first, to the degradation states $((M-1)_1, (M-1)_2, (M-1)_3)$ or to the state of catastrophic failure (state F), due to random shock. When the system reaches the first state of degradation, it can either remain in this state or go to the second degradation state $((M-2)_1, (M-2)_2, (M-2)_3)$, or it passes to state F. The same process is repeated at each degradation stage with the exception of the states $(0_1, 0_2, 0_3)$

Assumptions:

1. The system occupies $(M+2)$ states, where 0 and F are the states of failure, state i is a degradation state, $1 < i < M$;
2. No repair or maintenance is carried out on the system;
3. $Y_i(t);\ i = 1, 2, 3;$ is a not decreasing and not negative function. With respect to time t, it corresponds to an irreversible accumulation of damage;
4. $Y_i(t);\ i = 1, 2, 3;$ and $D(t)$ is statistically independent implying that the state of one process will have no effect on the other state;
5. At time $t = 0$, the system is at state M;
6. The system may fail due to;

- Degradation process if: $Y_i(t) > G_i$ for $i = 1, 2, 3$;
- Random shock process (the system passes to the condition of the catastrophic failure state F), if: $D(t) = \sum_{t=1}^{N(t)} Xt > S$; with G_i and S are critical level of degradation and shocks, respectively.

The reliability function is defined as follows:

$$\begin{aligned} R_M(t) &= P(state \geq 1) \\ &= \sum_{i=1}^{M} P_i(t) \\ &= P\{Y_1 \leq G_1, Y_2 \leq G_2, Y_3 \leq G_3, D(t) \leq S\} \end{aligned} \tag{22}$$

The system will fail if any of the degradation rates exceeds the critical level G_i; $i = 1, 2, 3$ or the process of shock also exceeds the critical level S.

The system subject to three displacement processes is defined by:

1. The process of increasing degradation representing the wear of the contacts of the circuit breaker is denoted by $Y_1(t)$;

2. The process of increasing degradation representing the aging of oil insulating circuit of the circuit breaker is denoted by $Y_2(t)$;

3. The degradation process of bus bars supports is denoted by $Y_3(t)$;

4. A random process of cumulative shock damages is given by $D(t) = \sum_{t=1}^{N(t)} Xt$.

We obtain a system with four competing degradation processes. For the wear of the contacts: $Y_1\ (t)$, $M_1 = 2$; for the aging of the oils $Y_2\ (t)$: $M_2 = 2$; and for the degradation of the supports: $Y_3\ (t)$, $M_3 = 3$; the system fails if the process of degradation $Y_i(t)$ exceeds a level G_i; $i = 1, 2, 3$ or the process of damages cumulates shocks, and D (t) exceeds the level S. It is assumed that the state spaces associated to Y_1, Y_2 and Y_3 are $\Omega_1 = \{2_1, 1_1, 0_1\}$, $\Omega_2 = \{2_2, 1_2, 0_2\}$, $\Omega_3 = \{3_3, 2_3, 1_3, 0_3\}$, respectively. Consequently, the space of the system is defined as follows: $\Omega = \{3; 2; 1; 0; F\}$. Thus, the function f is defined as being the Cartesian product of three sets following $f : R = \Omega_1 \times \Omega_2 \times \Omega_3 \rightarrow \Omega = \{3, 2, 1\}$, and the result is illustrated by the three-dimensional matrix given in **Figure 5**.

The implementation of the abovementioned models under the MATLAB software which has given the results of the probabilities of the sojourns in different states and the system reliability changing is shown in **Figure 6**. The system (oil circuit breaker, bus bars) is in good condition for 3 years (1100 days) with a probability greater than 0.9. After this period, the latter decreases exponentially to reach the zero value after 4 years (1490 days) without any maintenance actions.

The evolution of the probability of the system in degradation state 2 is complementary to that of the probability of the degradation state 3. Indeed, during 3 years of operation, the probability of being in state 2 is zero. Then, it increases exponentially to reach its maximum value of 0.41 up to 4 years (1280 days). As this state is transient, its probability function decreases to 0 after 5 years. For the system reliability, we note that during the first 8 years, it is expected to decrease by 20% due to the random shock process that governs the system during this period.

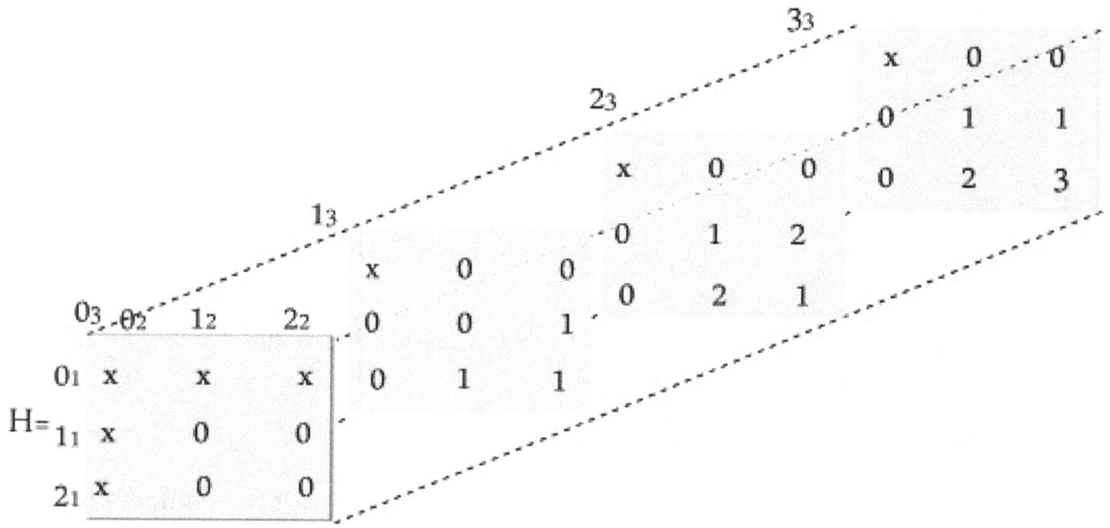

Figure 5. The three-dimensional matrix of the studied system.

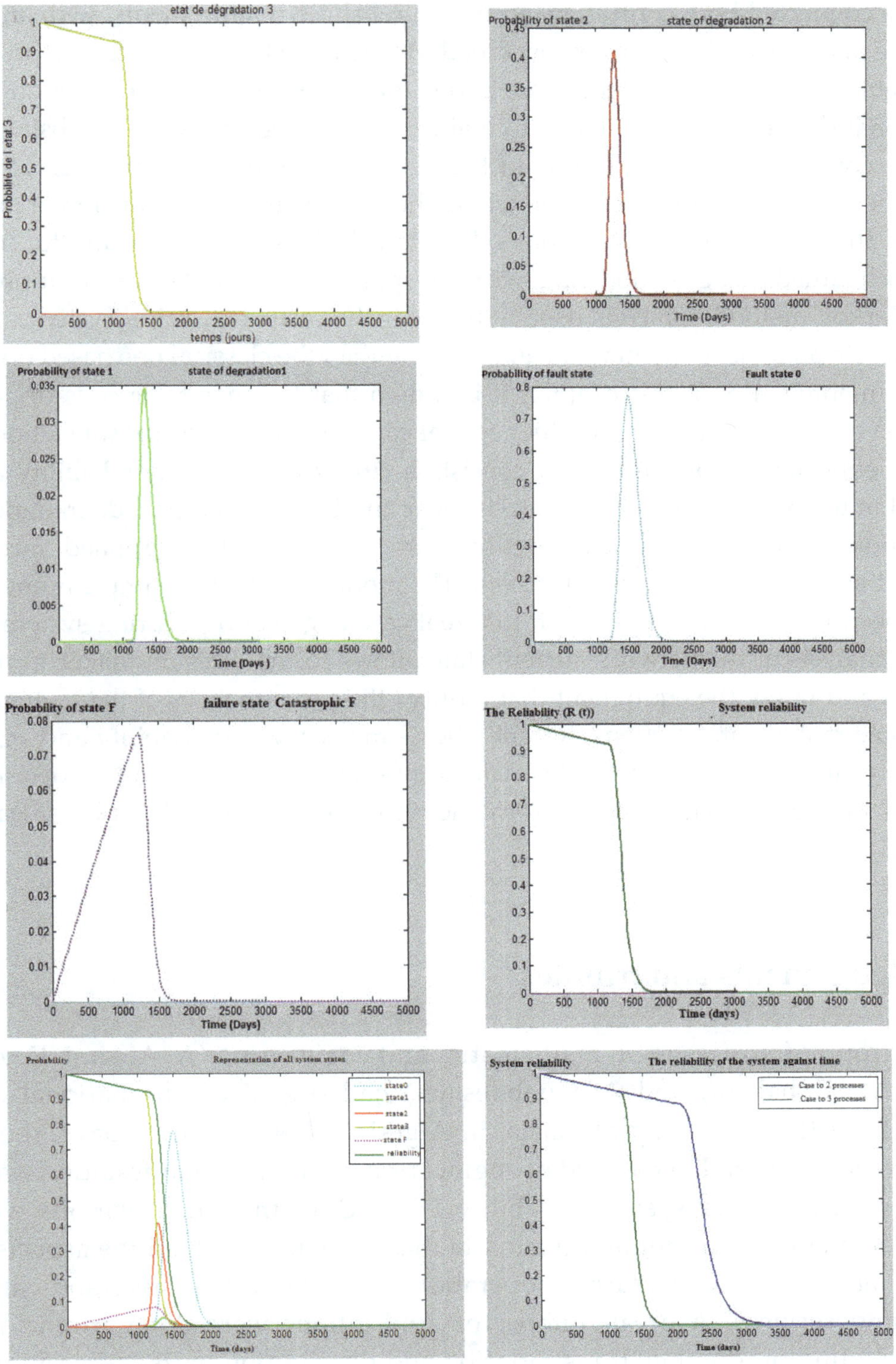

Figure 6. Probabilities of the states and the system reliability changing.

5. Conclusion and discussions

Recently researchers in electrical systems have proposed differentiated electricity service based on reliability and have shown some inconveniences to apply it into a real case. In the same location are connected both consumers with high reliability requirements, with an agreement to pay more and others who are not concerned. Because the technical measure proposed is to add a reliable feeder, the differentiation is not quite possible. Our proposition consists of organizational measure and is oriented to maintenance actions on MV/LV substations, which are directly connected to the end users of the network. The differentiated service is directly related to the reliability of the substation where the improvement is a function of maintenance actions and the frequency of interventions. For statistical considerations and for interruption (forced or scheduled outages) modeling, we have applied the Weibull-Markov approach rather than the Markov method, which is usually used for the case of electrical systems. It has been proven that it is possible to maintain in another way than the classical one based on systematic preventive maintenance. In this chapter, it is shown that the maintenance is decided on the reliability level and benefit bases. Another critical component of electrical substation is studied using competing failure processes and consists a circuit breaker. The reliability aspects are formulated in the bases of oil aging, contacts wear and bus bars support degradation. Investigations conducted by Pham in a theoretical framework have been applied successfully to complement system such as electrical system. The models applied on simple numerical examples have been validated by application to a real case engineering. During system operation, the results analysis of the network current state allows to the decision maker to reach better information and target the equipment that reduces the performances of the system and practicing suitable maintenance actions. Recent studies in energy sustainability and smart energy grid have revealed that reliability is the main criterion taken into account by decision makers in electricity market behavior and a performance index for electric utilities classifications.

Acknowledgements and tributes

In 1988, a group of professors in mathematics and engineering (A. Aissani, D. Aissani, K. D. Haim, A. Boubakeur, and A. Ouabdeslam) had organized the national conference named MFSI at the University of Bejaia, in Algeria, where during 2 days, the notion of reliability was vulgarized in the field of engineering. From that manifestation was born the group of work in power system reliability FSE2. Over than 1500 various works were conducted in engineer, master and doctorate theses dealing with all the aspects of power systems reliability and with a large cooperation with other universities and various manufactures and services. A great number of applications were done around the power systems including production, transportation and distribution parts. In recent years, a lot of novelties were developed compared to what is done over the word, such as the Weibull-Markov modeling in data analysis, nonparametric distributions in switching components behavior, Box and Jenkins models in blackouts forecasting and reliability aspects in smart grids development and multicriteria optimization. The results were valorized in a great number of international conference proceedings and in valuable international journals. This chapter dealing with power system reliability constitutes an interesting opportunity to express our acknowledgments and tributes to these pioneers of reliability in Algeria for what they have given for research.

Author details

Rabah Medjoudj[1]*, Hassiba Bediaf[2] and Djamil Aissani[3]

*Address all correspondence to: r.medjoudj66@gmail.com

1 Lamos Research Unit, Faculty of Technology, Electrical Engineering Department, University of Bejaia, Bejaia, Algeria

2 SOPERIE, Electrical Engineering Society, Bejaia, Algeria

3 Lamos Research Unit, Faculty of Exact Sciences, Operational Research Department, University of Bejaia, Bejaia, Algeria

References

[1] Medjoudj R, Aissani D, Haim KD. Power customer satisfaction and profitability analysis using multi-criteria decision making methods. International Journal of Electrical Power and Energy Systems (Elsevier). 2013;**45**(1):331-339. DOI: 10.1016/j.ijepes.2012.08.062

[2] Van Castaren J. Reliability assessment in electrical power systems: The Weibull-Markov stochastic model. IEEE Transactions on Industry Applications. 2000;**36**(6)

[3] Medjoudj R, Aissani D, Boubakeur A, Haim KD. Interruption modeling in electrical power distribution systems using Weibull-Markov model. Proceedings of the Institution of Mechanical Engineers (IMEchE), Part O: Journal of Risk and Reliability. 01-06-2009;**223**:145-157. DOI: 10.1243/1748006XJRR215

[4] Iberraken F, Medjoudj R, Medjoudj R, Aissani D. Combining reliability attributes to maintenance policies to improve high-voltage oil circuit breaker performances in the case of competing risks. Proceedings of the Institution of Mechanical Engineers, Part O: Journal of Risk and Reliability 1748006X15578572. First published on 02-04-2015. DOI: 10.1177/1748006X15578572

[5] Kuech JD, Kirby BJ, Overholt PN and Markel LC. Measurement practices for reliability and power quality. U.S Department of energy. ORNL/TM-2004/91

[6] Nourelfeth M, Ait-Kadi D. Optimization of series-parallel multi-state systems under maintenance policies. Reliability Engineering and System Safety. 2007;**92**:1620-1626

[7] Taboada HA, Espiritu JF, Coit DW. Design allocation of multistate series-parallel systems for power systems planning: A multiple objective evolutionary approach. Proceedings IMEchE, Part O: Journal of Risk and Reliability. 2008:381-391

[8] Levitin G, Lisnianski A. Multi-state system reliability analysis and optimization (UGF and GAA). In: Pham H, editor. Chapter 4, Handbook of Reliability Engineering. Springer; 2003

[9] Medjoudj R. Reliability Aspects of Electrical Systems: Semi-Markovian Modeling and Maintenance Optimization. Doctorate thesis: University of Bejaia, Algeria; 2009

[10] Iberraken F, Medjoudj R, Aissani D. Mathematical models to support the issue of electrical blackouts in the context of smart grids. Lecture Notes on Information Theory, Engineering and Technology Publishing. December 2014;**2**(4). DOI: 10.12720/lnit.2.4.295-301 (LNIT,ISSN: 2301-3788, www.lnit.org), 2014

[11] Endrenyi J. Reliability Modeling in Electric Power Systems. Toranto: Wiley & Interscience; 1978

[12] Brown ER, Marshall MW. The Cost of Reliability. ABB consulting report. 2001

[13] Pievatolo A. The downtime distribution after a failure of a system with multistate independent components. IMATI-CNR Technical report, Presented at the Mathematical Methods on Reliability conference, MMR, Glasgow, 2007

[14] Tsai YT, Wang KS, Tsai LC. A study of availability centred preventive maintenance for multi-component systems. Reliability Engineering and System Safety. 2004;**84**:261-269

[15] Dhilon BS. Design Reliability, Fundamentals and Applications. USA: CRC press LLC; 1999

[16] Billinton R, Goel R. An analytical approach to evaluate probability distributions associated with the reliability indices of electric distribution systems. IEEE Transactions on Power Delivery. 1986;**1**(3):245-251

[17] Li W, Pham H. Reliability modeling of multi-state degraded systems with multi-competing failures and random shocks. IEEE Transactions on Reliability. 2005;**54**(2):297-230

5

A Decision Support System for Planning and Operation of Maintenance and Customer Services in Electric Power Distribution Systems

Carlos Henrique Barriquello, Vinícius Jacques Garcia, Magdiel Schmitz, Daniel Pinheiro Bernardon and Júlio Schenato Fonini

Abstract

This chapter aims to present the design and development of a decision support system (DSS) for the analysis, simulation, planning, and operation of maintenance and customer services in electric power distribution system (EPDS). The main objective of the DSS is to improve the decision-making processes through visualization tools and simulation of real cases in the EPDS, in order to allow better planning in the short, medium, and long term. Therefore, the DSS helps managers and decision-makers to reduce maintenance and operational costs, to improve system reliability, and to analyze new scenarios and conditions for system expansion planning. First, we introduce the key challenges faced by the decision-makers in the planning and operation of maintenance and customer services in EPDS. Next, we discuss the benefits and the requirements for the DSS design and development, including use cases modeling and the software architecture. Afterwards, we present the capabilities of the DSS and discuss important decisions made during the implementation phases. We conclude the chapter with a discussion about the obtained results, pointing out the possible enhancements of the DSS, future extensions, and new use cases that may be addressed.

Keywords: decision support system, system reliability, visualization tools, maintenance, customer services, electric power distribution

1. Introduction

A reliable and efficient power grid plays a central role in the infrastructure of a country, with a huge impact to its economic and social progress. An uninterrupted, secure, and efficient power flow since generators to consumers is not only desirable but an important requirement for a reliable power grid, where electricity must be produced and delivered in the right amount at the right time.

Although electrical energy can be stored on energy storage systems (EESs), like batteries, to be available at the consumer side in case of a network fault, this is yet not an economically viable solution for bulky quantities [1]. Albeit rapid progress has been observed in the development of EES in the last years, mainly driven by the requirements imposed by the introduction of renewable energy resources into the electric power system, several drawbacks are still present. Most EES technologies are not yet mature, have very limited lifetime, low power/energy density and/or low storage capacity, and always have a round-trip efficiency penalty due to the required charge/discharge cycles [2].

Even in case of higher EES availability at the consumer side due to the current trend of technology evolution and price reduction, there is also the increasing penetration of renewable generation sources near to the consumer, known as distributed generation (DG). As such, DG sources are being adopted at a growing rate by the consumers; they also bring more challenges to the system operators and to the distribution infrastructure, imposing a higher demand for network reliability, mainly at distribution level [3].

This situation is even more challenging when considering the aging infrastructure of distribution systems, uncertainties and regulatory changes, the entry of new players into the energy market (e.g. prosumer), the constant pressure for cost reduction, and the more stringent requirements of reliability. On the contrary, the modernization of the distribution system with the deployment of information and communication technologies (ICTs) and the integration with operational technologies (OTs) is enabling a smarter grid and helping the distribution system operators (DSOs) to face those challenges in a better condition. Nevertheless, the deployment of those smart grid technologies, such as advanced metering infrastructure (AMI) and supervisory control and data acquisition systems (SCADA), also brings new decision variables to the DSOs and, consequently, the need of decision support systems (DSSs) to help them in their decision-making processes [4].

In the literature, several works have been published with proposals of DSS targeted to support the decision-making of the different players in the electric power systems, such as electric utilities, producers, consumers, and market regulators. In Refs. [5] and [6], DSSs were proposed for assisting the competitors in their trading strategies for the electricity market. Also, for the utility companies, there are proposals of DSS for the planning and expansion of the distribution system network based on geographic information system (GIS) [7, 8], for power grid operation and planning considering meteorological conditions [9], for analyzing disturbances of electrical power distribution transformers to diagnosing failures [10], and for assisting utilities to comply with the limits of sulfur dioxide (SO_2) emissions [11]. Additionally, some DSSs have been proposed for the producers and consumers, including support for the calculation of the solar energy potential of surfaces in urban landscapes [12], for managing energy-efficient buildings [13], and also for assisting consumers to participate in demand response programs [14]. There are still proposals of DSS for assisting regulators in deciding about the installation of DG facilities [15] and about the deregulation of the electricity market [1].

Different from prior works, the main contribution of the DSS proposed here is in supporting the distribution system operators in the planning and the execution of customer and maintenance services, which are introduced in the next section. In the sequel, the design and the development of the DSS are reported in the third section. The evaluation of the DSS is left for the fourth section, followed by our concluding remarks at the end of the chapter.

2. Customer and maintenance services in the electric power distribution system

The electric utilities are responsible for maintenance services required to maintain the distribution network operating satisfactorily. Accordingly, the utility shall allocate its material and human resources (e.g. maintenance crews) in order to attend its customers and to fix the network as soon as possible in case of emergency (e.g. equipment failures and blackouts). However, due to the resources constraints, there is a need to prioritize emergency orders over normal maintenance orders, while taking the best decision possible regarding the available maintenance teams and the required response time for each order in the service queue.

The decision problem that must be solved by system operators working at the utility company is called the emergency dispatching and routing problem (EDRP) and can be stated as follows. Given pending emergency work orders and normal work orders in a service queue, and a set of available maintenance teams, to allocate the emergency orders to the crews in order to minimize the waiting time, which is defined as the sum of the travel time and the execution times of normal orders under execution or scheduled to be executed before the pending emergency orders.

It is important to note that the ERDP is dynamic by nature, given that new emergency orders may arrive in an unpredictable manner, changing the inputs of the problem and, thus, requiring a real time and continuous decision-making process. Therefore, a new dispatching solution must be found during the execution of the previously defined routes, possibly requiring changes to the routes assigned to the teams and the insertion of new emergency orders into their service queue. On the other hand, the decision process must take into account several conflicting criteria, such as the minimization of the required time to complete the emergency orders and the reduction of the cost incurred by the changes in the assigned routes. Note that the routing cost includes the delays of the previously assigned services and the possible decrease of the work efficiency, which is the ratio of effective service time to the total spent time (i.e. service times plus travel times). The relationship among the decision variables and decision outputs is represented in **Figure 1**, while the overall decision process is illustrated in **Figure 2**.

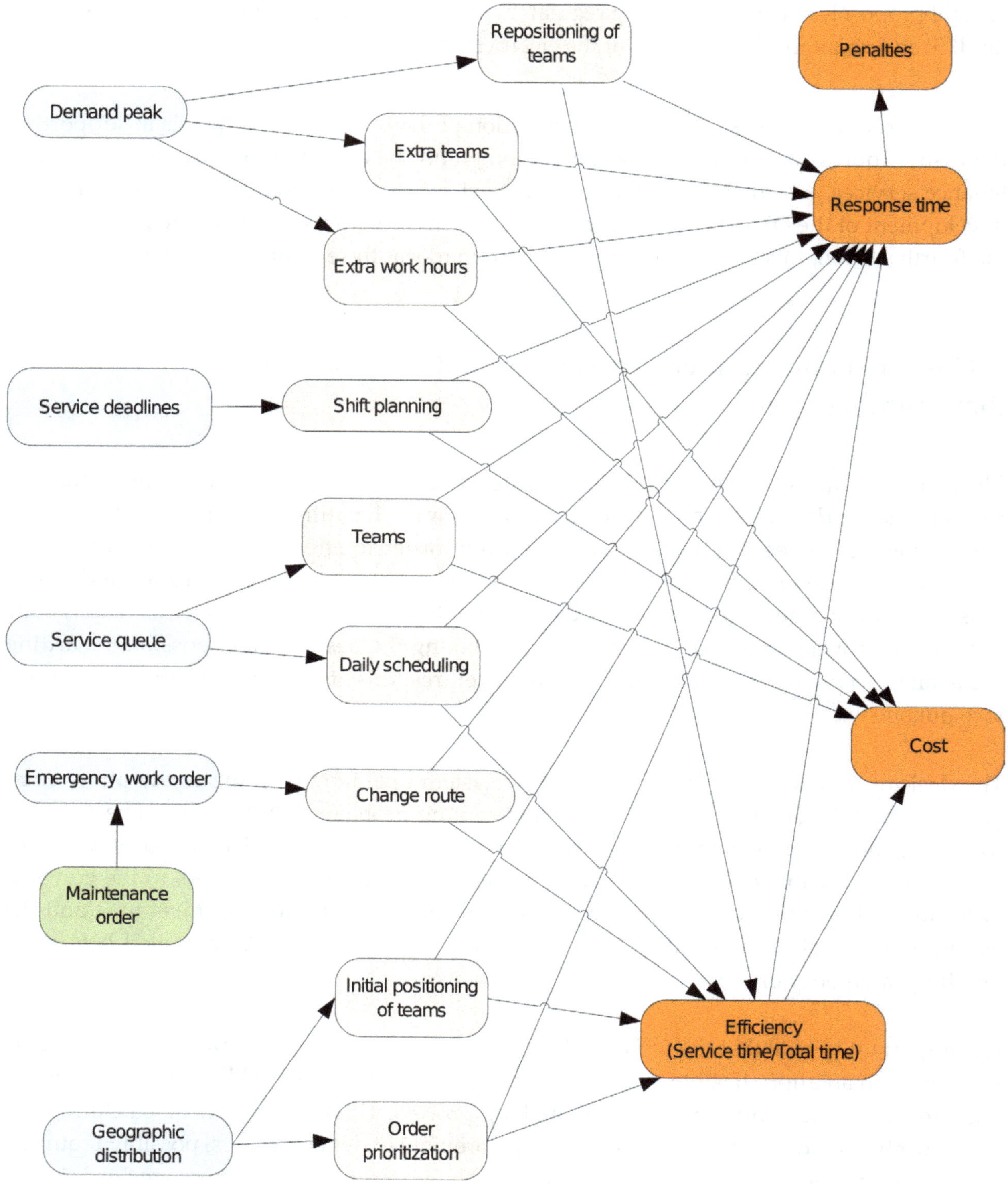

Figure 1. Diagram of the relationship among decision variables.

As shown in **Figure 1**, the main goals of the decision-maker are the increasing of the work efficiency and the reducing of the operational cost, the response time, and, consequently, the penalties incurred. Yet, the response time influences the calculation of the electric power distribution system (EPDS) reliability indices, such that the reduction of the response time has

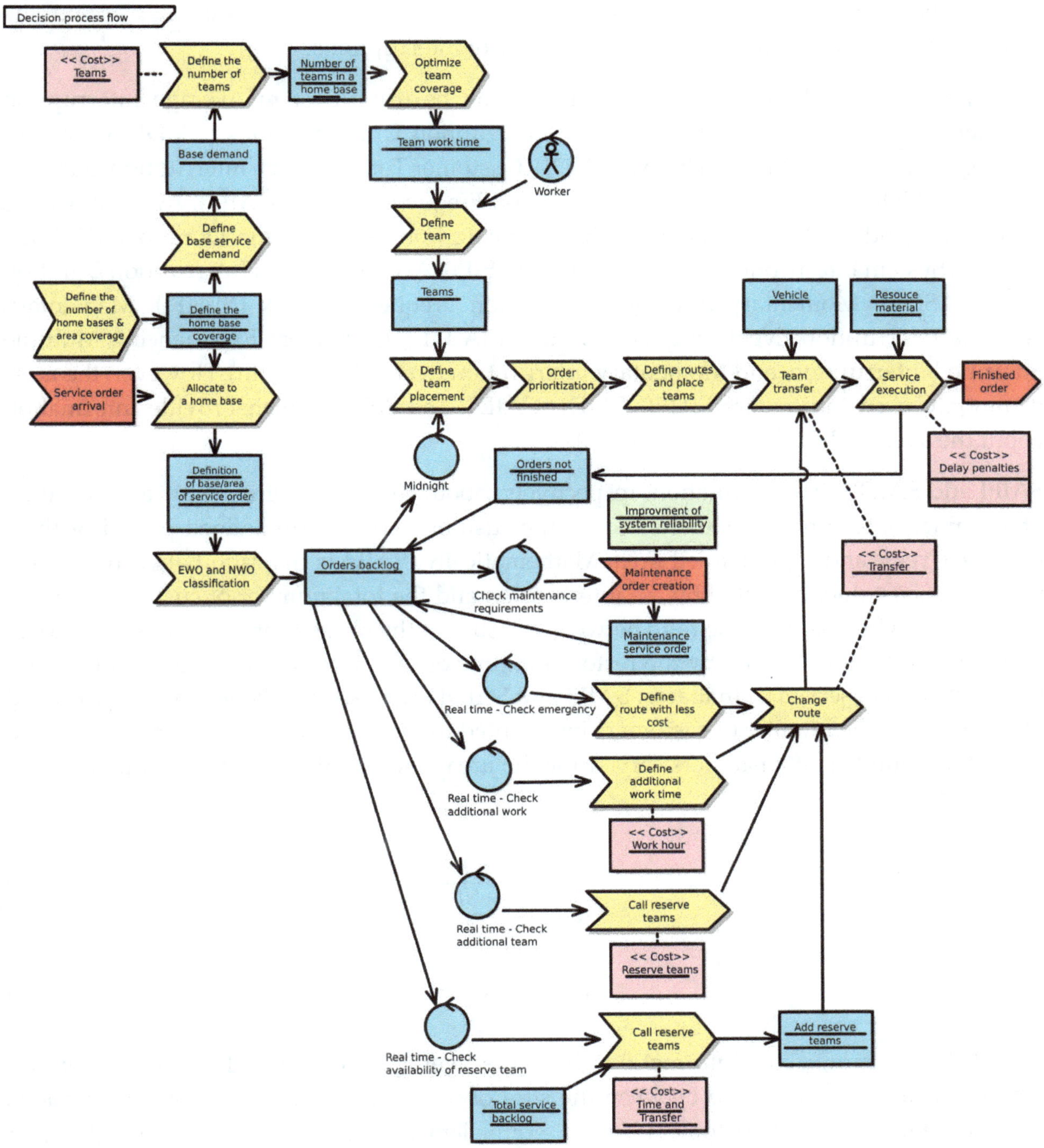

Figure 2. Decision process flowchart.

a positive impact in the improvement of the indices. Therefore, in fact, the decision-making process affects ultimately the EPDS reliability.

The assessment of the EPDS reliability is done by a set of reliability indices, whose definitions and forms of calculation can be found in Ref. [16]. In general, the reliability indices are measurements of the system performance and represent the frequency, duration, and magnitude of the interruptions of the electric energy supply. To this end, their calculations may take into account the number of customers, the connected load, the duration of the interruption, the amount of power interrupted, and the frequency of interruptions.

In summary, the following reliability indices can be used: System Average Interruption Frequency Index (SAIFI), System Average Interruption Duration Index (SAIDI), Customer Average Interruption Duration Index (CAIDI), Customer Total Average Interruption Duration Index (CTAIDI), Customer Average Interruption Frequency Index (CAIFI), Average Service Availability Index (ASAI), Customers Experiencing Long Interruption Durations (CELID), Average System Interruption Frequency Index (ASIFI), Average System Interruption Duration Index (ASIDI), Momentary Average Interruption Frequency Index (MAIFI), Momentary Average Interruption Event Frequency Index (MAIFIE), Customers Experiencing Multiple Sustained Interruption, and Momentary Interruption Events (CEMSMIn). However, the most commonly used indices are SAIFI, SAIDI, CAIDI, and ASAI, which provide information about the average EPDS performance [16].

SAIFI and SAIDI give information, respectively, about the average frequency and duration of sustained interruptions experienced by the customers over a predefined period of time, usually 1 year, and a predefined area. Mathematically, SAIFI is the ratio between the total number of customers affected by the interruption and the total number of customers served in that area, while SAIFI is the ratio between the sum of the number of customers affected by each interruption multiplied by the restoration time of each interruption and the total number of customers served in that area. SAIFI and SAIDI are, respectively, given in Eqs. (1) and (2), where Ni is the number of customers interrupted for each sustained interruption event, Nt is the total number of customers served for the area, and ti is the restoration time for each interruption

$$SAIFI = \frac{\sum Ni}{Nt} \tag{1}$$

$$SAII = \frac{\sum ti \times Ni}{Nt} \tag{2}$$

CAIDI, by its turn, gives information about the average time needed to restore service. Mathematically, it is the ratio between the sum of the durations of customer interruptions and the total number of customers served. Equivalently, it is the ratio between SAIDI and SAIFI, as given in Eq. (3)

$$CAIDI = \frac{SAIDI}{SAIFI} \tag{3}$$

Finally, ASAI represents a fraction of time of the defined reporting period, usually 8760 h (equivalent to 1 year), which the average customer has received power. It is given in Eq. (4), where Nhy is the number of hours in 1 year and is equal to 8760 h in a non-leap year and 8784 h in a leap year

$$ASAI = \frac{Nt \times Nhy - \sum ti \times Ni}{Nt \times Nhy} = 1 - \frac{SAIDI}{Nhy} \tag{4}$$

3. Decision support system design and development

The proposed DSS, named Pladin (a contraction of "dynamic planner"), was designed using a three-tier architecture based on a client/server model, as shown in **Figure 3**. The client/server model was chosen to allow the DSS to support its deployment in a cloud infrastructure. According to Ref. [17], cloud-based systems are attractive for utility companies due to the business model (i.e. memory/computation elasticity, scalability, economies of scale, pay-as-you-go model, etc.), beyond offering more efficiency, productivity, and anywhere access. Moreover, the deployment of a DSS in cloud is also interesting due to the possibly large variability of the required computing resources (e.g. memory and computing power) and the complete integration with web-based GIS portals when required [12].

In order to implement the proposed architecture, modern web technologies have been selected and employed according to the requirements of each layer of the DSS, as shown in **Figure 4**. On the server side, the decision models, the database access, and the middleware were implemented in Java language, where the middleware follows a Spring MVC [18] design pattern and the database access is managed by the Hibernate ORM tool [19]. On the client side, the user interface was implemented in Javascript language, following the React JS [20] and the Redux JS [21] frameworks. Additionally, the client/server data exchange is implemented as a web service Application Programming Interface (API) based on the Representational State Transfer (REST) architecture [22] using JavaScript Object Notation (JSON) data format [23].

The next step in the development process of the DSS was the design of the user interface, where special attention has been given to provide to the user visualization tools, such as maps and graphs, which are valuable for assisting her/him in the decision-making process given the spatial and temporal complexity of the problem at hand. Thus, the user interface has been prototyped and developed, according to the modeled use cases, which are shown in **Figure 5**.

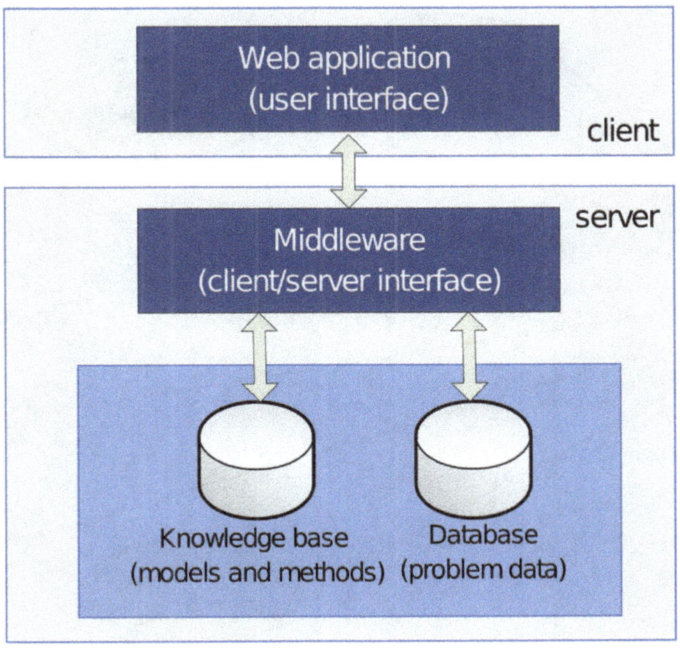

Figure 3. DSS three-layer architecture based on a client/server model.

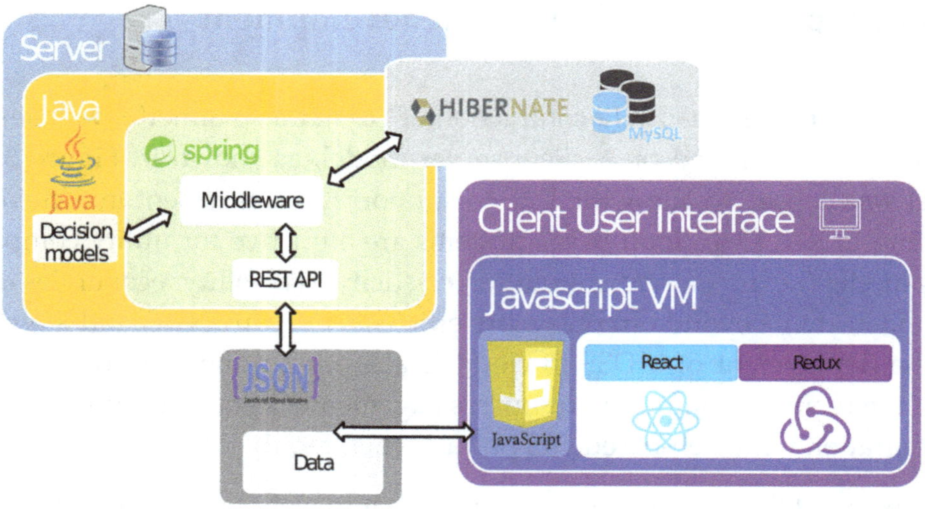

Figure 4. Web technologies used for client and server developments.

Due to the foreseen complexity of the user interface and the evolving requirements, it was decided to develop it as single-page application (SPA). Single-page applications are web applications written in Javascript that almost fully execute on the web browser at the client side, thus similar to desktop applications (i.e. applications that run natively). Some of the advantages of SPAs are flexibility, easy maintenance and deployment, rich visual experience, and a clear separation of concerns.

When developing an SPA, a choice of some architecture can ease the development and the maintenance of the application. There are basically four main architectures used in SPAs: model-view-controller (MVC), model-view-presenter (MVP), model-view-view-model (MVVM), and Flux. Flux is the most recent one and brings several advantages over the other architectures, mainly compared to the well-known MVC.

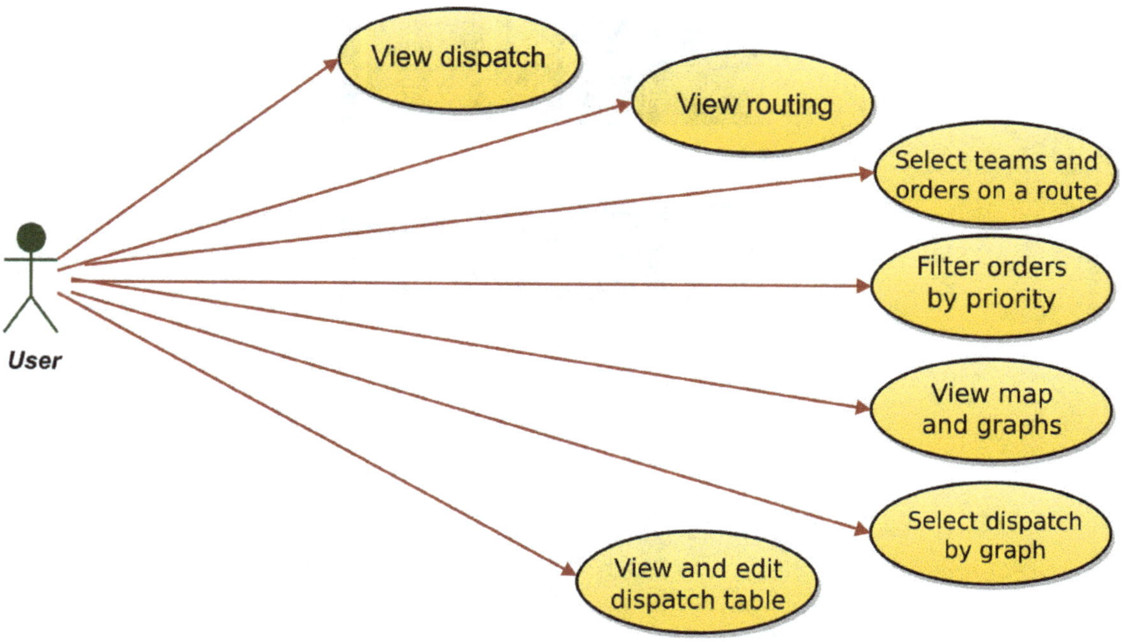

Figure 5. DSS use cases modeling.

In the MVC architecture, as shown in **Figure 6**, the model is responsible for maintaining and updating the application data, the view displays the data to the user (in some visual form) and receives his/her inputs, and the controller is in charge of taking the user inputs to update the model and to notify the view of the updates of the mode. The main principle of the MVC is the separation between the three components, which allows the implementation of different views for the same model and improves testability of the components. However, the MVC architecture is not easily scalable due to the need of adding more instances of the components (models, views, and controllers) as the application grows and the increasing difficulty of reasoning about the logic of the application due to the bidirectional data flows among the components.

The lack of the scalability of the MVC motivated the introduction of the Flux architecture, which adheres to the philosophy of unidirectional data flow. In Flux, as shown in **Figure 7**, there are four components: the action creators, the dispatcher, the stores, and the views. The dispatcher, the stores, and the views are components with independent inputs and outputs, while the action creators are components that create, as the name suggests, special objects known as actions, which carry the data of the application (payload) and a unique property that identifies the action type.

The dispatcher is a central hub that manages all the data flow in the application by maintaining a registry of callbacks into the stores. These callbacks are used as a mechanism for distributing the actions to the stores, thus inverting the control of the application. Actions are generated in the application usually in response to the user interaction with the views or may come from the server. Then, each generated action goes to the stores via the dispatcher and is processed in the stores. The stores contain the application state (data) and logic, and are somewhat similar to the models in MVC. However, the stores can manage the state of many objects and not a single one as models in MVC.

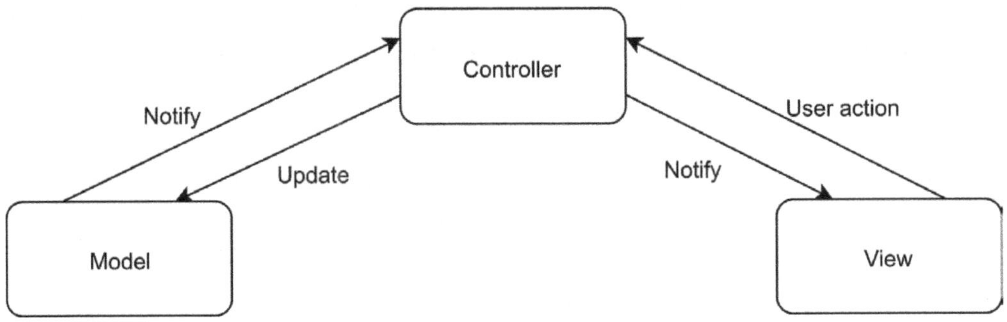

Figure 6. MVC architecture.

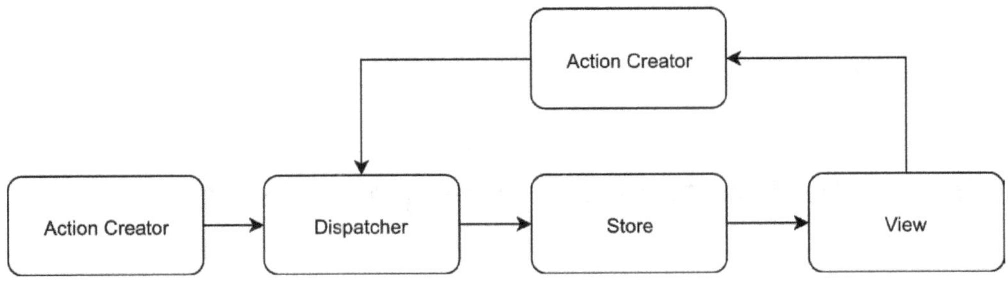

Figure 7. Flux architecture.

After receiving an action, a store interprets it according to the action's type using a switch statement. Then, the store updates its state and broadcasts an event declaring that its state has changed. In turn, special kinds of view, which are known as controller views, listen for events broadcast by the stores that they depend on. These controller views sit on top of a hierarchy of views and are able to obtain the data from the store and pass it down to their descendant views. Usually, the entire state of store is passed down the chain of views, so that views can take the part of the state they need and update themselves accordingly. Finally, views can create new actions based on the user interaction, closing the flow loop.

Basically, the Flux architecture avoids the complexities of the MVC architecture, mainly due to the restriction of unidirectional data flow in the application, which make easy for reasoning about the application workings. However, it is possible to simplify the Flux architecture even further in order for using a single store in the application. This concept is behind of Redux, which is an implementation of the Flux architecture, with added constraints in order to make it a predictable state container for the application, as shown in **Figure 8**.

Redux follows three fundamental principles: single source of truth, read-only state, and changes made with pure functions [21]. The first principle means that state of the whole application is stored in a single store as an object tree, which allows for easier debugging of the application. The second principle implies that it is only possible to change the state emitting actions, which are processed one by one in a strict ordering, avoiding race conditions and simplifying logging and debugging. And the third principle means that a new state of the application is created by a reducer, which is a function that takes as inputs the previous state and the action, and returns the next state without mutating the previous state. Therefore, because the state transitioning in Redux is done by functions (i.e. reducers) instead of event emitters as in the original Flux architecture, the Redux architecture does not have the concept of a dispatcher and allows easier composing of reducers. In fact, due to the several qualities of Redux compared to other alternatives architectures, such as simplicity, scalability, and predictability, it has been chosen as the basis architecture of the proposed DSS.

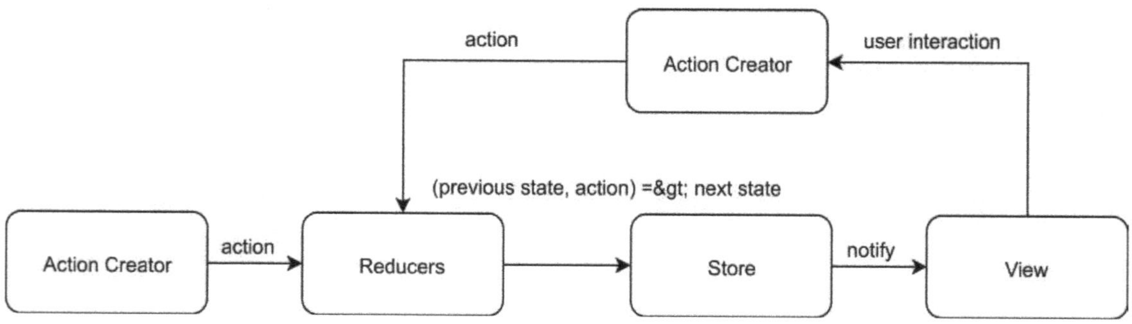

Figure 8. Redux architecture.

4. Evaluation of the developed decision support system

In this section, implemented use cases are presented and discussed. The first one is the "View dispatch" case, where the user can have an overview of a dispatch solution for a given instance of the dispatching problem. The instance is based on the user selection of the position of an operating base and the target date. Then, the user can select from a list one of the available dispatching solutions for the chosen day in order to have a graphical overview of its characteristics, such as the number of crews and orders, the classification and the priority assignment of the orders and the distribution of the execution and travel times by orders and by crews. A sample screen of this use case is shown in **Figure 9**.

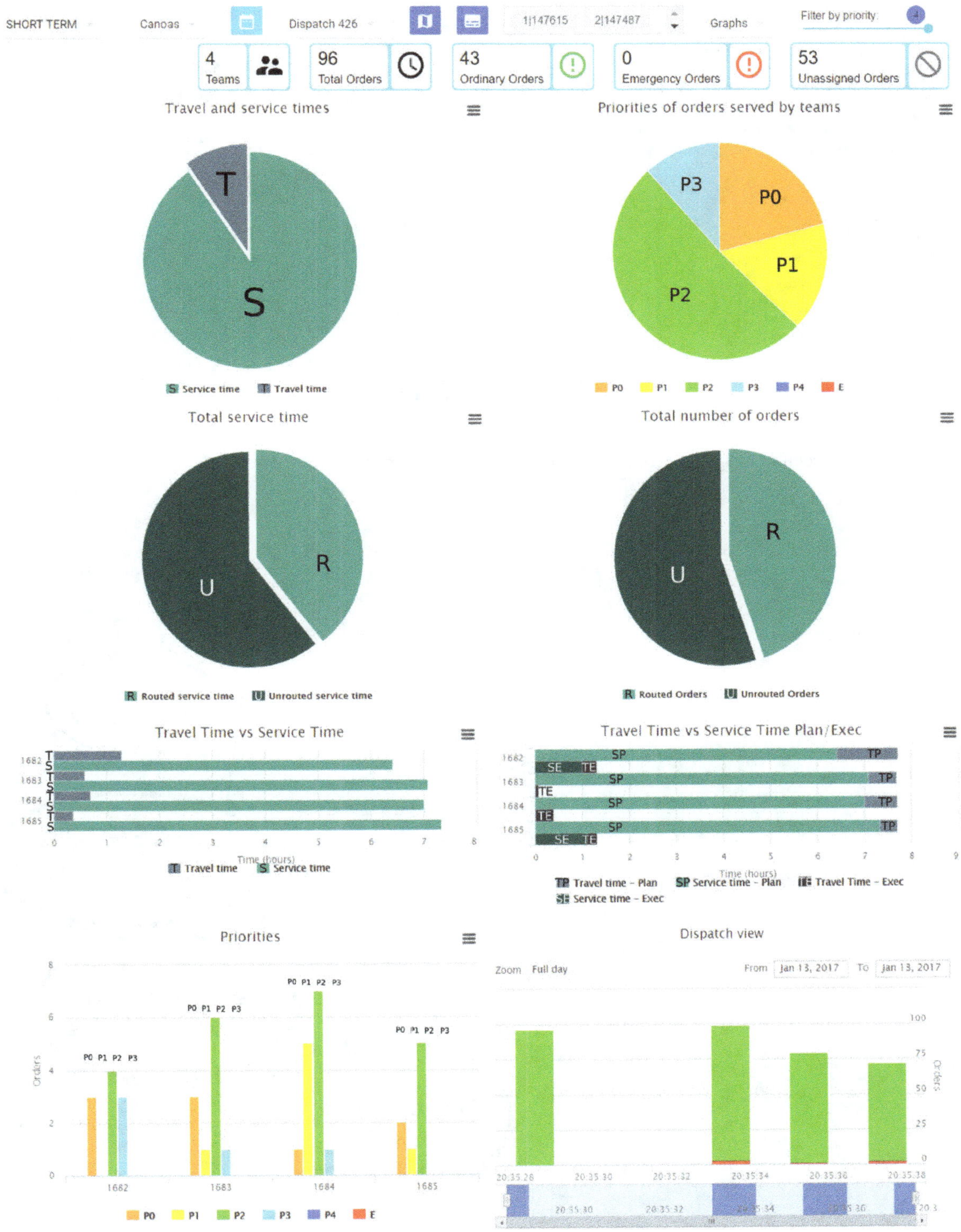

Figure 9. Sample screen of "View dispatch" use case of the Pladin DSS.

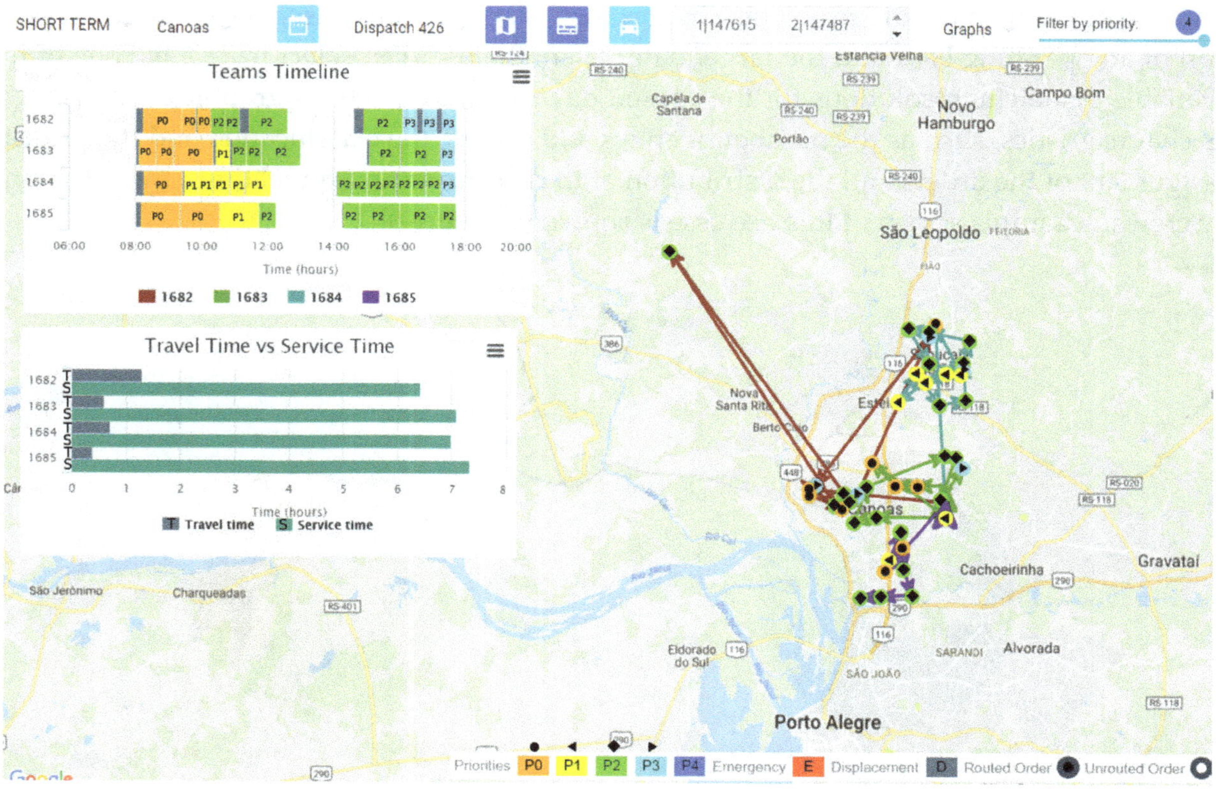

Figure 10. Sample screen of the "View routing" and "Show map and graphs" use cases.

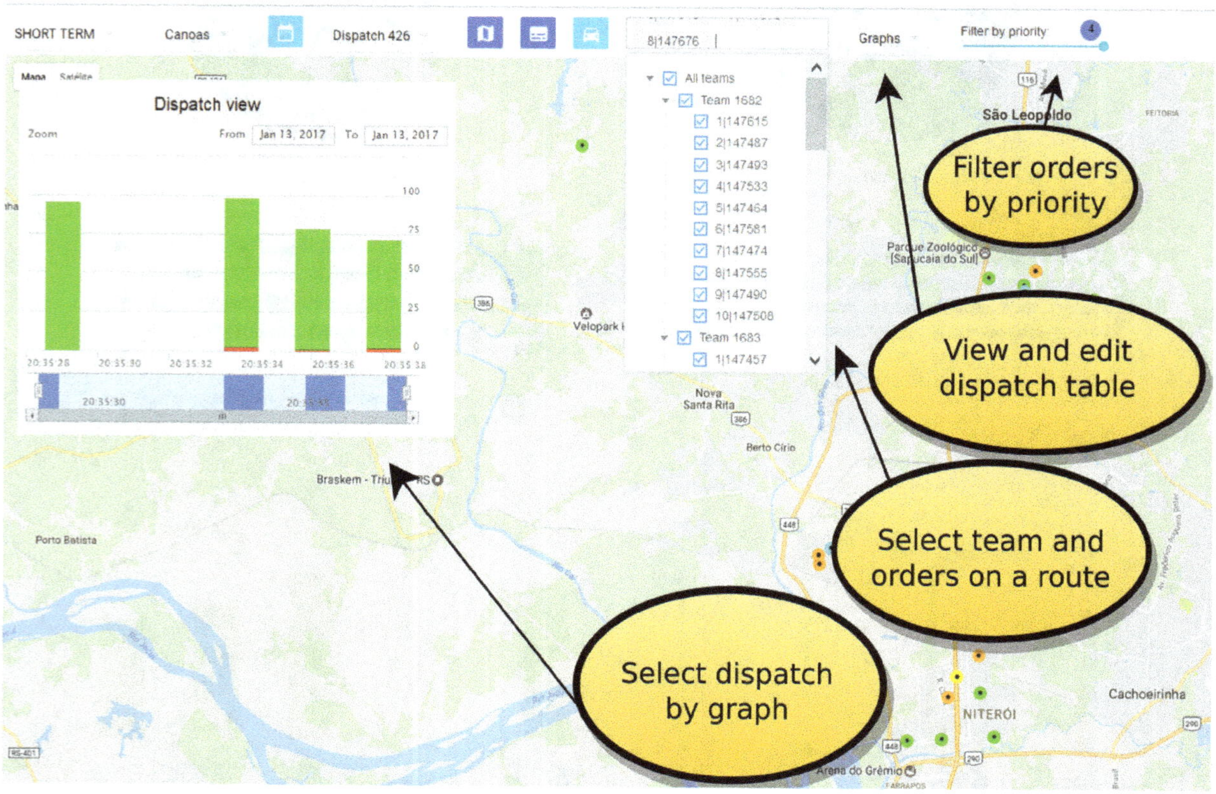

Figure 11. Sample screen highlighting the use cases "Filter orders by priority" and "Select teams and orders on a route," "Select a dispatch by graph," and "View and edit dispatch table".

After the user has selected a dispatching solution, he/she also can view the routes taken for each crew, according to the "View routing" use case. This use case allows the user to analyze the quality of the proposed dispatching solution with assistance of graphical tools, which are provided according to the use case "View map and graphs." A sample screen including both use cases is shown in **Figure 10**.

From the screen shown in **Figure 10**, the user also can access four other functionalities, which are implementations for the use cases "Filter orders by priority," "Select teams and orders on a route," "Select dispatch by graph," and "View and edit dispatch table." These use cases are highlighted in **Figure 11**.

Using the Pladin DSS, the user can better take his/her decisions for the planning and the execution of customer and maintenance services in the electric power distribution system. Therefore, he/she can appropriately allocate the available resources (e.g. teams) for improving the efficiency of the teams by reducing travel and response times. Thus, leading ultimately to reduce operational costs and improving the distribution system reliability.

5. Conclusions

In this chapter, we have introduced a web-based decision support system aimed to aid distribution system operators for planning and executing customer and maintenance services in the electric power distribution system. The DSS provides visualization tools for supporting the user decision to better allocating the available resources in order to solve the emergency dispatching and routing problem in real time.

In addition, the user interface of the DSS also allows for tweaking the input parameters, in order to change the solution for improving the work efficiency and reducing operational costs. Moreover, the user has access to facilities for analyzing what-if scenarios and studying the influence of the decision variables in the solution provided by the DSS.

Finally, due to the simplicity and the scalability of the chosen application architecture, it is expected that future extensions and enhancements will be added for improving it. Currently, the DSS is being used by the system operators working at a utility company located at the South Brazil, and thus the enhancements will be added after receiving the feedback of the current users. Possible enhancements and new use cases include the addition of more visualization tools and analytics services, artificial intelligence techniques, and the integration of more information sources.

Acknowledgements

The authors would like to thank the technical and financial support of RGE Sul Power Utility by project "Planejamento Dinâmico de Operações" (P&D/ANEEL), Coordination for the Improvement of High Level Personnel (CAPES), and the National Center of Scientific and Technological Development (CNPq).

Author details

Carlos Henrique Barriquello[1*], Vinícius Jacques Garcia[1], Magdiel Schmitz[1], Daniel Pinheiro Bernardon[1] and Júlio Schenato Fonini[2]

*Address all correspondence to: barriquello@gmail.com

1 Federal University of Santa Maria, Santa Maria, Brazil

2 RGE Sul, Brazil

References

[1] Bergey PK, Ragsdale CT, Hoskote M. A decision support system for the electrical power districting problem. Decision Support Systems. 2003;**36**(1):1-17. DOI: http://dx.doi.org/10.1016/S1344-6223(02)00033-0

[2] Aneke M, Wang M. Energy storage technologies and real life applications – A state of the art review. Applied Energy. 1 October 2016;**179**:350-377. ISSN 0306-2619. Available from: http://dx.doi.org/10.1016/j.apenergy.2016.06.097

[3] Ipakchi A, Albuyeh F. Grid of the future. IEEE Power and Energy Magazine, March–April 2009;**7**(2):52-62. DOI: 10.1109/MPE.2008.931384

[4] Doğdu E et al. Ontology-centric data modelling and decision support in smart grid applications a distribution service operator perspective. In: Proceedings of the IEEE International Conference on Intelligent Energy and Power Systems (IEPS), 2-6 June 2014; Kiev. New York: IEEE; 2014. pp. 198-204. DOI: 10.1109/IEPS.2014.6874179

[5] Sancho J, Sánchez-Soriano J, Chazarra JA, Aparicio J. Design and implementation of a decision support system for competitive electricity markets. Decision Support Systems. March 2008;**44**(4):765-784. ISSN 0167-9236, http://dx.doi.org/10.1016/j.dss.2007.09.008

[6] Sueyoshi T, Tadiparthi GR. An agent-based decision support system for wholesale electricity market. Decision Support Systems. January 2008;**44**(2):425-446. ISSN 0167-9236, http://dx.doi.org/10.1016/j.dss.2007.05.007

[7] Yao FS, Zhang XQ, Zhang Y and Wang TH. Computer decision-making support system for power distribution network planning based on geographical information system. In: Proceedings of the International Conference on Electricity Distribution, 10-13 December 2008; Guangzhou. New York: IEEE; 2009. pp. 1-6. doi: 10.1109/CICED.2008.5211676

[8] Luo F et al. A practical GIS-based decision-making support system for urban distribution network expansion planning. In: Proceedings of the International Conference on Sustainable Power Generation and Supply; 6-7 April 2009; Nanjing. New York: IEEE; 2009. pp. 1-6. DOI: 10.1109/SUPERGEN.2009.5348306

[9] Li L, Yao-qiang X, Xing-zhi W, Kai W. Study on analysis and decision support system of power grid operation considering meteorological environment based on big data and GIS. In: Proceedings of the International Conference on Electricity Distribution (CICED); 10-13 Aug. 2016; Xi'an. New York: IEEE; 2016, pp. 1-6. DOI: 10.1109/CICED.2016.7576045

[10] Nina DLF, Neto JVdF, Ferreira EFM, Santos AMd. Hybrid support system for decision making based on MLP-ANN, IED and SCADA for disturbances analysis of electrical power distribution transformers. In: 2013 UKSim 15th International Conference on Computer Modelling and Simulation; Cambridge; 2013. pp. 12-20. DOI: 10.1109/UKSim.2013.147

[11] Ghandforoush P, Sen TK, Wander M. A decision support system for electric utilities: Compliance with clean air act. Decision Support Systems. October 1999;**26**(4):261-273. ISSN 0167-9236. Available from: http://dx.doi.org/10.1016/S0167-9236(99)00056-1

[12] Boulmier A, White J, Abdennadher N. Towards a Cloud Based Decision Support System for Solar Map Generation. In: Proceedings of the IEEE International Conference on Cloud Computing Technology and Science (CloudCom); 12-15 Dec. 2016; Luxembourg City,.New York: IEEE; 2017, pp. 230-236. DOI: 10.1109/CloudCom.2016.0047

[13] Perea E et al. Decision support system for distributed energy resources and efficient utilisation of energy in buildings. In: Proceedings of the 22nd International Conference and Exhibition on Electricity Distribution (CIRED 2013); 10-13 June 213; Stockholm. New York: IEEE; 2013, pp. 1-4. DOI: 10.1049/cp.2013.0557

[14] Sianaki OA, Hussain O, Dillon T, Tabesh AR. Intelligent Decision Support System for Including Consumers' Preferences in Residential Energy Consumption in Smart Grid. In: Proceedings of the Second International Conference on Computational Intelligence, Modelling and Simulation; 28-30 Sept. 2010; Bali. New York: IEEE; 2011, pp. 154-159. DOI: 10.1109/CIMSiM.2010.84

[15] Monteiro C et al. Spatial decision support system for site permitting of distributed generation facilities. In: Proceedings of the IEEE Porto Power Tech Proceedings (Cat. No.01EX502); 10-13 Sept. 2001; Porto. New York: IEEE; 2002. pp. 6 pp. vol.3. DOI: 10.1109/PTC.2001.964892

[16] IEEE Guide for Electric Power Distribution Reliability Indices. In: IEEE Std 1366-2012 (Revision of IEEE Std 1366-2003), May 31 2012. New York: IEEE; 2012. vol., no., pp.1-43. DOI: 10.1109/IEEESTD.2012.6209381

[17] Qin YB, Housell J, Rodero I. Cloud-Based Data Analytics Framework for Autonomic Smart Grid Management. In: Proceedings of the International Conference on Cloud and Autonomic Computing; 8-12 Sept. 2014; London. New York: IEEE; 2015, pp. 97-100. DOI: 10.1109/ICCAC.2014.39

[18] Spring MVC official website [Internet]. 2017. Available from: spring.io [Accessed: 2017-02-19]

[19] Hibernate ORM official website [Internet]. 2017. Available from: hibernate.org/orm [Accessed: 2017-02-19]

[20] React JS official website [Internet]. 2017. Available from: facebook.github.io/react [Accessed: 2017-02-19]

[21] Redux JS official website [Internet]. 2015. Available from: redux.js.org [Accessed: 2017-03-10]

[22] Fielding RT. Chapter 5: Representational State Transfer (REST). Architectural styles and the design of network-based software architectures [Ph.D.]. Irvine: University of California; 2000

[23] Bray T. RFC7159—The JavaScript Object Notation (JSON) Data Interchange Format [Internet]. March 2014. Available from: tools.ietf.org/html/rfc7159

6

Down Time Terms and Information Used for Assessment of Equipment Reliability and Maintenance Performance

Jon T. Selvik and Eric P. Ford

Abstract

Reliability and maintenance data is important for predictive analysis related to equipment downtime in the oil and gas industry. For example, downtime data together with equipment reliability data is vital for improving system designs, for optimizing maintenance and in estimating the potential for hazardous events that could harm both people and the environment. The quality is largely influenced by the repair time taxonomy, such as the measures used to define downtime linked to equipment failures. However, although it is important to achieve high quality from maintenance operations as part of this picture, these often seem to receive less focus compared to reliability aspects. Literature and experiences from, e.g., the OREDA project suggest several challenging issues, which we discuss in this chapter, e.g., for the interpretation of "MTTR." Another challenge relates to the duration of maintenance activities. For example, while performing corrective maintenance on an item, one could also be working on several other items while being on site. This provides an opening for different ways of recording the mobilization time and repair time, which may then influence the data used for predictive analysis. Some relevant examples are included to illustrate some of the challenges posed, and some remedial actions are proposed.

Keywords: reliability, maintenance, data collection, downtime, MTTR, detection time

1. Introduction

Equipment reliability and maintenance (RM) data are widely collected in the oil and gas industry and are needed for predictive analysis, e.g., for oil and gas production systems, to achieve safe and cost-efficient solutions. An important benefit of such activity is optimized maintenance [1], for example, when finding the optimal inspection intervals for pipelines (see e.g., [2–4]) or when deciding upon the appropriate testing modes for safety instrumented systems (see e.g., [5]).

In the current chapter, we focus on the maintenance part of the RM data collection and mainly on the measures linked to maintenance activities associated with failed items, i.e., equipment with downtime. Thus, attention in this chapter is primarily on the time interval from when a failure of a reparable item occurs to the time when it is back in an upstate.

Updated information about equipment downtime is important and relevant for various types of analysis within the oil and gas industry to inform decision-making. In particular, the measure "mean time to repair," commonly abbreviated as "MTTR" is widely used within this industry for the purpose of reliability, availability and maintainability analysis (see e.g., [6–9]). It is also widely used in, for example, design planning (e.g., [10, 11]) and also in relation to safety integrity level verification for safety instrumented systems (e.g., [5, 12, 13]).

However, there are several obstacles related to the collection of such data. One of the challenges is simply to get sufficiently high quality in the data for downtime predictions, for example, for assessment of the expected repairing time, i.e., the "MTTR." The problem is typically twofold; the relevant population is too small (e.g., technology is changing, and old data, which has taken years to collect, may not be relevant anymore), and the taxonomy used for the data collection may not be sufficiently clear, which may give room for interpreting information differently.

This is the reason why the international standard ISO 14224 [14] is such an important document. The document is partly a result of industry field feedback implemented for more than 25 years within the OREDA project (see [1, 15]), which has led to the ISO 14224 standard [16]. It represents a main guidance document for collection of RM data within the oil and gas industry, applicable in par with the dependability standard IEC 60300-3-2 [17], and outlines principles and taxonomy for how to achieve quality data. It is a way of ensuring data quality, e.g., through a consistent taxonomy being used (see e.g., [18, 19]).

A much-ignored issue when collecting downtime data is the difficulty to measure the time to detect the failure, i.e., the exact time of failure. This time is normally referred to as the "detection time" (TD) (see e.g., [14]). It is a key value needed, for example, when assessing the expected time to achieve the repair of the failed item, i.e., the so-called mean time to repair, widely referred to by the abbreviation "MTTR." The main problem is that, in situations where failures are hidden and these are not evident before some demand occurs, it is most difficult for a data collector to assess or specify the exact time of the failures. Besides, it may not be possible to confirm whether in fact these values are true or not. Often, rather one attempts to ignore the issue by claiming that the precision of this value is negligible, as $TD \ll MTTR$, or $MTTR \ll MTTF$, where MTTF is the common abbreviation for the "mean time to failure." In other words, the numeric value of TD is assumed to be of low importance to the predictive analysis.

In addition, when using the data on repair times from a database, one normally mixes together situations where the failure is detected at once (being in continuous mode) and those situations where the failure is hidden (or dormant) for a while until a demand occurs (e.g., revealed from a functional test of the equipment). Consequently, to guide for more consistent data, the newly issued [14] has tried to limit the use of MTTR, despite the strong position and frequent use of this measure in the oil and gas industry.

Similarly, the time to mobilize or carry out the repair may also be subject to uncertainty, depending on how the data collector interprets these measures. Having a consistent interpretation of the key terms used to predict downtime is crucial in order to obtain high-quality maintenance data.

An objective of this chapter is to critically examine key terms and associated measures used in maintenance data collection, by studying the repair time taxonomy defined in Ref. [14], i.e., the different terms used in data collection to define downtime in relation to failed items (see Section 2).

The remaining part of the chapter is structured as follows. Section 2 gives a brief description and definitions of key measures, mostly based on the terminology in [14, 20], such as the MTTR. Section 3 provides two example cases illustrating the effects of using different interpretations of the MTTR. Section 4 links data collection to decision-making and presents findings from data collection experiences identified from published literature, addressing several challenging issues that could compromise the maintenance data quality and use. Then Section 5 provides a discussion on to what extent the industry is in line with ISO 14224 taxonomy and how to handle the issues identified. In Section 6 we give some conclusions.

2. Key downtime terms and measures in maintenance data collection

As indicated in the previous section, ISO 14224 [14] gives guidance both on how to collect and how to analyze downtime data and is a document strongly linked to ISO/TR 12489 [21] on reliability modeling and also the two dependability documents [17, 20].

The recently issued [14] specifies four different maintenance data categories; see **Table 1**, where details are given on what is the minimum data that must be collected. For example, any corrective maintenance action shall be paired with the associated failure event (i.e., failure record). Downtime data is also labeled as minimum, meaning that the data collector must specify the total length of the downtime interval.

It is common to split the downtime interval in three main parts, i.e., the active repair time, and the activities before and after (such as waiting, delay, start-up). Obtaining accurate information regarding these down time segments is challenging. One such example is already mentioned, i.e., the detection time.

Another example is the "active repair time," which could easily be confused with the "active maintenance time." [14] distinguishes between similar terms such as the "active maintenance time," "active repair time" and "overall repairing time" and uses these when addressing

Data category	Examples	Minimum data
Identification	Equipment tag number, failure reference	Unique maintenance identification, equipment identification/location, failure record
Maintenance data	Maintenance category = preventive or corrective	Date of maintenance, maintenance category
Maintenance resources	Type of main resource(s) and number of days used, e.g., drilling rig, diving vessel, service vessel	(No data specified as minimum)
Maintenance times	Time duration for active maintenance work being done on the equipment	Active maintenance time, downtime

Table 1. Maintenance data categories (based on ([14], Table 8)).

measures, i.e., expected values, such as the "mean active repair time," "mean overall repairing time," "mean time to restoration" and "mean time to repair." We will further address and discuss the meaning of these terms and measures in the following sections.

2.1. Mobilization time

The mobilization time is normally a main part of the repair preparations. It includes mobilization of all types of resources required, such as vessels, personnel and ROVs. It includes all activities carried out to get the necessary resources available to execute the active repair of the failed items.

In Ref. [14], there is also a relevant note to the entry associated to the definition, which states that "time spent before starting the maintenance is dependent on access to resources, e.g., spare parts, tools, personnel, subsea intervention and support vessels." The mobilization time is therefore sometimes difficult to distinguish from delays caused by manufacturing time and transportation.

In practical terms, the mobilization of intervention vessels is often described using the term "opportunity maintenance," meaning that the intervention vessel is on site or called for in relation to other activities. For example, the vessel could already be on site when the failure is detected, or some critical failure is somehow making the mobilization more urgent and prioritized. To deal with such situations, it is important to have clear procedures for how to collect the mobilization data. Typically, the mobilization time T_M is specified as $T_M = 0$, if other items are mainly responsible for the vessel order. However, for analysis purposes, it is important to be aware of which maintenance activities are included in the order and which are not.

Information about an intervention vessel can also be found in long-term schedules, where the time when the vessel is on site largely depends on the planned route. For example, the vessel could plan for 30 days at a "Site A," then 30 days at a "Site B" (which could then be operated by a different company), then at "Site C," etc. Mobilization will then depend on both whether the item is critical for production and safety and whether the intervention vessel by chance is at the site or soon is coming to this site.

Furthermore, it is not straightforward how to cope with the issue of multiple maintenance and mobilization activities. Often, several maintenance calls are issued, and the allocation of a maintenance vessel needs to be cost-efficient. Thus, one may find that there is a need to remobilize the vessel, as it was not able to complete the maintenance actions on site before moving to the next location.

2.2. Active maintenance time

The active maintenance time is, as defined in Ref. [20], "Duration of the maintenance action, excluding logistic delays." Per the definition, other delays such as technical delays are thus included in the active maintenance time. The measure is often referred to as "the effective time to repair" (see e.g., [12]). This is regardless of whether the repair is performed in one go.

However, there is nothing stating that the maintenance activity must be completed during the downtime. Hence, parts of the maintenance, or in some situations (and especially for preventive maintenance) most of the maintenance, relate to an upstate of the equipment. As mentioned, several separate activities may be performed before the maintenance is completed. For example, one may try different options and run checks on whether the performance is satisfying, or one might have to wait for equipment parts to arrive, etc. However, the time used to run-down or start-up the equipment is considered as part of the uptime and not the downtime.

In practice, the active maintenance time could include various testing of the equipment, to check its condition. Depending on the urgency of getting the item back into operation, the duration of this activity might be significant. This also relates to, e.g., wells that are alternating in production, which usually makes the maintenance activity less urgent and allows for more "experimentation." The most cost-efficient solution is then perhaps not the most time-efficient one. For example, one could opt to use time-demanding repairing tools with lower cost and lower efficiency.

The active maintenance time is, by including technical delay, interpreted in Ref. [14] as, "the calendar time during which maintenance work on the item is actually performed." Although normally not the case, it is possible for the active maintenance time interval to be larger than the total downtime. This would be true in situations where the equipment is running in operation while the maintenance activity is ongoing.

2.3. Active repair time and mean active repair time

The effective time needed to achieve the repair of an item is called the "active repair time" (see [14]) and is a key part of the downtime as shown in **Table 2**, in Phase No. 3. It consists of, per the IEC 60050-192 [20], three possible activities, i.e., the fault localization time, the fault correction time and the function checkout time. See also ([21], **Figure 5**), which compares repair time taxonomies provided in [20–22] (currently also available in the International Association of Drilling Contractors (IADC) Lexicon definition of "mean active repair time" [23].

The active repair time is different from the active maintenance time, as the preparations and delays are not included (Phase No. 2 and No. 4). But, it is not necessarily the same as the number

Phase No.	Description	State of equipment
1	Time of failure and then run-down	Uptime
2	Preparations and delays	Downtime
3	Active repair time	Downtime
4	Waiting and delays	Downtime
5	Start-up	Uptime

Table 2. Equipment state categorization.

of man-hours needed to achieve the repair. The number of man-hours also relates to the amount of personnel working with the repair and is thus not directly linked to the active repair time. Similarly, the time of resource use may only capture a part of, and give a misleading picture of, the overall active repair time. For example, an ROV is often used for subsea maintenance operations, but the use of the actual ROV time may be split on several simultaneous maintenance operations. There may be several other ROV activities carried out, while the ROV is subsea.

The expectation of the active repair time is a relevant key performance indicator (KPI) for downtime, i.e., the mean active repair time (MART). It is listed in ([14], Annex E), to have purpose and value through "indication of the productivity and work content of repair activities." It is noted that if one is also interested in the preparation and delay times, then the mean downtime (MDT), comprising Phase Nos. 2–4, is a relevant KPI or measure.

2.4. Overall repairing time and mean overall repairing time

ISO/TR 12489:2013 defines the mean overall repairing time (MRT) as the "expected time to achieve the following actions:

- The time spent before starting the repair; and,
- The effective time to repair; and,
- The time before the item is made available to be put back into operation."

[12] gives the same understanding of the elements included but instead refers to MRT as the "mean repair time," which although somewhat similar introduces a variation of the term that is found in practice but which is distinct from the "mean time to repair" (see Section 2.6).

By including the time spent before starting the repair and the time to prepare the item for operation, the measure is synonymous with the MDT when fault detection time is equal to zero. The MDT is simply the expectation of the downtime, i.e., the mean time the equipment is not in a standby or operating mode (upstate).

2.5. Detection time

Detection time is the period from when a fault occurs to the time when this is detected, where fault in this context refers to the equipment being unable to perform as required due to some internal state, in line with the definition in Ref. [20]. The term "fault detection time" is sometimes used for more specificity, as, e.g., in Ref. [21]. However, this term is not the same as the "fault localization time," i.e., the time taken to complete fault localization, although the two terms appear similar. Fault localization takes place after the fault is detected and during the period of corrective maintenance action. Fault localization often includes the activity of diagnosing at what time the fault occurred (see e.g., [24]).

Fault detection may be achieved through manual or automatic operations, depending on modes of operation and system characteristics. Faults of safety systems with possible long detection time, e.g., revealed through functional testing, often make it challenging to identify the exact detection time. Assumptions and estimates are then normally made based on the testing intervals.

The expectation of the fault detection time is called the "mean fault detection time" and is abbreviated as the "MFDT" (see e.g., [21]). For immediately revealed failures, the value of MFDT is equal to zero and negligible for failures with short detection time. Otherwise, this value strongly depends on the test policy for the equipment. For hidden failures, the detection time may represent the main part of the downtime.

Sometimes, for assessment of reliability and maintenance performance, the abbreviation "MFDT" is also used for the "mean fractional dead time." This term has a completely different meaning, i.e., a measure for the average unavailability expressing the expected fraction of time in a nonfunctional condition (refer to use in, e.g., ([25], p. 428, [26])). Obviously, the two terms should not be mixed.

2.6. Restoration time, mean time to restoration and mean time to repair

Restoration time (or time to restoration; see [14]) includes also the fault detection time in addition to the elements comprised by the overall repairing time (see above). The expectation of this value, i.e., the mean time to restoration (MTTRes), thus includes the full picture of:

- Fault detection time.
- Preparation and delays (administrative, logistic and technical delays).
- Active repair time.
- Delays after the item is repaired (mainly administrative).

The variation in the meaning of MTTR as the 'mean time to restoration' versus the 'mean time to repair' makes it unclear whether all the four elements are captured, i.e. the full picture above. It is often challenging to separate what definition is used in practice by only looking at the abbreviation "MTTR." The change of the meaning of MTTR in 1999 (see [27]) has the engineering population still divided between those using the present time definition, "mean time to restoration" and those keeping with the old definition, "mean time to repair" and a reluctance to change [28].

[14] also defines the mean time to repair (MTTR) as "expected time to achieve repair of a failed item." The problem with this definition is the fault detection time, which is either zero when the failure is revealed immediately, or it is unknown. If it is possible to include the detection time, the MTTR is equal to the MTTRes; otherwise MRT is considered a more appropriate

measure. When using the MTTR, the meaning could be all of the three measures above, as illustrated in **Figure 1**, depending on the length of fault detection, active repair time and preparations and delays, which is causing unnecessary confusion to data collection and analysis. [21] therefore avoids the use of the term MTTR.

2.7. Availability measures: intrinsic availability

The length of downtime and associated measures are important in computations of availability, where availability is often estimated (e.g., by Monte Carlo simulations) by the use of terms such as the MTTR and MTTF. This is the case when calculating the intrinsic (or inherent) availability for some component (A_I), where one considers the corrective maintenance downtime of the system:

$$A_I = MTTF/(MTTF + MTTR) \tag{1}$$

where MTTF is the mean time to failure. However, as already mentioned [14] notes that it is more appropriate to refer to the active maintenance time observed in the field, i.e., the MTTRes, which represents a more meaningful term compared with, e.g., the MTTR. The formula should therefore be instead.

$$A_I = MTTF/(MTTF + MTTRes) \tag{2}$$

The MTTRes here is not the same as the mean downtime, MDT, although the two measures may have the same value. Replacing MDT with MTTRes, and MTTF with mean uptime (MUT), would give the operational availability instead of the A_I and is perhaps a more relevant measure from a maintenance perspective. Both are used to express proportion of time that the equipment or system is in an upstate, but the MDT is generally not considered an intrinsic property. The duration of MDT could in practice include a variety of delays (e.g., detection, isolation, spare parts, standby, repair duration, reinstatement, etc.; see ([14], Annex C)). The mean up- and downtime are measures that depend strongly on the system performance, i.e.,

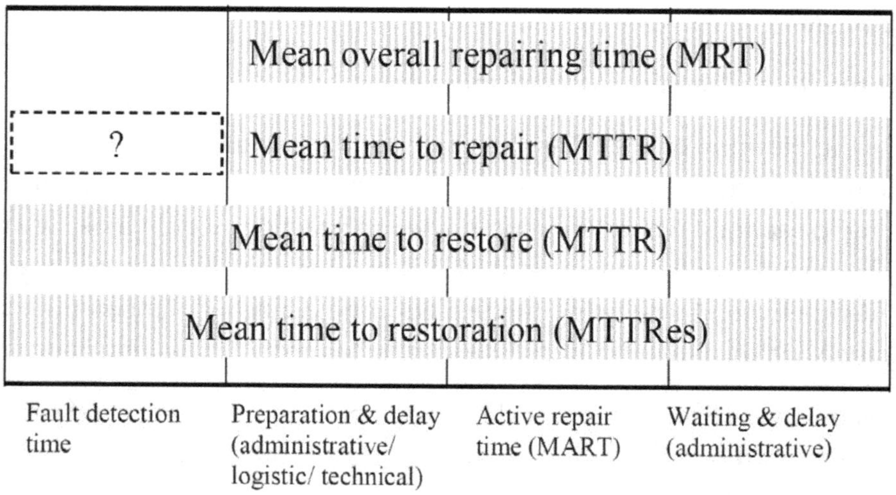

Figure 1. Different measures that could represent the mean time to repair (MTTR).

reliability, maintainability and maintenance support, and therefore dependent on the context in which they are used. The MTTRes, by focusing on the maintenance resources and disregarding external resources, thus allows for a more intrinsic analysis.

3. Significance of taxonomical differences

To illustrate the effects of different interpretations of maintenance times, two example cases are provided that illustrate the challenges related to situations where failures are detected with short-time versus long-time intervals. In the first case, we use a data set for subsea control modules (SCMs) obtained from the OREDA database. In the second case, we have constructed a data set for downhole safety valves (DHSVs) based on reliability data collected annually by the Petroleum Safety Authority Norway (PSA) to analyze and map the risk level on the Norwegian Continental Shelf, the so-called RNNP project (see e.g., [29]). Both data sets have been randomized and anonymized for confidentially reasons and are thus solely for illustration purposes.

3.1. Example I: short detection time

The data set consists of 375 SCMs, with a total time in service of 1.44×10^7 hour. During this period, a total of 255 failures have occurred (counting any type of failure severity, i.e., critical, degraded or incipient). An estimator for the mean time to failure, MTTF, is thus given as $1.44 \times 10^7/255 = 56{,}471$ hour. We now consider the implications of varying definitions of the repair time, in this context referred to simply as MTTR. Let us assume that there is uncertainty on all of the components that together form the MTTRes, i.e., fault detection time, administrative delay prior to repair, logistic delay, technical delay, active repair time and administrative delay post repair. For simplicity, we will assume that each of these parameters is represented by a triangular distribution denoted T1:

- T1 (1; 8; 20) hours, i.e., minimum = 1 hour, peak value = 8 hour and maximum value = 20 hour.
- All of the components of the MTTRes are assumed to be independent of each other.

This distribution is not based on any real data set, but delays and repair times on an SCM could realistically be at least 20 hour.

Recalling from the previous chapter we have:

- MART = active repair time (fault localization + correction + function checkout).
- MRT (overall repair time) = MART + administrative delay prior to repair + logistical delay + technical delay + administrative delay post repair.
- MTTRes (time to restoration) = MRT + fault detection time.

Thus, we can establish three definitions for the intrinsic (inherent, technical) system availability, A:

$$A_1 = \text{MTTF}/(\text{MTTF} + \text{MTTRes}) \tag{3}$$

$$A_2 = \text{MTTF}/(\text{MTTF} + \text{MRT}) \tag{4}$$

$$A_3 = \text{MTTF}/(\text{MTTF} + \text{MART}) \tag{5}$$

For now, we assume that the MTTF is deterministic, and thus all uncertainty is placed on the MTTR component. Running a Monte Carlo simulation using N = 10,000 gives the distributions for A_1, A_2 and A_3 shown in **Figure 2**.

As **Figure 2** shows, widening the definition of repair time to include delays and fault detection will lower the system availability and increase the standard deviation, as there is obviously more uncertainty. The deviations between the three definitions of repair time are not significant in this particular case, as the MTTF is relatively much larger. Consider however **Figure 3**, where a sensitivity analysis is run for the MTTF, showing the expected absolute and relative difference between A_1 and A_3 (blue line) and A_2 and A_3 (red line), where A_3 is a main element of both A_2 and A_3, i.e., the differences thus indicating the contribution of detection time. As the MTTF becomes lower, the significance in the varying interpretation of repair time becomes as expected greater.

Figure 3 also shows the expected relative difference between the two measures, showing a decreasing value in the approximate MTTF interval = [500; 10,000]. Indicating that especially when the MTTF is higher than 10,000 hours, i.e., around 1 year, it is fully acceptable to ignore the contribution of detection time in such calculations.

While there is the obvious point that MTTR generally plays a more important part of availability the lower the MTTF is, there are also other reasons why the MTTR generally is skewed

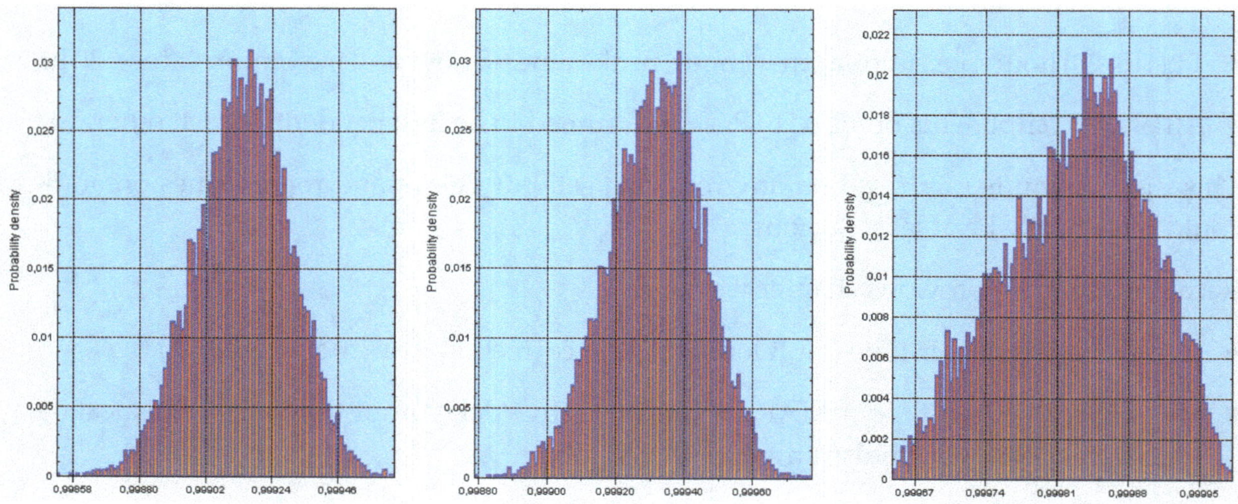

Figure 2. Intrinsic availability distribution for A_1, (left) using MTTR = MTTRes. Mean = 99.915% and stdev = 1.5×10^{-4}. A_2, (center) using MTTR = MRT. Mean = 99.932% and stdev = 1.4×10^{-4}. A_3, (right) using MTTR = MART. Mean = 99.983% and stdev = 7.0×10^{-5}.

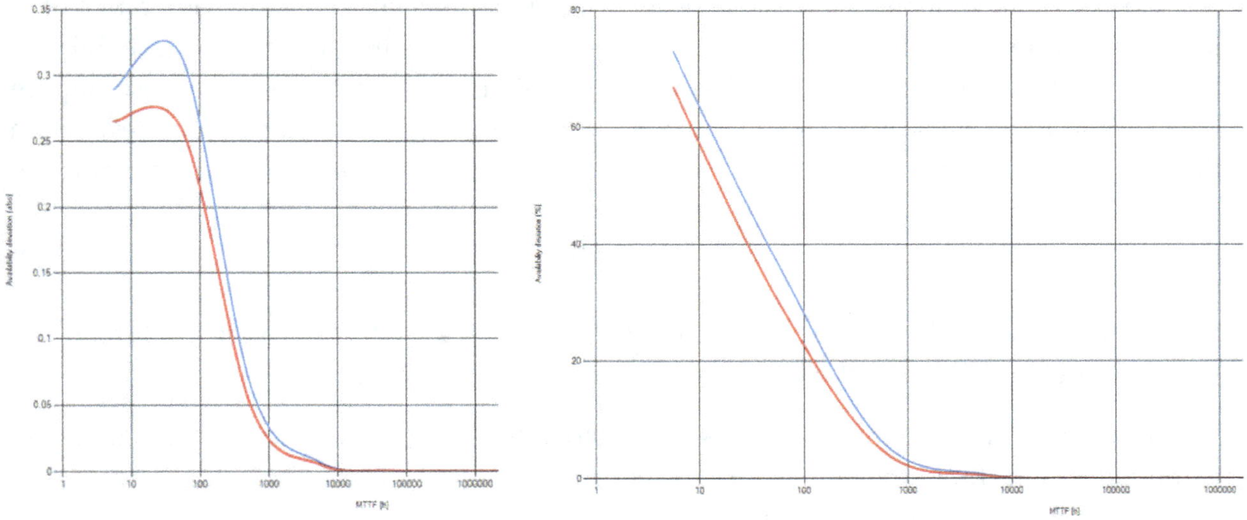

Figure 3. Expected absolute (left) and expected relative (right) difference between A_1 and A_3 (upper line) and A_2 and A_3 (lower line), as a function of varying MTTF [hours].

towards the right of **Figure 2**. When collecting subsea data, for example, for OREDA, this process is often time-demanding and costly; see e.g., [15] when done manually, which is often the case. For this reason, the priority is often to collect failures and any maintenance related directly to these. This comes at the cost of sometimes disregarding opportunity maintenance or other types of no-failure maintenance, where the equipment in question is actually in a downtime state, thus overestimating the total time in service and consequently also the MTTF. Essentially, the actual MTTF is bound to be lower than what is often used as an estimate, and thus the difference in repair time definition becomes greater. In addition to component-related maintenance, there are also at times planned shutdowns of wells or even fields, which are not always captured for the same reasons and which also emphasize the point made. Furthermore, as mentioned in Section 2.1, there is the difficulty of distinguishing mobilization time from delays caused by manufacturing time and transportation, meaning there is a possibility of potentially both longer or shorter times added to, or subtracted from, extended definitions of MTTR. According to Ref. [15], quality checks on various equipment items (not necessarily subsea items) yielded wrong interpretations or coding used during data collection in 39% of the cases. Such errors could also swing both the MTTF and the MTTR in either direction but will certainly give rise to variations in the expected system availability.

3.2. Example II: hidden failures

A main objective of this second example case is to address the important issue of hidden failures (also called dormant failures). These are failures that, according to Ref. [14], are not immediately evident to operations and maintenance personnel. This means that it may take some time before detection, as it is not possible to detect these failures unless some specific action, such as a periodic test, is performed.

For the second data set, we refer to a population of 8714 DHSV tests collected from 73 facilities operating at the Norwegian Continental Shelf during the time period in 2012–2016 from the

RNNP project [29], which in contrast to the SCM population is taken from a wide range of oil and gas operators. The availability requirements refer to an industry standard (see **Table 3**) based on the requirements set by the Norwegian oil and gas operator Statoil. These requirements are also referred to as the "failure fraction," FF, the ratio between number of safety critical failures revealed from periodic testing, x, and the corresponding number of tests performed, n:

$$\mathrm{FF} = x/n \tag{6}$$

This measure may further be used to estimate the MTTF (see [30, 31]), which will then also depend on the time between the tests (the periodic testing interval).

For simplicity, we assume that the valves are tested at the maximum interval, i.e., twice a year (as defined in Ref. [32]), meaning that if a failure is detected from testing, the valve failure has occurred at some point in time within the interval = [0, 6] months. The data set then corresponds to an estimated 4357 DHSVs. These valves are associated with a total number of 200 failures recorded from this population.

Furthermore, we assume that, except for the detection time TD, each of the parameters is representable by the triangle distribution defined in Example I, T1 (1; 8; 20) hours. In addition, another triangular probability distribution is made for the time to detect the failure, denoted T2 (1; 2160; 4320) hours, i.e., a peak (mean) value of 3 months. This corresponds to a total downtime of 4.32×10^5 hours.

Although the various delay times may be correlated, we assume the parameters are independent of each other. For the time to failure, we calculate this based on the RNNP data in **Table 3**, showing the percentage of critical failures, i.e., exceeding the acceptance criteria of the barrier testing. The calculations based on PSA data [29] yield a total time in service of 7.49×10^7 hours. An estimator for the mean time to failure, MTTF, is thus given as $7.49 \times 10^7/200 = 3.74 \times 10^5$ hours. We now consider the implications of varying definitions of the repair time, in this context referred to simply as MTTR. We assume that there is uncertainty on all of the components which together form the MTTRes.

Figure 4 shows, similarly to **Figure 2** for Example I, that widening the definition of repair time to include delays and fault detection will lower the system availability and increase the standard

Barrier element: DHSV	
Number of facilities where tests were performed in 2016	73
Average number of tests for facilities where tests were performed in 2016	119
Number of facilities with percentage failures in 2016 greater than the industry standard	25
Total (mean) percentage of failures in 2016	0.023 (0.026)
Total (mean) percentage failures 2012–2016	0.021 (0.021)
Industry standard for availability (Statoil value)	0.02

Table 3. General calculations and comparison with industry standards [29].

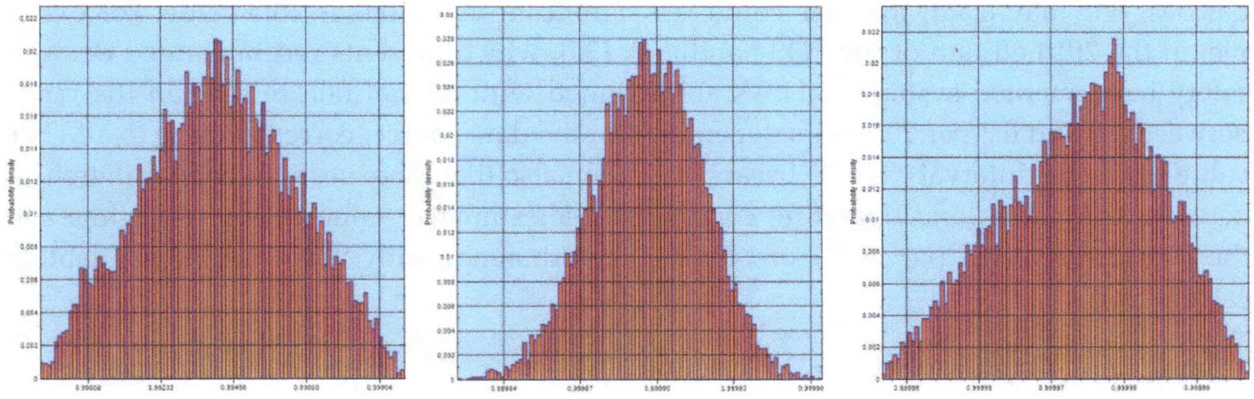

Figure 4. Intrinsic availability distribution for A_1, (left) using MTTR = MTTRes. Mean = 99.416 and stdev = 2.3×10^{-3}. A_2, (center) using MTTR = MRT. Mean = 99.990% and stdev = 0.2×10^{-4}. A_3, (right) using MTTR = MART. Mean = 99.997% and stdev = 0.1×10^{-4}.

deviation. For this example, in contrast to the previous, the detection time is significant. Also for this example, the MTTF is relatively high; however, now the differences between using MTTRes and MRT are much greater, where the deviations in the mean availability between the A_1 (i.e., MTTR = MTTRes) and A_2 (i.e., MTTR = MRT) is equal to 0.547% versus a deviation of 0.061% given for Example I. And although there is significant uncertainty related to the value of MTTF, the relevance of detection time is significant.

Consider also the value of the MTTF in **Figure 5**, where a sensitivity analysis is run, showing both the expected absolute and relative difference between A_1 and A_3 (blue line) and A_2 and A_3 (red line). As the MTTF becomes lower, the significance in the varying interpretation of repair time becomes as expected greater also in this example.

Example II indicates that the MTTF must be significantly higher before the availability deviations converge towards zero, at least a factor of 10 higher than the SCM situation (Example I). Only when the MTTF is in the region of 500,000 hours (60 years or more) can the contribution from detection time be considered negligible in this example.

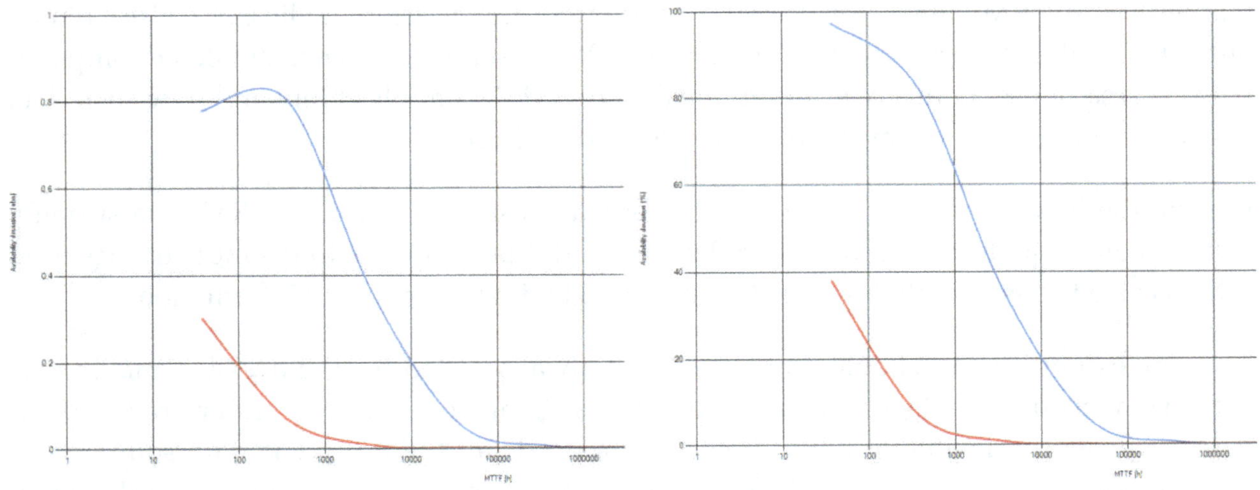

Figure 5. Expected absolute (figure to the left) and expected relative (figure to the right) difference between A_1 and A_3 (upper line) and A_2 and A_3 (lower line), as a function of varying MTTF [hours].

Besides, one may claim that one could see significant variations in this value. For example, in the 2010 edition of the PDS handbook [30], which presents recommended data for safety instrumented systems, DHSVs are assigned with overall failure rates in the range between 2.0 and 6.7 per 10^6 hours depending on the data source, corresponding to a MTTF value inside the interval = [58, 17] years. Knowing also that there are significant differences between companies and operating conditions, the estimated value of detection time and the measure selected (i.e., MRT or MTTRes) could significantly influence the availability calculations.

4. Maintenance data collection link to decision-making and challenges

Information about downtime is highly important for decision-making in the oil and gas industry, including for subsea systems. For example, such data is needed to track maintenance KPIs and achieve the so-called maintenance excellence; refer to references [33, 34].

In general, having useful information about RM is important for high-quality decision-making, and it is one of the six key dimensions that are used to evaluate decision quality [35]:

1. Helpful frame (what is it that I am deciding?)
2. Creative alternatives (what are my choices?)
3. Useful information (what do I know?)
4. Clear values (what consequences do I care about?)
5. Sound reasoning (am I thinking straight about this?)
6. Commitment to follow through (will I really take action?)

The above six dimensions can be visualized as a chain of decision links (see also ([36], p. 55)), where the decision is not stronger than its weakest link, which simply means that poor information (i.e., the RM data in this context) deteriorates the decision quality. It is also pointed out in Ref. [35] that the information or data should be "useful" and hence should be compared with its area of use, in this case within the area of RM data applicability and how such data may create business value by contributing to good decisions.

Despite the link to business value, and the broad consensus that use of RM data strongly depends on its quality, collecting such data about RM performance in the industry has typically received considerably less attention compared with the use of the information.

To some extent it is inevitable that data is not always suited for the decision-making, as the system requirements, design and operations change over time. Neither are the databases sufficiently flexible to adopt for the data needs all the time. The data sources are, in many situations, unmanageable at the time when they are needed. For example, when analyzing and predicting downtime for some subsea safety valve, one typically uses the data already collected and at the time available from some database or source, such as

OREDA (see e.g., [1, 37]), WellMaster (see e.g., [38]) or some internal database. There may be at the time limited room for collecting new and more relevant information. Data collection requires time, personnel and software resources. It could take years to build a quality database, and strategic decisions should be taken about what equipment data and format are relevant in the future. At the time of analysis and decision-making, there may not be sufficient resources (e.g., time) to collect additional information.

RM data collection is often an issue of resources, such as cost. In Ref. [14], it is mentioned that "collecting RM data is costly and therefore it is necessary that this effort is balanced against the intended use and benefits." The RM data collection activity is considered an investment, but it is an issue that strongly depends on both how and why the data are collected. Besides, it is not always clear what time perspective one is considering. Several years of data collection could be required for the data to have significant population and value in decision-making.

A similar point is made in Ref. [39], with reference to the Well Integrity Management System (WIMS). This is a software application for data collection, and [39] claim that it is important that both accurate and reliable information are achieved, at least from a user perspective. Nowadays data is shared with, and should be compatible with, other systems used by the oil and gas companies, such as, the data management system SAP, to synchronize data quality. The taxonomy used in the RM database should to a large extent also be reflected in the sources where the data is first recorded, such that the transfer of failure data does not have to go through several interpretation steps where essential information is missing.

Another issue is cost and ownership, which is a main reason why the data collection process is often delayed. Experience shows that often the data collection is performed at a quite late stage compared to when the actual failure occurred. The main information is typically captured by, e.g., reports from the maintenance provider. Then it is later the task of some data collector to transfer the relevant information into a RM database, including identification of missing and low-quality information. For example, the time to mobilize is normally given with low accuracy in such reports. The same is the situation when studying data from typical replacements, for example, choke replacements. The number of hours to replace such an item is too often given as an estimate, for example, 24 hours, as the detailed information may not be available from the maintenance reports.

The information provided by the MTTR, which is a common and much used KPI in the oil and gas industry (see e.g., [40, 41]), is also challenging. The KPI has a strong position within this industry, especially for availability calculations, and it is highly difficult to avoid the use of this term by replacing it with MRT and MTTRes, as suggested in Ref. [21]. A quick search in "Google scholar" (January 2016), for the period 2013–2016, confirms that the term "mean time to repair" (MTTR) is used significantly and way more than the term "mean time to restoration" (MTTRes). The search results confirm that on several accounts, the term used is MTTR, while the actual basis for estimating this value is equivalent to "mean time to restoration," which is the specific term and abbreviation used in the two IEC documents [20, 22] but then with the meaning of MTTRes. The abbreviation is also used for the mean time to recover (or recovery), synonymous with the "mean time to repair."

5. Discussion

RM data is used for a vast array of different applications, both operational and engineering. Data use today is as relevant as it was 30 years ago, for example, according to Ref. [42], reliability data is used to design the operational phase in terms of evaluating the operational performance of equipment, adjusting maintenance intervals, optimizing test intervals, establishing failure probability distributions, optimizing spare parts and logistics and job priority scheduling. Of similar importance to the engineering of equipment components, reliability data provide input to analysis such as safety integrity level analyses, RAM studies, required maintainability, selection of equipment and parts based on reliability experience, choice of maintenance strategy and qualification testing.

The OREDA database, which has a taxonomy based directly on ISO 14224, contains more than 39,000 failures and 73,000 maintenance records, as well as over 2000 years of operating experience from subsea fields. Its current member list includes BP, Engie E&P Norge, Eni, Gassco, Petrobras, Shell, Statoil and Total. Expert members from the OREDA project are also involved in the development of ISO 14224, to make sure that the challenges experienced in practice are captured when revising the standard. When the latest edition was issued [14], it was possible to include several definitions of key measures relevant for downtime predictions, such as, the new definition of the mean time to restoration (MTTRes).

Considering the cost of subsea maintenance operations, downtime and mobilization of vessels, the range of affected decisions, operations and procedures and not least the extent to which safety systems are used in today's industry, the importance of RM data quality cannot be emphasized enough. We will next discuss some challenges relating to the practical use of the ISO 14224 taxonomy, what impacts this may have on decisions, operations and design and some suggestions to improvements.

5.1. Data separation

One of the key challenges when collecting RM data is related to the fact that the RM database in many cases exists as a separate entity which may only partially communicate or not communicate at all, with other relevant systems (e.g., SAP). It is not rare to observe that data stored in an RM system such as OREDA or WellMaster is essentially data extracted from another system or database, converted into an appropriate format and then re-entered. This creates several challenges relating to both data integrity and quality:

(1) Creates an unnecessary overhead; data could simply be stored in one place, possibly supplemented by automatic or manual conversion if a special format or interpretation was required, as is often the case for standards that use a specific taxonomy.

(2) Requiring access to multiple data systems, means spending more time interviewing and more time spent on data interpretation, since it is likely that the different systems have different data formats, requiring multiple "reformatting" efforts to reach taxonomical compliance. The data are not even necessarily found within the same company, but may be located with suppliers or sub-suppliers.

(3) Since crucial data, such as maintenance or vessel data, may not be stored directly into a RM system, then there is also the risk that original records do not contain all required fields, meaning that if the persons responsible for the operations cannot be contacted or do not recall specifics, then required data will be impaired, severely inaccurate or lost forever.

5.2. Mobilization time

The mobilization time, as previously mentioned, includes all resources, including personnel. The source for obtaining such data varies. Sometimes, such as in the event of unexpected, significant failures, specific detailed maintenance reports may outline the mobilization time of resources used. In other cases, such data are available through data stored in vessel log reports, and the data stored into the RM database is typically an interpretation of this. These reports however seldom store records of other resources than the vessels themselves, such as personnel. Maintenance records for which no records exist in neither vessel logs nor other systems have unknown mobilization times and must therefore be estimated. This is typically done based on expected duration to bring the vessel to its destination. Such estimates will thus be optimistic figures in cases where there were delays, where there were shortages in spare parts or where there were other delays that should have been considered, but were not. Conversely, in some cases, due to lack of accurate maintenance records or other source documents, the time allocated to mobilization becomes too large if the same maintenance campaign covers several maintenance jobs but where mobilization is not split across all maintenance records. There is also a fair chance that data collection will not necessarily capture any remobilization related to maintenance jobs, if these occur at a point after the maintenance data was collected or if this remobilization is not logically mapped to the original mobilization activity.

It may be challenging to use statistical data to estimate the mobilization time due to the issues discussed above. In many situations, one must rely on experts having insights into the planning of maintenance activities, to assess and recommend representative values of mobilization time. For example, the mobilization is often highly area and company sensitive, where the mobilization time could be significantly shorter within one specific geographic area, e.g., due to a higher number of cooperating vessels operating within the area.

5.3. Detection time

The abbreviation for "mean fault detection time" (i.e., MFDT) is, as already mentioned in Section 2.5, an abbreviation that may have two distinct meanings, both of which are relevant for assessment of reliability and maintenance performance. However, in practice, this discrepancy is not widely problematic, as the specific meaning of both terms is considered well known.

In many situations, it may be reasonable to ignore the contribution of detection time when making assessments involving downtime measures. However, as we may see from the two examples given in the previous section, the significance partly depends on the equipment dealt with and obviously the type of assessment. It is especially important to distinguish between the two types of "MTTR," i.e., "MRT" and "MTTRes," when the failure rate is high and the time of failure is uncertain.

The detection time may naturally be subject to high uncertainty when dealing with hidden failures. However, that is not always the situation. Detection time is closely linked to equipment degradation modeling, such as, the use of so-called P-F interval models (see e.g., ([25], p. 394)). A P-F interval is the time from when there is a potential for failure (TP) due to the equipment being in a condition where it is possible to reveal some fault (e.g., from periodic functional testing, condition monitoring or inspection) to the time when the failure occurs (TF); see **Figure 6.** If the item is subject to periodic testing, and the test interval is shorter than the P-F interval, then one may have a situation where a fault is always detected from testing before any failure occurs. By using such models, it is possible to assess the time from when the fault is revealed to the time of the fault by comparing the condition of the item and the P-F interval model.

Detection time should also be seen in relation to the probability of detection, which relates to both the quality of testing and inspection activities, and to the incentives of accepting versus failing tests. For functional testing of equipment exposed to hidden failures, there may be strong incentives of passing tests that are close to the acceptance criterion. For example, if failing a valve leakage test leads to a more frequent test schedule, documentation work, etc., then the test personnel might pass a test even though some initial result shows that the leakage is 2.1 bar and thus just over the acceptance criterion of 2.0 bar. One could find it convenient to extend the test or make some adjustments on site to achieve a time interval where the leakage result is found acceptable. The incentives of failing or accepting tests is likely to both influence testing schedules and the statistics concerning how many of the results in the area around the acceptance criterion are reported as "failures" and thus also the estimated MTTF value.

5.4. Mean time to repair (MTTR)

As with MFDT, MTTR is also an abbreviation that may have several different meanings. Within the oil and gas industry, the letters used in the abbreviation could refer to several terms which all make sense within the area of analysis, and may thus cause analytic ambiguity. The

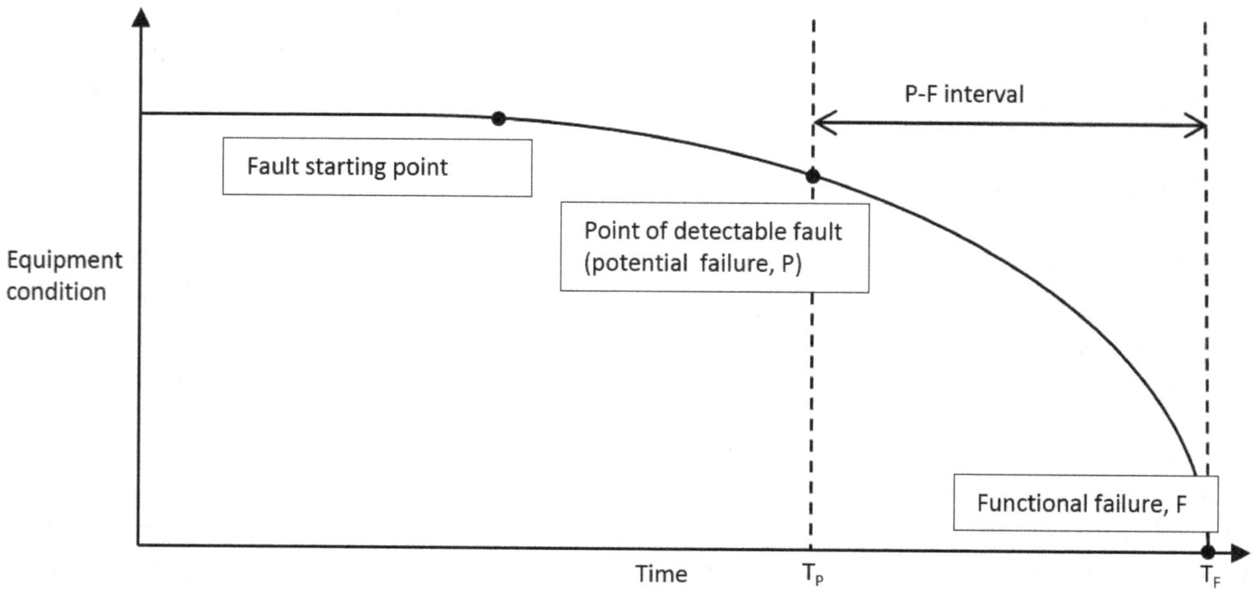

Figure 6. P-F interval model (equipment degradation model).

"TT" term consistently refers to "time to." However, the two other letters are found highly inconsistent in use. "M" could refer to "minimum," "mean" or "maximum" and likewise the "R" term could also refer to different meaningful words, such as "repair," "recovery," "respond" or "restore/restoration." It is therefore sometimes confusing when the abbreviation is not explicitly defined.

The use of the term "mean time to restoration," with the abbreviation MTTRes, as suggested in ISO 14224 [14] clearly reduces the chance of making wrong assumptions regarding use of MTTR.

The use of MTTR very much is an issue of quality. By comprising both MTTRes and MRT, it represents a measure that fails to give consistent information. In some situations, MTTR will provide the MTTRes information, and in some situations, it will provide a combination of the two. To use the MTTRes makes it clear that the intended meaning is captured. By keeping the MTTR as a useful term, one must ensure that the "R" refers to "restore" and not "repair." For decision-making purposes, the MTTR measure is better avoided, as also recommended in the international standard [14].

5.5. Failure fraction (FF)

The failure fraction, while being a measure that is simple to understand and use, also has some limitations with respect to time aspects. Using the FF as an availability measure, for example, makes it difficult to draw conclusions when the population varies significantly in detection time and number of demands. The value expressed from using this measure does not separate between equipment tested monthly or tested once a year, which makes it difficult to use the values for estimating the MTTF.

Failure fraction is typically used for equipment linked to hidden failures. If assuming that the fault does not occur during the testing, the measure provides relevant information about the number of tests that reveal hidden failures. However, such an assumption strongly depends on the testing interval, as there is a possibility that the fraction of test inducted faults may be high, and one could then argue that the FF, and consequently also the MTTF, would be far higher if the testing interval was reduced.

A main challenge is that the failure fraction ignores the number of demands or faults that occurred since the last test, which is important information regarding the availability of the system. Hence, it is possible that instead of one failure, there should have been recorded, e.g., two failures, where a shorter test interval could have detected the two distinct faults initiated at different points in time instead of only the one that was registered at the point of testing. Furthermore, some of the failures could be observed from a demand and maintained prior to the functional testing and thus not be included in the FF statistics.

5.6. Decision-making quality

Typically, the quality of data influences the quality of decision-making. The process also comprises other elements, as listed in Section 4. Part of this quality relates to consistency in the use of the downtime measures for, e.g., availability calculations, which we exemplify in the current chapter.

Based on the above discussion, the use of downtime measures is found ambiguous in the sense that they can be given different interpretations, such as the situation now appears to be with the term "mean time to repair," and, in particular, the abbreviation "MTTR," and may thus contribute to reduced data and decision-making quality. The attempt of [20] to reduce ambiguity by defining "MTTR" as "mean time to restore" and thereby include the detection time is welcome in that sense (see also [27]). However, as the "MTTR" is still being widely used with previous definitions, the "mean time to repair" and the "mean time to restore" could be difficult to separate. The ISO 14224 [14] term "mean time to restoration" with the abbreviation "MTTRes" is considered less ambiguous.

6. Conclusions

In this chapter, we have looked at different terms used to describe downtime. Different terms are used in data collection to define downtime, where some may be questioned to not provide adequate quality needed for associated analyzes and decision-making where the RM data is used.

The 'mean time to repair' is a term that is well-established and widely used for e.g. availability calculations. Although the meaning of this term has shifted over time, we find examples where MTTR refers to different meanings and thus is a challenging term to use. The solution proposed in Ref. [14], to avoid the use of this term, and instead use MTTRes and MRT, is considered an acceptable way to deal with this challenge. Data collection experience indicates that to complete a change of the MTTR meaning is difficult, as the term has such a strong position within the oil and gas industry.

Another challenge discussed in this chapter relates to the mobilization of maintenance resources. Experience shows that it is difficult to both interpret and quantify mobilization times in practice. Part of this problem is that resources may be linked to several maintenance activities on site, which provides an opening for different ways of recording the actual time used to mobilize and to repair the specific item. It becomes an interpretation issue, which may influence the values used for prediction of time needed to achieve repair of a failed item. Limited guidance, except for adequate definition, is given in Ref. [14] on this issue.

In general, we recommend that data collection is given higher attention compared with the situation today. Typically, investments are focused on building models and using the data rather than obtaining them and ensuring high quality. Especially, more focus should be on achieving high-quality data from maintenance operations.

Acknowledgements

The current book chapter is based on the paper: "Maintenance data collection for subsea systems: A critical look at terms and information used for prediction of down time" [43], presented at the European Safety and Reliability (ESREL) conference in Portorož, Slovenia, June 18–22, 2017.

Author details

Jon T. Selvik[1]* and Eric P. Ford[2]

* Address all correspondence to: jon.t.selvik@uis.no

1 University of Stavanger, Stavanger, Norway

2 International Research Institute of Stavanger (IRIS), Stavanger, Norway

References

[1] Langseth H, Haugen K, Sandtorv H. Analysis of OREDA data for maintenance optimisation. Reliability Engineering and System Safety. 1998;**60**(2):103-110

[2] Dawotola AW, van Gelder PHAJM, Vrijling JK. Integrity maintenance of petroleum pipelines. Oil and Gas Facilities. SPE 162873. December 2012 https://www.onepetro.org/journal-paper/SPE-162873-PA

[3] Dawotola AW, Trafalis TB, Mustaffa Z, van Gelder PHAJM, Vrijling JK. Data-driven risk based maintenance optimization of petroleum pipelines subject to corrosion. In: Proceedings of the Twenty-first (2011) International Offshore and Polar Engineering Conference (ISOPE), Maui, Hawaii, USA, June 19-24, 2011

[4] Nezamian A, Rodriques C, Nezamian M. Asset integrity management and life extension of gas distribution facilities and pipeline network. SPE 180889-MS. In: Proceedings of the SPE Trinidad and Tobago Section Energy Resources Conference, Port of Spain, Trinidad and Tobago, 13-15 June 2016; https://www.onepetro.org/conference-paper/SPE-180889-MS

[5] Longhi AEB, Pessoa AA, de Almada Garcia PA. Multiobjective optimization of strategies for operation and testing of low-demand safety instrumented systems using a generic algorithm and fault trees. Reliability Engineering and System Safety. 2015;**142**:525-538

[6] Atwood CL, Engelhardt M. Bayesian estimation of unavailability. Reliability Engineering and System Safety. 2004;**84**:225-239

[7] Bevilacqua M, Ciarapica FE, Giacchetta G, Marchetti B. Evaluation of MTBF and MTTR trends for a selected equipment in an oil refinery. Article 18122. In: Proceedings of the 18th ISSAT International Conference on Reliability and Quality in Design. Boston, Massachusetts, U.S.A. International Society of Science and Applied Technologies. July 26-28, 2012

[8] Rajiv SK, Poja S. Computing ram indices for reliable operation of production systems. Advances in Production Engineering & Management. 2012;**7**(4):245-254

[9] Corvaro F, Giacchetta G, Marchetti B, Recanti M. Reliability, availability, maintainability (RAM) study, on reciprocating compressors API 618. Petroleum. 2017;**3**:266-272

[10] Gupta P, Gupta S, Gandhi OP. 2013. Modelling and evaluation of mean time to repair at

product design stage based on contextual criteria. Journal of Engineering Design. 2013;**24**(7): 499-523

[11] Coulibaly A, Houssin R, Mutel B. Maintainability and safety indicators at design stage for mechanical products. Computers in Industry. 2008;**59**(5):438-449

[12] Ding L, Wang H, Kang K, Wang K. A novel method for SIL verification based on system degradation using reliability block diagram. Reliability Engineering and System Safety. 2014;**132**:36-45

[13] Jigar AA, Liu Y, Lundteigen MA. Spurious activation analysis of safety-instrumented systems. Reliability Engineering and System Safety. 2016;**156**:15-23

[14] Petroleum, Petrochemical and Natural Gas Industries—Collection and Exchange of Reli ability and Maintenance Data for Equipment. 3rd ed. Geneva, Switzerland: International Organization for Standardization (ISO); 2016

[15] Sandtorv HA, Hokstad P, Thompson DW. Practical experiences with a data collection project: The OREDA project. Reliability Engineering and System Safety. 1996;**51**:159-167

[16] Signoret J-P. Dependability & safety modeling and calculation: Petri nets. IFAC Proceedings Volumes. 2009;**42**(5):203-208

[17] IEC 60300-3-2: Dependability Management—Part 3-2: Application Guide—Collection of Dependability Data from the Field. Geneva, Switzerland: The International Electrotechnical Commission (IEC); 2004

[18] Brissaud F. Using field feedback to estimate failure rates of safety-related systems. Reliability Engineering and System Safety. 2017;**159**:206-213

[19] Selvik JT, Bellamy LJ. On the use of the international standard ISO 14224 on reliability data collection in the oil and gas industry: How to consider failure causes from a human error perspective. In: Walls L, Revie M, Bedford T, editors. Risk, Reliability and Safety: Innovating Theory and Practice. London: Taylor & Francis Group; 2017

[20] IEC 60050-192: International Electrotechnical Vocabulary – Part 192: Dependability. Geneva, Switzerland: The International Electrotechnical Commission (IEC); 2015

[21] ISO/TR 12489: Petroleum, Petrochemical and Natural Gas Industries—Reliability Model ling and Calculation of Safety Systems. Geneva Switzerland: The International Organi zation for Standardization (ISO); 2013

[22] IEC 61508: Functional Safety of Electrical/Electronic/Programmable Electronic Safety-Related Systems. Geneva, Switzerland: The International Electrotechnical Commission (IEC); 2010

[23] IADC—International Association of Drilling Contractors. Mean Active Repair Time. Lexicon. Available from: http://www.iadclexicon.org/mean-active-repair-time/: IADC; Accessed 01-09-2017

[24] Goot R, Levin I. Estimating the latent time of fault detection in finite automaton tested in real time. Automation and Remote Control. 2008;**69**(10):128-141

[25] Rausand M, Høyland A. System Reliability Theory – Models, Statistical Methods, and Applications. 2nd ed. Hoboken, New Jersey, USA: John Wiley & Sons, Inc.; 2004

[26] Ahmadi A, Ghodrati B, Garmabaki AHS, Kumar U. Optimum inspection interval for hidden functions during extended life. In: Proceedings of the 27th International Congress of Condition Monitoring and Diagnostic Engineering Management, 16-18 September 2014. Brisbane, Australia

[27] IEC 60050-191/ADM1: Amendment 1—International Electrotechnical Vocabulary. Chapter 191: Dependability and Quality of Service. Geneva Switzerland: The International Electrotechnical Commission (IEC); 1999

[28] Selvik JT, Signoret J-P. How to interpret safety critical failures in risk and reliability assessments. Reliability Engineering and System Safety. 2017;**161**:61-68

[29] PSA – Petroleum Safety Authority Norway: RNNP – Trends in risk level in the petroleum activity – Summary report 2016 – Norwegian Continental shelf. (2017). Available from: http://www.psa.no/getfile.php/1344338/PDF/RNNP%202016/ENG_summary_RNNP2016.pdf

[30] Hauge S, Onshus T. Reliability Data for Safety Instrumented Systems - PDS Data Handbook 2010 Edition. SINTEF Report no. A13502. SINTEF Technology and Society; 2010

[31] Selvik JT, Abrahamsen EB. A review of safety valve reliability using failure fraction information. In: Podofillini L et al., editors. Safety and Reliability of Complex Engineered Systems: ESREL 2015. London: Taylor & Francis Group; 2016

[32] NORSOK D-010: Well integrity in drilling and well operations. Edition 4:June 2013. Standards Norway. Available from: https://www.standard.no/en/sectors/energi-og-klima/petroleum/norsok-standard-categories/d-drilling/d-0104/

[33] Jansen M. Shell maintenance excellence. SPE 177964-MS: In: Proceedings of the Abu Dhabi International Petroleum Exhibition and Conference, 9-12 November 2015, Abu Dhabi, UAE; https://www.onepetro.org/conference-paper/SPE-177964-MS

[34] Smart K, Blakey K. 2014. Achieving maintenance excellence in Maersk Oil Qatar. International Petroleum Technology Conference; IPTC 17623; https://www.onepetro.org/conference-paper/IPTC-17623-MS

[35] Matheson D, Matheson J. The Smart Organization – Creating Value Through Strategic R&D, Boston. MA, USA: Harvard Business School Press; 1998

[36] Bratvold RB, Begg SH. Making Good Decisions. Richardson, TX, USA: Society of Petroleum Engineers; 2010

[37] Gjerstad T. OREDA—The reliability data reference for the offshore industry. SPE-14000-MS. In: Proceedings of Offshore Europe, 10-13 September 1985, Aberdeen, United Kingdom. https://www.onepetro.org/conference-paper/SPE-14000-MS

[38] Molnes E, Strand G-O. Application of a completion equipment database in decision-making. SPE 63112. In: Proceedings of the SPE Annual Technical Conference and Exhibition, 1-4 October 2000, Dallas, Texas. https://www.onepetro.org/conference-paper/SPE-63112-MS

[39] Corneliussen K, Sørli F, Brandanger Haga H, Tenold E, Menezes C, Grimbert B, Owren K. Well Integrity Management System (WIMS) – A systematic way of describing the actual and historic status of operational wells. SPE 110347. In: Proceedings of the SPE Annual Technical Conference and Exhibition, 11-14 November 2007, Anaheim, California, U.S.A. https://www.onepetro.org/conference-paper/SPE-110347-MS

[40] Reith CA, Lagstrom KB: Reliable subsea oil and gas transportation systems. In: Proceedings of the SNAME World Maritime Technology Conference 2015, Rhode Island, USA, 3-7 November 2015.

[41] Gustavsson F, Eisinger S, Kraggerud AG. Simulation of industrial systems. In: Proceedings of the ESREL 2003 Conference, Maastricht, The Netherlands, 15-18 June 2003.

[42] Lydersen S, Sandtorv H, Rausand M. Processing and Application of Reliability Data. Foundation for Scientific and Industrial Research at the Norwegian Institute of Technology. Trondheim, Norway: SINTEF Report; 1987

[43] Selvik JT, Ford EP. Maintenance data collection for subsea systems: a critical look at terms and information used for prediction of down time. In: Čepin M, Briš R, editors. Safety and Reliability – Theory and Applications. London: CRC Press/Taylor & Francis Group; 2017. p. 2989-2996

7

Techno-Economic Feasibility Study of Autonomous Hybrid AC/DC Microgrid System

Atanda K. Raji

Abstract

Distributed generation technology based on diesel generators often has been considered as a viable solution to providing power to remote areas, but the sky-rocketing of diesel fuel price and the increasing cost of delivery to such remote sites have called for providing a sustainable solution that is environmentally friendly, economical, affordable, and easily accessible. To this end, the use of locally available energy resources is accepted as a sustainable solution in providing electricity for rural and remote settlements. The system cost of wind and solar energy systems is continuously decreasing because of the increase in the acceptance and deployment of the energy systems based on these renewable energy resources. A standalone hybrid AC/DC electric power system is designed, modeled, simulated, and optimized in HOMER Pro. HOMER is a Hybrid Optimization Model of Electric Renewable that enables the comparison of electric and thermal power production technologies across an extensive variety of applications. Both cycle-charging and load-following dispatched strategies are investigated. Plausible selected system components ratings are chosen for the simulation to ensure that there is enough search space for HOMER Pro to obtain an optimal system configuration. Net present cost (NPC) is used as an economic metric to assess the optimal configuration that is technically feasible.

Keywords: HOMER, wind, solar PV, renewable energy system, electrochemical devices, net present cost, economics analysis, Sub-Sahara Africa

1. Introduction

Energy access and affordability is a very big concern for the majority of the Africa populace. With over 620 million people without access to modern electricity in the Sub-Saharan Africa alone [1], different innovative solutions are being considered. Grid extension is definitely not an option because of the sparse population distribution, which makes the economic viability of such action infeasible. Distributed generation technology based on diesel generators has often been considered as a viable solution but the sky-rocketing of diesel fuel price and the increasing cost of delivery to remote sites have called for providing a sustainable solution that is environmentally friendly, economical, affordable, and easily accessible. To this end, the use of locally available energy resource has been accepted as a sustainable solution in providing electricity for rural and remote settlements. Matching the consumer demand profile and renewable energy production always has been perceived as enormous problem but can be mitigated by incorporating adequate sized energy storage systems [2, 3].

An autonomous electric power system is one that is not connected to the utility for one or all of the reasons given in the last paragraph. Wind and solar resources are local energy resources that can be efficiently and economically harnessed to provide electric power to remote site electricity consumers. While the cost of such renewable resources are cost-free, but the system required for harnessing and processing the energy plus the balance of system (BOS) is expensive. However, the system cost of wind and solar energy systems have been continuously decreasing because of the increase in the acceptance and deployment of the energy systems based on renewable resources. For the many Africa people who do not have access to modern energy system, it has been proven that high-quality electric power can be delivered to remote areas using local energy resources such as wind, solar, biomass, and so on [4–6].

Many of the historical technological advantages that resulted in AC power system to dominate over DC power systems are not valid any longer. For example, in the electric power generation sector, many distributed energy resources, such as electrochemical storage systems, fuel cells system, and photovoltaic systems, produce efficient and economical DC power. Many of the modern electrical and electronic loads, as well as energy storage systems, are either internally DC power operated or worked equally well with DC power sources and connect to the AC systems through power electronic converters [7]. Overall improved system efficiency, reduced system cost, increased reliability, and reduced system footprint are attained when multiple conversion stages are eliminated in externally operated AC power system for internally operated DC power devices. Also, many power quality issues, such as harmonics and unbalances, are not present in DC power systems [8, 9]. Advancements in power electronic technology continue unabated, leading to more applications in modern systems such as VSC-HVDC, electric vehicles, wireless power transfer, fuel cells vehicles, and so on [10, 11].

The design of hybrid renewable energy systems has been extensively studied, and numerous optimization techniques, such as genetic algorithm (GA), linear programming, etc., have been implemented for optimum economic and technical feasibility [12–16]. Many software packages, such as HOMER Pro [9], RETScreen [17], and Hybrid2 [18], have been developed for the proper selection of appropriate generation technologies and capacity sizing. This software makes a multi-objective optimization process easy to carry out to arrive at a robust and reliable decision. HOMER Pro is user-friendly software produced by the national renewable energy laboratory by the department of energy in the USA. It uses hourly data for the assessment of hybrid renewable energy systems and performs optimization based on net present cost (NPC). HOMER Pro is used in this study because of its proven ability to design, model, simulate, and perform sensitivity analysis of complex mix of different energy technologies, thermal and electrical load, and storage devices. Both cycle-charging and load-following dispatch strategies are available to choose from or use simultaneously as utilized in this study [19, 20].

The rest of the chapter is organized as follows. A brief discussion of two different dispatch strategies used in hybrid energy systems is provided in Section 2 of this chapter. All the energy system components of the studied autonomous hybrid AC/DC microgrid system are briefly discussed in Section 3. Section 4 explains the capability and functionalities of Hybrid Optimization Model of Electric Renewable (HOMER) as powerful software that is designed and optimized for modeling, simulating, optimizing, and performing sensitivity analysis of micropower systems. The system model and simulation process of the studied hybrid system is comprehensively detailed in Section 5 of this chapter. The system simulation results and discussion and conclusion are given in the penultimate and Section 7 of this chapter, respectively.

2. Cycle-charging and load-following dispatch strategy

Two system dispatch strategies are employed in HOMER Pro namely: cycle-charging and load-following dispatch strategy. Under the load-following dispatch strategy, a power generator produces only enough to serve the load and does not charge the battery bank. For the cycle-charging dispatch strategy, the generator run at full power to serve the load and any excess is used to charge the battery [7].

Tuglie and Torelli [21] developed a new load following dispatch algorithm that correlates system imbalances with unbalanced transaction in a bilateral transaction-based market. The developed algorithm is implemented on the IEEE 30-bus test system under a variety of operating conditions. In Bizon [22], a hybrid power source comprising wind energy conversion system, solar Photovoltaic system, and fuel cell system with energy storage system is designed and optimized for load-following dispatch strategy. The battery/supercapacitor hybrid energy storage system operates as an auxiliary source for supplying the power deficit based on dynamic balance strategy. An optimal energy management system for a stand-alone wind turbine/photovoltaic/hydrogen/battery hybrid system with supervisory control based on fuzzy logic is presented in Pablo et al [23]. The energy management system in Ref. [23] uses a fuzzy logic control to satisfy the energy demanded by the load and maintain the state-of-charge of the battery and the hydrogen tank level between desired target margins, while trying to optimize the utilization cost and lifetime of the energy storage system. A new method to model and control the aggregated power demand from a population of thermostatically controlled load, with the goal of delivering ancillary services such as frequency regulation and load following, is developed by Callaway [24]. It is shown in the paper that identified models perform only marginally better than the theoretical model. Subho and Sharma [25] developed a hybrid system with cycle charging dispatch strategy using swarm optimization algorithm for a remote area in India. A comparative study of particle swarm optimization (PSO), genetic algorithm, and HOMER is performed in Ref. [25]. PSO is judged to be better than GA and HOMER in terms of the minimum cost of electricity generation. In another paper by Subho and Sharma [26], different optimization algorithms and software packages are used to evaluate the economics of a hybrid renewable energy system. It is concluded that cycle charging strategy is most cost-effective compared to the load following and peak shaving dispatch strategies.

Both dispatch strategies are employed in this study to determine which is economically viable and technically feasible.

3. Energy system components

For the considered autonomous electric power system model in this study, all the components and energy resources involved are briefly discussed. The components related to this work are Solar PV, wind energy conversion system (WECS), electrochemical battery storage device, and bidirectional power electronics converter.

3.1. Solar PV energy system

Solar PV technology uses semiconductor material, such as silicon, to convert solar radiation energy to direct current (DC) electrical power. The generated DC power is regulated by charge regulator to charge electrochemical battery storage device. The electrochemical storage device supplies power directly to the electrical load if it is a DC load. If not, an inverter is used to convert the DC power supplied by the electrochemical storage device into an AC power for the AC load.

PV cell is usually modeled using a single diode or double diodes, but a single-diode model is shown in **Figure 1**. Series and shunt resistances represented in the figure are used to represent the leakage current and the inherent loss due to internal electrical connection, respectively [27–29].

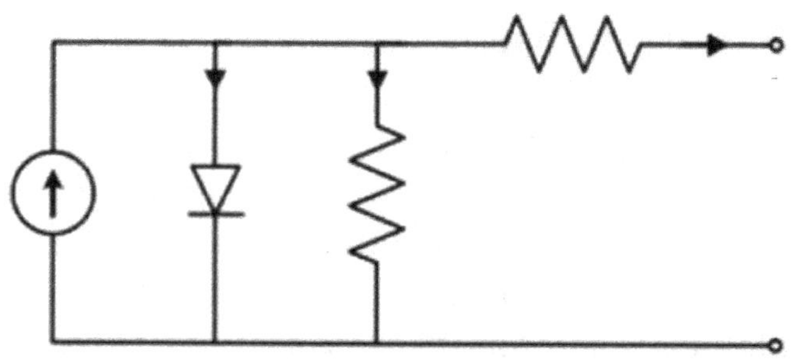

Figure 1. Single-diode model of solar cell.

The insolation intensity is strongly related to the solar cell current, and the cell operating temperature is inversely proportional to the output voltage across the solar cell. These factors basically limit the maximum power that can be generated by the solar cell [30].

Homer Pro calculates the power output of the PV array using equation 1 below.

$$P_{PV} = f_{PV} Y_{PV} \frac{I_T}{I_S} \quad (1)$$

Where f_{PV} is the PV derating factor, Y_{PV} is the rated capacity of the PV array (kW), I_T is the global solar radiation (beam plus diffuse) incident on the surface of the PV array (kW/m^2), and I_S is the standard amount of radiation used to rate the capacity of the PV array given as 1 kW/m^2.

A generic flat plate PV system is selected for the modeled hybrid microgrid from the HOMER Pro components library with a rated capacity of 40 kW. The capital cost of the PV is $3000/$kW_p$, replacement cost of $2000, and operation and maintenance cost of $10.00/year.

3.2. Wind energy conversion system

A blowing wind consists of kinetic energy carried by the atmospheric air. The moving air is converted to rotational kinetic energy by the rotor blade of the wind turbine system. The rotating rotor is coupled via a solid shaft to an electrical generator to generate AC electric power. The extraction of the power contained in the moving is limited by the aerodynamic efficiency of the rotor blade as illustrated in **Figure 2**.

The theoretically generated turbine power limit of which is 59% is known as the Betz limit. Most of the modern wind turbines are practically in the 30–40% range. Equation 1 below gives the relationship between the extracted power and other parameters and variable of a wind

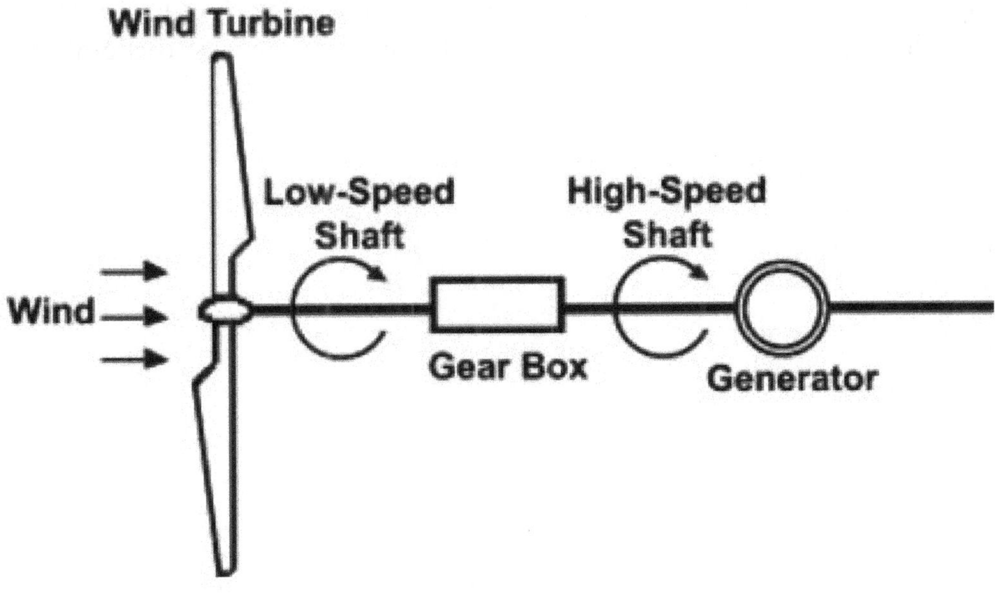

Figure 2. Wind energy conversion system (WECS) adopted from Ref. [31].

energy conversion system. The wind speed is strongly coupled to the available power, whereas the power coefficient depends on the tip speed ration (TSR) and the blade pitch angle [10, 16].

$$P = \frac{1}{2} A \rho C_p V^3 \quad (2)$$

A is the rotor swept area, ρ is the air density, C_p is the power coefficient that is dependent on the rotor blade pitch angle and tip speed ratio, and V is the average wind speed.

HOMER Pro determines the output power of the wind turbine in a four-step process. In the first step, it determines the average wind speed for the hour at the anemometer height by referring to the wind resource data. In the second step of the process, it determines the corresponding wind speed at the turbine's height using either the logarithmic law or the power law (the logarithmic law is used in this study). Thirdly, it refers to the turbine's power curve to calculate its output power value at that hourly wind speed assuming standard air density. Finally, it multiplies that output power value by the air density ratio, which is the ratio of the actual air density to the standard air density. HOMER Pro also assumes that air density ratio is constant throughout the year.

A generic wind turbine model of 10 kW rated power is selected in HOMER Pro for the study. The capital cost is $5000/kW, replacement cost of $5000/kW, operation, and maintenance cost of $500/year, operational lifetime of 20 years. The hub height of the generic wind turbine model is 24 m above sea level. The power curve of the selected turbine is shown in **Figure 3**.

3.3. Electrochemical battery energy storage system

A common electrochemical energy storage used in autonomous and emergency standby is Lead-Acid battery that has proven to be robust, cost-efficient, and cost-effective. However, many new battery technologies are being developed and utilized for practical applications with improved operational characteristics but are currently more expensive than the traditionally used battery technology [7]. With the interest in electric vehicle, lithium-ion battery technology is in the pole position to be the mainstream electrochemical technology

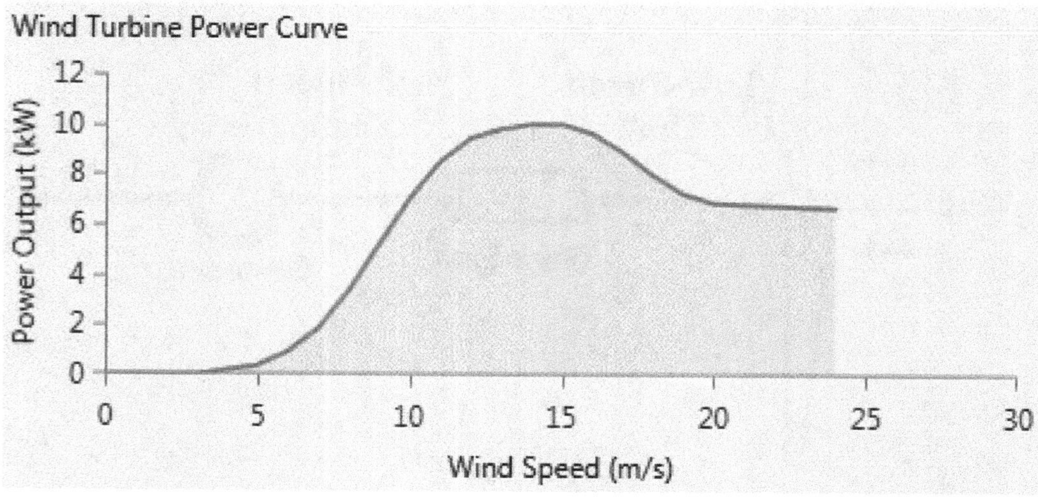

Figure 3. 10 kW generic wind turbine power curve (HOMER Pro).

for various applications. Among the operational parameters of Lead-Acid battery, maximum state of discharge (DOD), minimum state of charge (SOC), cycle times, round trip efficiency, maximum charge rate, maximum discharge rate are the most important. This battery technology also needs regular maintenance and should be well ventilated especially if it is not the sealed type [7].

A generic 1 kWh lead-acid battery is selected from the Homer Pro library for the hybrid microgrid presented in this study. The nominal voltage of the battery is 12 V configured in a string of 20 batteries (240 Vdc) to conform to the bidirectional rectifier rated input voltage. The maximum capacity and round trip efficiency of the battery are 83.4 Ah and 80% respectively. The capital cost is $300/kWh, replacement cost of $200/kWh, operation and maintenance cost of $20/unit/year.

3.4. Static power converters

Power electronics converters are needed for various power conversions and conditioning to optimally match electrical power source with electrical and electronic loads. Four main types of such converters are available: AC-AC, AC-DC, DC-DC, and DC-AC power converter [7]. A bidirectional power electronic converter is considered in this study because the wind energy conversion system is modeled as an AC power source, whereas the energy storage and solar PV are modeled as DC power source. Such power electronic converter allows power to flow bidirectionally between the AC and DC bus. Such functionality is possible by adequately

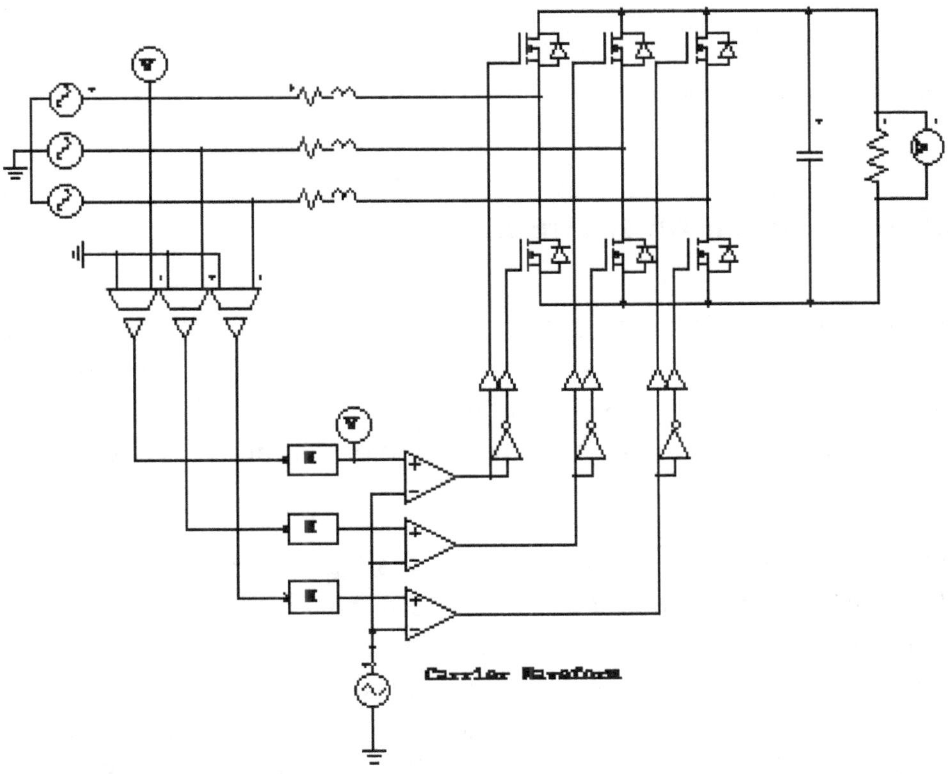

Figure 4. Three-phase pulse-width modulation (PWM) rectifier.

modulating the semiconductor switches that are embedded in the power converter topology [7, 32]. Such configuration that operates as an active end rectifier is shown in **Figure 4**.

HOMER Pro assumes that the inverter and rectifier capacities are not surge capacities that the device can tolerate for only short periods of time, but rather, constant capacities that the device can withstand for as long as needed [7].

A Leonics MTP41FP 25 kW 240V DC bidirectional power converter is selected for the studied hybrid AC/DC microgrid to interconnect the DC and AC bus. The capital cost of the bidirectional converter is $600/kW, replacement cost of $600, operation and maintenance cost of $0/year. The lifetime of the selected converter is 10 years and its inverter efficiency is 96% while the rectifier relative capacity is chosen to be 80% and its efficiency is 94%.

Table 1 shows the associated costs of all the system model components. These costs are used in Section 4 to perform the system simulation in HOMER Pro software.

System component	Capital cost per unit of rated value	Replacement cost per unit	Operating and maintenance cost per unit per year
PV panels	$3000	$2000	$10
Wind turbine	$5000	$5000	$500
Bidirectional power converter	$600	$600	$0
Electrochemical battery storage system	$300	$200	$20

Table 1. Costs of system model components [33].

4. HOMER pro simulation environment

HOMER Pro software is able to evaluate off-grid or grid connected power system design, choose the best system based on cost, technical requirements, or environmental considerations. It can also simulate many design configurations under market uncertainty and evaluate risk plus the ability of the modeler to choose the best addition or retrofit for an existing system [7].

HOMER Pro software is used in this research investigation. Wind energy conversion system, solar PV system, bidirectional converter, electrochemical batteries, AC primary load, and renewable resources are modeled and analyzed. Energy resources data, system components data, typical community load profile, and system components costs are some of the important parameters that are used as inputs in the HOMER Pro model. AC and DC bus are created in the HOMER Pro model to facilitate the connection of different types of locally available resources. HOMER Pro is able to optimize feasible electric power system configurations that are technically possible [7]. Sensitivity analysis of feasible configurations is performed to give an indication of the system robustness which is reported in Section 4 that follows.

5. System modeling and simulation

The investigated hybrid AC/DC microgrid system model architecture is shown in **Figure 5**. The system consists of two electrical buses: AC and DC bus. To ensure improved overall system efficiency, 25% of total electrical load is the DC load that is connected to the DC bus. The wind turbine and solar energy system are connected to the AC and DC bus, respectively. In modeling the battery, HOMER Pro includes in the battery model, a charge regulator to ensure that the specified operating condition is not violated via the SOC and DOD characteristic curves. The interlinking power converter connects the two buses together by acting as a rectifier or inverter depending on the power state of the hybrid system. In the rectifier mode, it transfers power from the AC bus to the DC bus whenever there is excess power in the AC bus and there is instant demand of power from the DC bus. Alternatively, in the inverter mode of operation, power is transferred from the DC bus to the AC bus whenever there is excess power on the DC and there is instant demand of power from the AC bus. In both modes of operation, the SOC of the energy storage system is to be taken into consideration. The dispatch strategy can be set in HOMER Pro. Load-following and cycle-charging dispatch strategies are available for selection in HOMER Pro. Under the load-following strategy, generators produce only enough power to serve the load, and do not charge the energy storage device. For the cycle-charging, the excess energy available after serving the load is used to charge the battery [2, 7, 34].

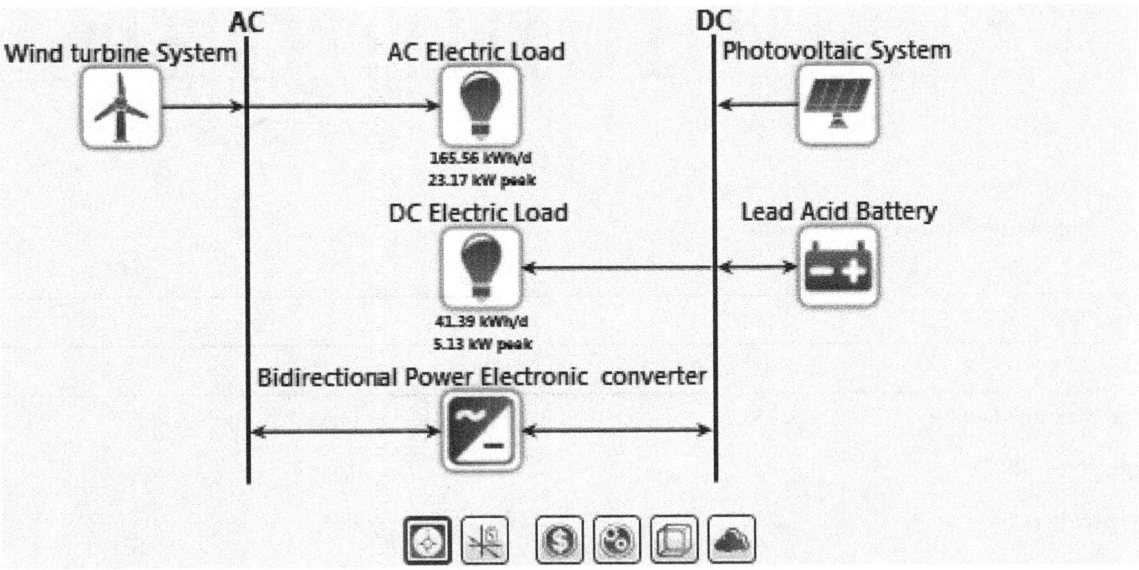

Figure 5. Autonomous AC/DC hybrid microgrid system.

The simulation process models the micropower system under investigation, and the optimization process determines the optimal system configuration that satisfies the modeler-specific constraints at the lowest total net present cost. The sensitivity analysis functionality of HOMER Pro allows energy planner, which energy technology combinations are optimal under different conditions [7, 9].

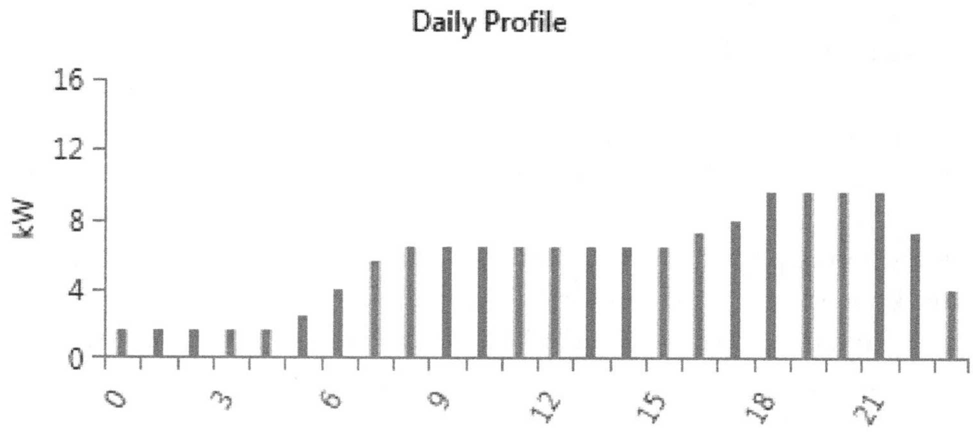

Figure 6. Community daily AC load profile.

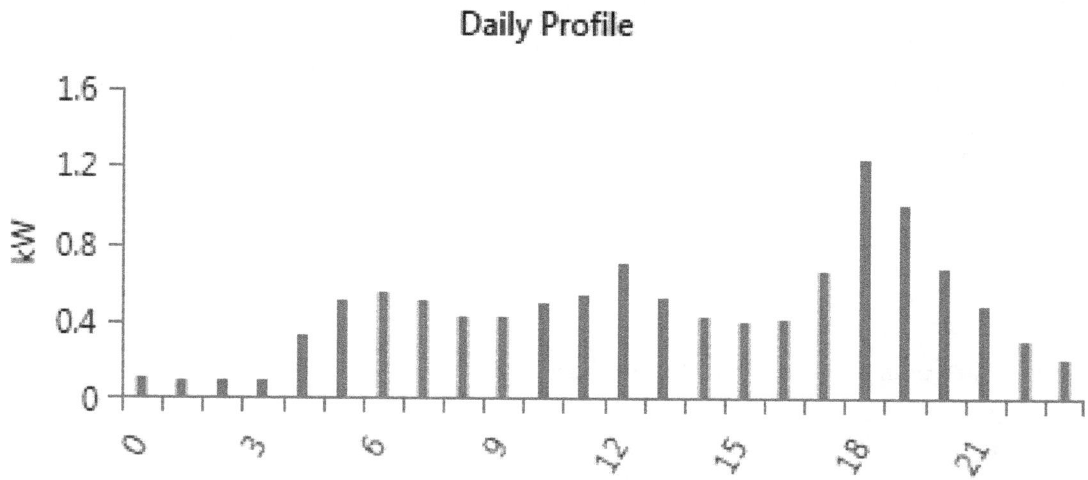

Figure 7. Community daily DC load profile.

Electrical loads		
	AC load	**DC load**
Average consumed energy	165.56	14.39
Average power (kW)	6.9	0.6
Peak power (kW)	23.17	2.67
Load factor	0.3	0.22

Table 2. Community load profile data.

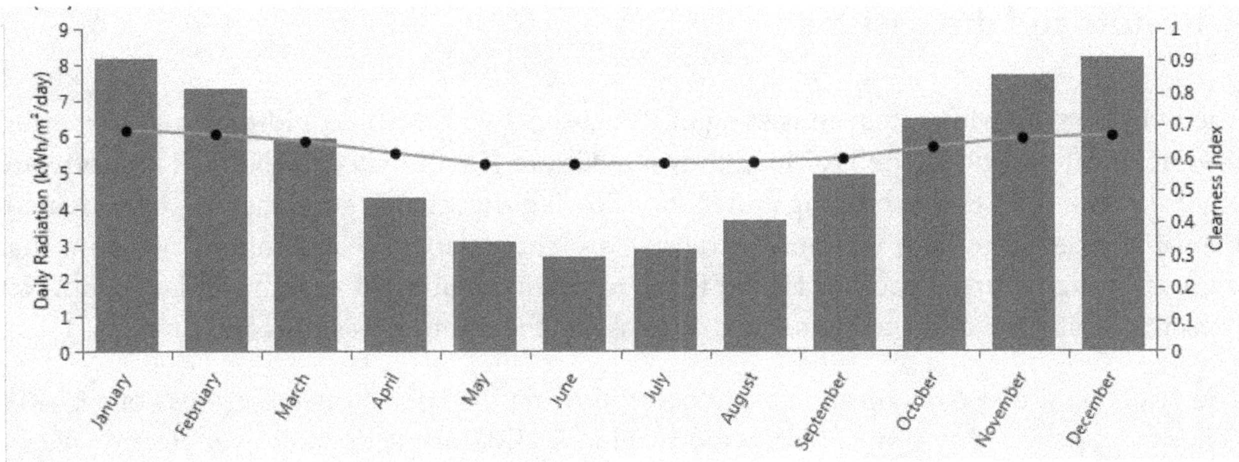

Figure 8. Monthly average solar global horizontal irradiance of Cape Town South Africa.

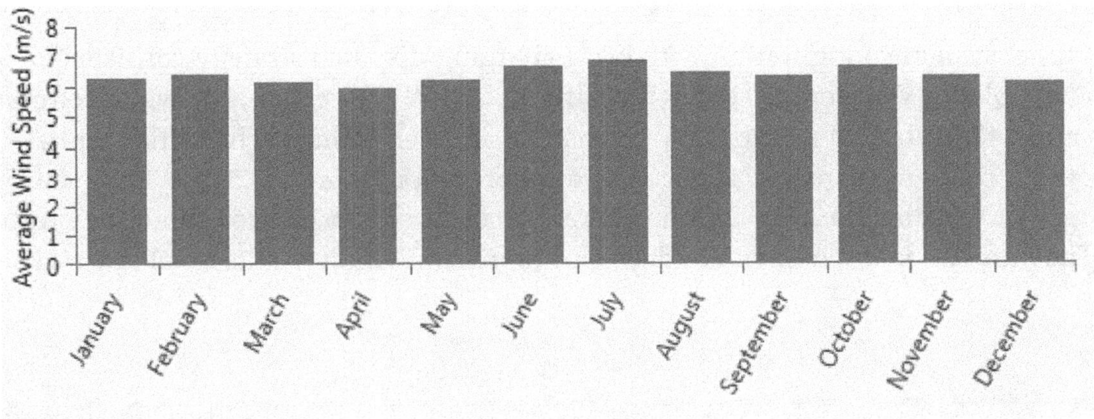

Figure 9. Average wind speed of Cape Town South Africa.

The AC load daily and DC load profiles are shown in **Figures 6** and **7**, respectively. The data used for this plot are shown in **Table 2**. The community load profile used in this study is the reference load date available in the HOMER Pro library [9].

The wind and solar resource data are obtained from the NASA Surface Metrological and Solar Energy database website. **Figure 8** depicts the monthly average solar global horizontal irradiance data. These monthly average data are averaged over 22-year period (July 1983–June 2005) [35]. The solar irradiance of Cape Town is low during the winter months of April, May, June, and July as shown in **Figure 8**. The wind speed data are also obtained from NASA Surface Metrological and Solar Energy database website. Wind speed at 50 m above the surface of the earth for terrain similar to airport, and monthly averaged values over 10 years period (July 1983–June 1993) are used in this work. These data are displayed in **Figure 9** (NASA Surface Meteorology and Solar Energy).

The Cape Town average wind speed as reported by the NASA database is nearly constant throughout the year, with average speed hovering around 6 m/s.

6. Results and discussions

The hybrid microgrid system consisting of PV system, wind turbine, bidirectional converter, lead-acid battery, and DC and AC load shown in **Figure 4** is simulated in HOMER Pro software environment. Both cycle-charging and load-following dispatched strategies are investigated. Plausible selected system component ratings are chosen for the simulation to ensure that there is enough search space for HOMER Pro to obtain an optimal system configuration. Net present cost is used as an economic metric to assess the optimal feasible configuration.

The optimal system configuration with load-following dispatch strategy consists of a 50-kW generic flat plate PV system, 20 kW wind turbine, 15 kW bidirectional power electronic converter, and 160 kWh lead-acid battery consisting of eight string. The initial capital cost of this optimal configuration is $307 000. The total present cost is $558,717, with levelized cost of energy (COE) being $0.3652, and the operational cost is of $11 363 throughout the project lifetime. The number of autonomy is 11 days. For more simulation results, see **Figure 10**.

The optimal system configuration with cycle-charging dispatch strategy consists of a 30-kW generic flat plate PV system, 20 kW wind turbine, 15 kW bidirectional power electronic converter and 140 kWh lead-acid battery consisting of seven strings. The initial capital cost of this optimal configuration is $241,000. The total present cost is $453,517, with levelized cost of energy (COE) being $0.2973, and the operation cost is of $9 593 throughout the project lifetime. The number of autonomy is 10 days. For more simulation results, see **Figure 11**.

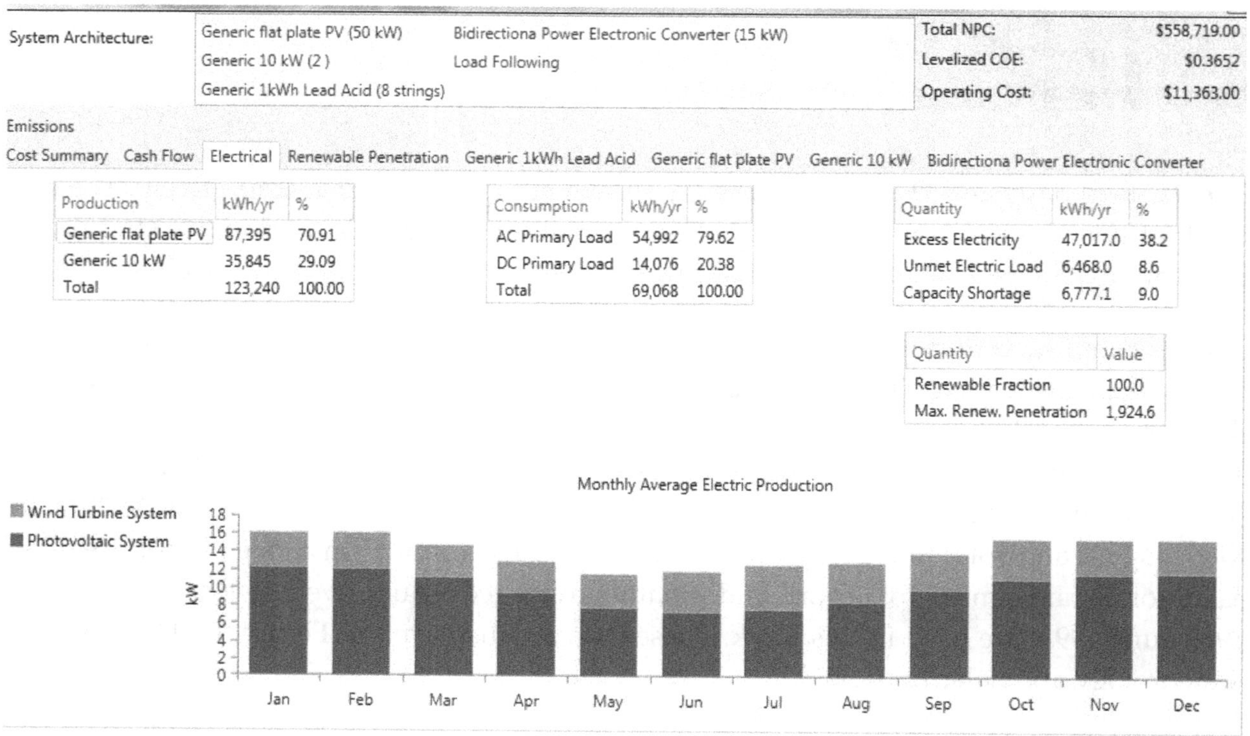

Figure 10. Simulation results of the optimal architecture with load-following dispatch strategy.

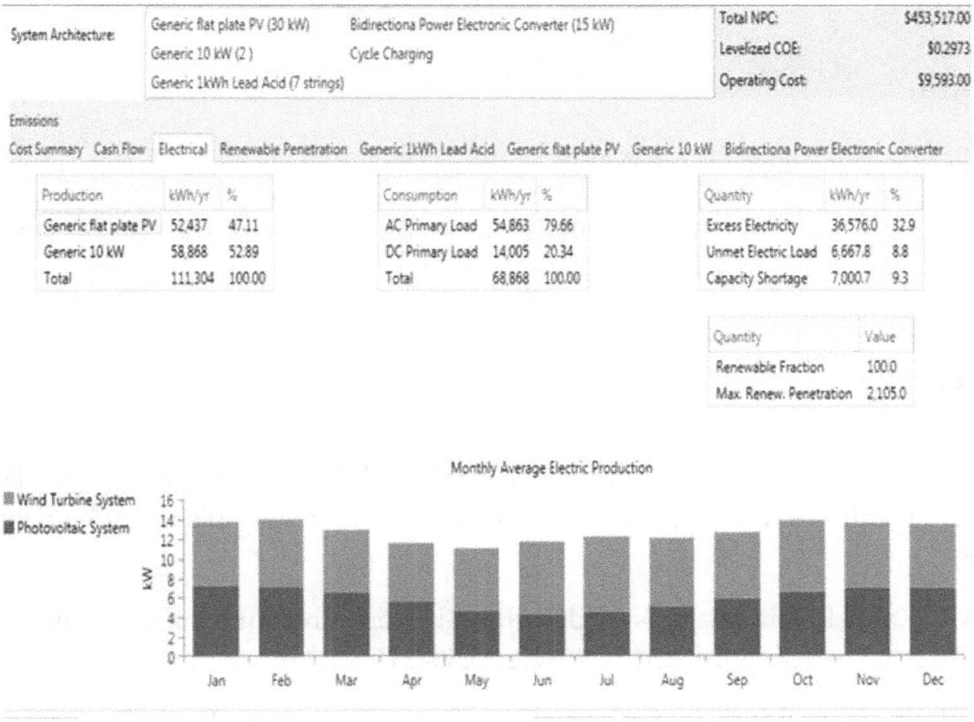

Figure 11. Simulation results of the optimal architecture with cycle-charging dispatch strategy.

For the same technically feasible configuration, the total net present cost and the initial capital cost of the load-following dispatch strategy are more expensive than the cycle-charging dispatch strategy. The levelized cost of energy of optimal configuration with load-following strategy is higher compared to the one with cycle-charging dispatch strategy. In the periods of extended cloudy season and low solar insolation, the configuration with load-following dispatch strategy offers a marginally extended period of autonomy.

7. Conclusion

The use of renewable energy as local energy resources enhances the accessibility and affordability of energy to unconnected areas around the world. Renewable energies, such as wind, solar, and biomass, have the potential to reduce the energy poverty of rural areas and sites very far from the grid. Being local resources, they also provide higher energy security to the populace.

The technical and economic analysis of the designed hybrid AC/DC micropower system studied in this work reveals the optimal electric power system configuration that is technically and economically viable using the net present cost as the econometric index. It is concluded that it is marginally better to dispatch load-following strategy for higher net present cost and autonomy compared to cycle-charging strategy dispatch. For lower levelized cost of energy (COE) and initial capital cost, it is marginally better to consider a cycle-charging dispatch strategy compared to a load-following dispatch strategy.

Author details

Atanda K. Raji

Address all correspondence to: rajia@cput.ac.za

Centre for Distributed Power and Electronic Systems, Cape Peninsula University of Technology, Bellville Campus, Cape Town, South Africa

References

[1] Cha J. Electrifying Sub-Saharan Africa Sustainably with Renewable Energy. SAIS Perspectives; April 2011. http://www.saisperspectives.com/2017issue/renewable-energy-africa (Accessed 8-06-2017)

[2] Raji AK, Kahn MTE. Analysis of distributed energy resources for domestic electricity users. Journal of Southern Africa. 2012;**2**(1):50-55

[3] Shaahid SM, Elhadidy MA. Opportunities for utilization of stand-alone hybrid (photovoltaic + diesel + battery) power systems in hot climates. Renewable Energy. 2003;**28**(11): 1741-1753

[4] Bokanga GM, Raji A, Kahn MTE. Design of a low voltage DC microgrid system for rural electrification in South Africa. Journal of Energy in Southern Africa. 2014;**25**(4):9-14

[5] Schmidt K, Patterson DJ. Benefits of load management applied to an optimally dimensioned wind/photovoltaic/diesel/battery hybrid power system. Proceedings of solar, Australian and New Zealand Solar Energy Society. 1997; Paper 139. 1-6

[6] Kellogg WD, Nehrir MH, Venkataramanan G, Gerez V. Generation unit sizing and cost analysis for stand-alone wind, photovoltaic and hybrid wind/PV systems. IEEE Transactions on Energy Conversion. 1998;**13**(1):70-75

[7] Farret FA, Godoy SM. Integration of Alternative Sources of Energy. New Jersey: John Wiley & Sons; 2006. p. 471

[8] Venter C, Raji A, Adonis M. The DC house for low power households-DC-DC converter analysis. In: Proceedings of Domestic Use of Energy International Conference (DUE); 30 March–1 April 2015; Cape Town, South Africa. 143-146

[9] HOMER (The Hybrid Optimization Model for Electric Renewables). http://www.nrel/gov/HOMER (Accessed 10-02-2015)

[10] Raji AK, Kahn MTE. Fuel cell systems as an efficient domestic distributed generation unit. In: Proceedings of Domestic Use of Energy International Conference (DUE); 2-4 April, 2013; Cape Town, South Africa. 91-94

[11] Borowy B, Salameh Z. Methodology for optimally sizing the combination of a battery bank and PV Array in a Wind/PV Hybrid system. IEEE Transactions on Energy Conversion. 1995;**11**(2):367-375

[12] Kusakana K, Vermaak HJ, Numbi BP. Optimal sizing of a hybrid renewable energy plant using linear programming. In: Proceedings of IEEE PES Power Africa 2012 Conference and Exposition; 09-13 July 2012; Johannesburg, South Africa. http://ieeexplore.ieee.org/stamp/stamp.jsp?arnumber=6498608 (Accessed 23-04-2016)

[13] Connoly D, Lund H, Mathiesen BV, Leahy M. A review of computer tools for analysing the integration of renewable energy into various energy systems. Applied Energy. 2010;**87**:1059-1082

[14] Bagul AD, Salameh ZM, Borowy B. Sizing of a stand-alone hybrid wind photovoltaic system using a three-event probability density approximation. Solar Energy. 1996;**56**(4)

[15] Yang HX, Lu L, Zhou W. A novel optimization sizing model for hybrid solar– wind power generation system. Solar Energy. 2007;**81**(1):76-84

[16] Mellit A, Kalogirou SA, Hontoria L, Shaari S. Artificial intelligence techniques for sizing photovoltaic systems: a review. Renewable and Sustainable Energy Reviews. 2009; **13**(2):406-419

[17] Retscreen Expert. https://www.nrcan.gc.ca/energy/software-tools/7465 (Accessed 11-02-2015)

[18] Hybrid2. http://www.umass.edu/windenergy/research/topics/tools/software/hybrid2 (Accessed 10-02-2015)

[19] Choi B-Y, Noh Y-S, HyokJi Y, Lee B-K, Won C-Y. Battery-integrated power optimizer for PV-battery. In: Proceedings of IEEE Vehicle Power and Propulsion Conference; 9-12 October 2012; Seoul, Korea. http://ieeexplore.ieee.org/stamp.jsp?arnumber=6422686 (Accessed 20-01-2017)

[20] Li C-H, Zhu X-J, Cao G-Y, Sui S, Hu M-R. Dynamic modeling and sizing optimization of stand-alone photovoltaic power systems using hybrid energy storage technology. Renewable Energy. 2009;**32**(3):815-826

[21] Tuglie ED, Torelli F. Load following control schemes for deregulated energy markets. IEEE Transaction on Power Systems. 2006;**21**(4):1691-1697

[22] Bizon N. Load-following mode control of a standalone renewable/fuel cell hybrid power source. Energy Conversion and Management. 2014;**77**:763-772

[23] Garcia P, Torreglosa JP, Fernandez LM, Jurado F. Optimal energy management system for stand-alone wind turbine/photovoltaic/hydrogen/battery hybrid system with supervisory control based on fuzzy logic. International Journal of Hydrogen Energy. 2013;**38**:14146-14158

[24] Callaway DS. Tapping the energy storage potential in electric loads to deliver load following and regulation, with application to wind energy. Energy Conversion and Management. 2009;**50**:1389-1400

[25] Subho U, Sharma MP. Development of hybrid energy system with cycle charging strategy using particle swarm optimization for a remote area in India. Renewable Energy. 2015;**77**:586-598

[26] Subho U, Sharma MP. Selection of a suitable energy management strategy for a hybrid energy system in a remote rural area of India. Energy. 2016;**94**:352-366

[27] McGowan JG, Manwell JF. Hybrid wind/PV/diesel system experiences. Renewable Energy 1999;**16**(1-4):928-933

[28] Chedid R, Saliba Y. Optimization and control of autonomous renewable energy systems. International Journal of Energy Research. 1996;**20**(7):609-624

[29] Protogeropoulos C, Brinkworth BJ, Marshall RH. Sizing and techno-economical optimization for hybrid solar photovoltaic/wind power systems with battery storage. International Journal of Energy Research. 1998;**21**(6):465-479

[30] Sissine F. Renewable energy: Background and issues. In: the 110th Congress Resources 2008. http://nationalaglawcenter.org/wp-content/uploads/assets/crs/RL343162.pdf (Accessed 15-05-2017)

[31] Javashri R, Kumudini Devi RP. Effect of tuned unified power flow controller to mitigate the rotor speed instability of fixed-speed wind turbines. Renewable Energy. 2009; **34**(3):591-596

[32] Kellogg WD, Nehrir MH. Venkataramanan G, Gerez V. Generation unit sizing and cost analysis for stand-alone wind photovoltaic and hybrid wind/PV systems. IEEE Transactions on Energy Conversion 1998;**13**(1):70-75

[33] Online Eco Store Online. Available online http://www.sustainable.co.za

[34] Raji AK. Impacts of electricity rates on the economics of rooftop PV system. International Journal of Advanced Research in Engineering. 2016;**2**(2):1-6

[35] NASA Surface Meteorology and Solar Energy. http://eosweb.larc.nasa.gov/sse/ (Accessed 10-02-2017)

8

Complex System Reliability Analysis Method: Goal-Oriented Methodology

Yi Xiao-Jian, Shi Jian and Hou Peng

Abstract

Goal-oriented (GO) methodology is a success-oriented method for complex system reliability analysis based on modeling the normal operating sequence of a system and all possible system states. Recently, GO method has been applied in reliability and safety analysis of a number of systems, spanning defense, transportation, and power systems. This chapter provides a new approach for reliability analysis of complex systems, first, by providing its development history, its engineering applications, and the future directions. Then, the basic theory of GO method is expounded. Finally, the comparison of GO method, fault tree analysis and Monte-Carlo simulation is discussed.

Keywords: system reliability analysis, complex system, GO method, reliability model

1. Introduction

Quality and reliability are key attributes of economic success of a system because they result in an increase in productivity at low cost and vital for business growth and enhanced competitive position. The recent advances in electronics, computing, communication, control, and networking have resulted in integrated systems that are: (i) complex in structure, (ii) large in scope and scale, (iii) characterized by multimode operation, (iii) capable of working in varied working conditions, and (iv) hierarchically organized. The reliability of such complex systems is a critical factor of their fitness for their intended use and hence is vital for in design and manufacturing. Reliability analysis method of complex systems is conducive to prevent defects in the first place in all aspects when they do occur in operation, in order to improve their reliability and reduce their life-cycle cost.

Fault tree analysis (FTA) and Monte-Carlo simulation (MCS) are now the standard reliability and safety analysis methods. Different from them, goal-oriented (GO) methodology [1] is a success-oriented method for system reliability analysis based on modeling the normal operating sequence of a system and all possible system states. It is especially suitable for complex systems having time-sequence of operation, multiple states, and so on, and the quantitative analysis and the qualitative analysis of the GO method are conducted by the GO operation according to the GO model. The keys of the GO method are the GO model and the GO operation. Although the GO method was introduced in 1976 [1], it was largely unknown until recently. The GO method has become increasingly popular in recent years because of its advantages in terms of its ease of creating a model and of its representational and analysis power [2, 3]. This chapter is an attempt at providing a basic theory of GO method in terms of GO model, GO operation, and comparison with FTA and MCS.

1.1. Development of GO method

The major application of GO method is in establishing a system reliability model, and its quantitative and qualitative analysis. The chronological development of the GO method can be broadly divided into two periods of growth: (1) 1970s~the early 1990s, the basic model and theory of the GO method, its comparison with FTA method, and its GO operator type and function were explained in the research reports by the Electric Power Research Institute in US [4–8]; (2) After the late 1990s, the GO method has attracted more attention again, in particular in the People's Republic of China. Perhaps, the initial application of the GO method was in reliability and safety analysis of missile and weapons systems. Recently, the GO method has also been applied in the reliability analysis of defense systems, water, oil, and gas supply systems, manufacturing systems, transportation systems, power systems, and logistics management systems [2, 3]. Furthermore, the theory of GO method for a complex system with complex correlations (dependencies), closed-loop feedback, multiple functions, multiple fault modes, etc., has been developed from three aspects, which are GO model, basic GO algorithm, and GO method for complex systems with various characteristics [9–17]. Literature [2, 3] gave the overview of development of GO method in detail.

1.2. Further of GO method

The GO method has been widely used. There are a number of areas where the powerful advantages of the GO model and its reliability analysis methodology can be exploited.

Although there is a great deal of interest in the design of complex systems for reliability, research on deigning these systems for both functional and structural reliability and life-cycle cost is needed. To go further, there is still a substantial research gap in the optimal design of systems for reliability and life-cycle cost, taking into account the issues of structure, function, behavior, and other characteristics, such as active or cold standby redundancy, and fault-tolerant mechanisms. In terms of modeling, analysis, and software tools, there are a number of research issues to consider. These include: (i) How to integrate the product structure, behavior, and functions in a reliability model; (ii) How to conduct the reliability analysis of complex systems accurately, thoroughly, and quickly; (iii) How to optimally allocate reliability among subsystems, taking into consideration the structural and functional hierarchy, as well as redundancy management techniques under resource constraints; and (iv) How to develop software tools to support design of complex systems for reliability that are intuitive and support collaborative design. Considering the advantages of GO model and its reliability analysis method, the GO method not only can solve the existing problems above, but also it can further develop system reliability theory and application. Meanwhile, the corresponding software will also have a more extensive application prospects and important value.

Another important direction for research is the application of the GO methodology to related areas of quality control, fault diagnosis and prognosis, and condition-based maintenance.

2. GO model

The GO model is a key element of GO method. It is developed directly using product schematic diagrams, its structure, and its functional hierarchy. According to the GO model, the reliability analysis is conducted by GO operation. The GO model composed of GO operator and signal flow.

2.1. GO operator

GO operator contains function operator and logical operator, which represents the unit itself or logical relationship, respectively. Its data, type, and GO operation formula are the basic attributes of GO operator. There are 17 standard GO operators in basic GO theory, and their signs and description are shown in **Figure 1** and **Table 1**, respectively. In **Figure 1**, S, C, and R are the input signal, GO operator itself, and output signal, respectively.

2.1.1. Standard GO operator

In this section, six frequently-used standard GO operators are illustrated from aspects of description, operation rule table, and GO operation formula, respectively. In operation rule table, V_S, V_C, and V_R are the state of the input signal, GO operator itself, and output signal, respectively, and $0, \cdots, N$ represent their state values. In GO operation formula, $P_S(i)$, $P_R(i)$, and $P_C(i)$ are the state probability of the input signal, GO operator itself, and output signal, respectively. The state cumulative probability of input signal $A_S(i)$ and output signal $A_R(i)$ are defined as

$$\begin{cases} A_S(i) = \sum_{j=0}^{i} P_S(j) \quad i = 0, \cdots, N-1; A_S(N) = 1 \\ A_R(i) = \sum_{j=0}^{i} P_R(j) \quad i = 0, \cdots, N-1; A_R(N) = 1 \end{cases} \tag{1}$$

1. Type 1 operator

- **Description:** It describes the unit with two states, which are success state (enable signal flow pass) and failure state (stop signal flow passing). For example, electric resistance, switch, valve, and pipeline.

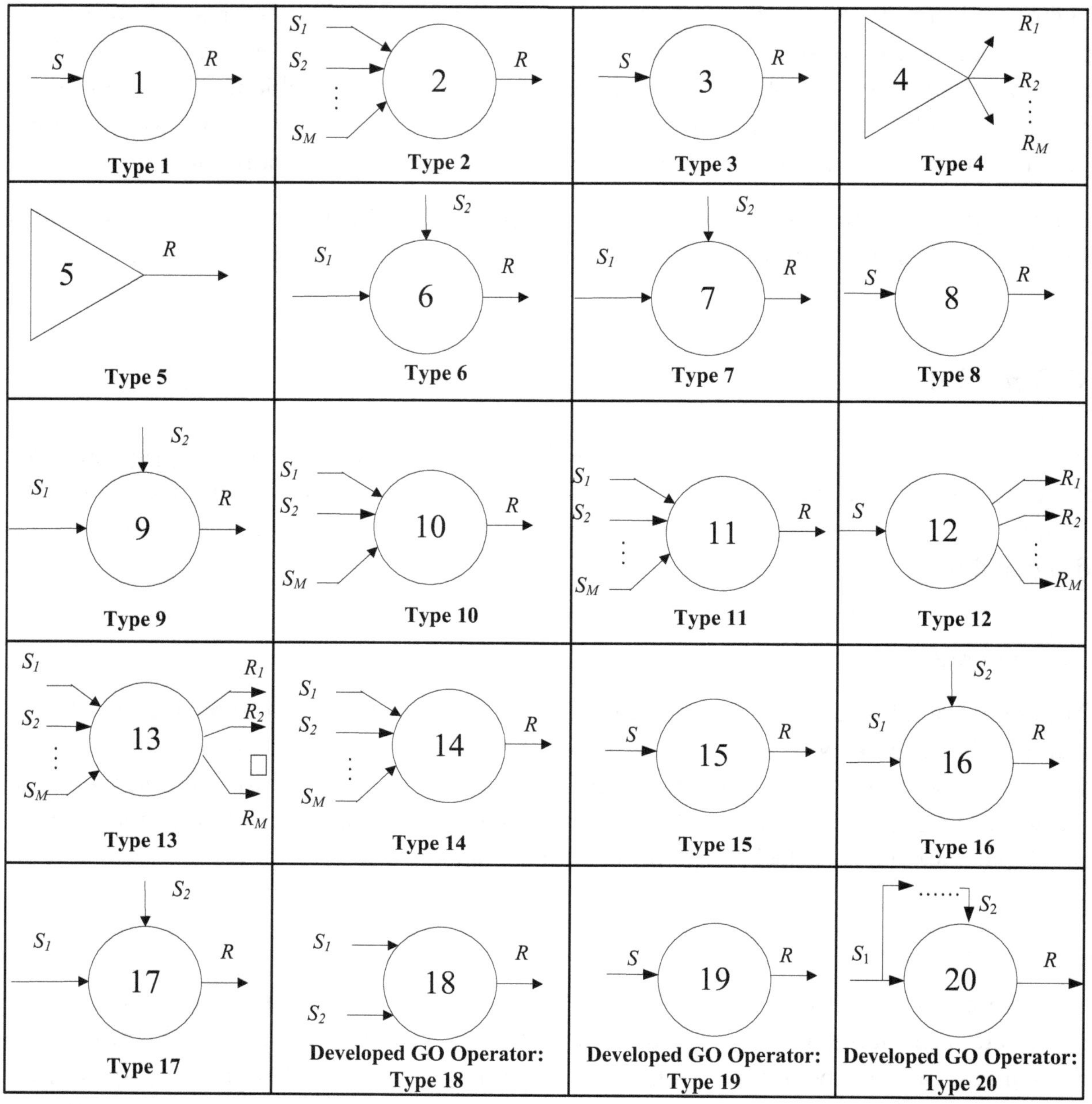

Figure 1. Signs of standard GO operators and three developed GO operators.

- **Operation rule table:**

V_S	V_C	V_R
0, …, N−1	1	0, …, N−1
N	1	N
0, …, N	2	N

- **GO operation formula:**

$$\begin{cases} P_R(i) = P_S(i) \cdot P_C(1), \quad i = 0, 1, \cdots, N-1 \\ P_R(N) = P_S(N) \cdot P_C(1) + P_C(2) \end{cases} \tag{2}$$

2. Type 2 operator

- **Description:** It describes the logical relationship OR among some inputs signal and one output signal.
- **Operation rule table:**

V_{S1}	V_{S2}	…	V_{SM}	V_R
0, …, N	0, …, N	0, …, N	0, …, N	MIN{V_{S1}, V_{S1},…, V_{SM}}

- **GO operation formula:**

$$\begin{cases} P_R(0) = A_R(0) \\ P_R(i) = A_R(i) - A_R(i-1), i = 1, \cdots, N \end{cases} \tag{3}$$

3. Type 3 operator

- **Description:** It describes unit with failure state, operating state, and operating ahead state. For example, control system, contactor coil, and so on.
- **Operation rule table:**

V_S	V_C	V_R
0, …, N	0	0
0, …, N–1	1	0, …, N–1
N	1	N
0, …, N	2	N

- **GO operation formula:**

$$\begin{cases} P_R(0) = P_C(0) + P_S(0) \cdot P_C(1) \\ P_R(i) = P_S(i) \cdot P_C(1), i = 1, \cdots, N-1 \\ P_R(N) = P_S(N) \cdot P_C(1) + P_C(2) \end{cases} \tag{4}$$

4. Type 5 operator

- **Description**: It describes the single input unit, which as system input. For example, battery, water source, and so on.
- **GO operation formula:**

$$P_R(i) = \begin{cases} P_j, i = I_j \\ 0, i \neq I_j \end{cases}, j = 1, \cdots, L, i = 0, \cdots, N \tag{5}$$

5. Type 6 operator

- **Description:** It describes the unit receiving signal to turn on. For example, electric water pump, contactor, and so on.

Operator type	Functional description
Type 1	Unit with failure state and operating state
Type 2	Logical relation "OR"
Type 3	Unit with failure state, operating state, and operating ahead state
Type 4	Multi-signal input unit
Type 5	Single input unit
Type 6	Unit receiving signal to turn on
Type 7	Unit receiving signal to turn off
Type 8	Unit with delayed response
Type 9	Output signal decided by the state of two input signal
Type 10	Logical relation "AND"
Type 11	Logical relation "k-out-of-m"
Type 12	Input signal can choose the output path
Type 13	Unit with multiple input signals and output signals
Type 14	Linear relation between multiple input signals and one output signal
Type 15	Logical relation of output signal affected by the probability event of input signal
Type 16	Unit requested to resume OFF-state
Type 17	Unit requested to resume ON-state

Table 1. Functional description of standard GO operators.

- **Operation rule table:**

V_{S1}	V_{S2}	V_C	V_R
$I_1(0, \ldots, N)$	$I_2(0, \ldots, N)$	0	I_1
$I_1(0, \ldots, N)$	$I_2(0, \ldots, N)$	1	$MAX\{I_1, I_2\}$
$I_1(0, \ldots, N)$	$I_2(0, \ldots, N)$	2	N

- **GO operation formula:**

$$\begin{cases} A_R(i) = A_{S1}(i)[P_C(0) + A_{S2}(i) \cdot P_C(1), \ i = 0, \cdots, N-1 \\ P_R(N) = P_{S1}(N) + A_{S1}(N-1)[P_C(2) + P_{S2}(N) \cdot P_C(1)] \\ P_R(0) = A_R(0) \\ P_R(i) = A_R(i) - A_R(i-1) \ i = 1, \cdots, N-1 \end{cases} \tag{6}$$

6. Type 10 operator

- **Description**: It describes the logical relationship AND among some inputs signal and one output signal.

- **Operation rule table**:

V_{S1}	V_{S2}	...	V_{SM}	V_R
0, ..., N	0, ..., N	0, ..., N	0, ..., N	MAX{V_{S1}, V_{S1},..., V_{SM}}

- **GO operation formula**:

$$\begin{cases} A_R(i) = \prod_{j=1}^{M} A_{Sj}(i),\ i = 0, \cdots, N-1 \\ P_R(N) = 1 - A_R(N-1) \\ P_R(0) = A_R(0) \\ P_R(i) = A_R(i) - A_R(i-1)\ i = 1, \cdots, N-1 \end{cases} \tag{7}$$

2.1.2. Developed GO operator

In this section, three developed GO operators are illustrated from aspects of description, operation rule table, and GO operation formula, respectively.

1. Type 18 operator

- **Description:** It describes the logical relation of standby mode, which is combination of primary equipment group *CG* and standby equipment group *CBG* working under condition of primary equipment group faulting. The input signals and output signal of Type 18 operator are denoted as L_1, $L_{2,}$ and *R*, respectively. The signal flow L_1 represents primary equipment group working. The signal flow L_2 represents standby equipment group working under condition of primary equipment group faulting, and the signal flow L_2 is also the output signal of GO operator, which represents the standby equipment group. And the signal flow *R* represents standby structure working. L_1, $L_{2,}$ and *R* have two states, which are state 1: success state and state 2: fault state.

- **Operation rule table:** The combination composed of a Type 18 operator and a Type 20 operator is often used to represent standby structure at any place. S_1, $S_{2,}$ and S_0 represent the input signal flows and output signal flow of Type 20 operator, respectively, and S_2represents also the input signal flow of Type 18 operator, i.e., L_1.

V_{CG}	V_{CBG}	V_{S1}	V_{S0}	V_{S2}	V_{L2}	V_R
1	1	1	2	1	2	1
1	1	2	2	2	2	2
1	2	1	2	1	2	1
1	2	2	2	2	2	2
2	1	1	1	2	1	1
2	1	2	2	2	2	2
2	2	1	1	2	2	2
2	2	2	2	2	2	2

- **GO operation formula:**

$$\begin{cases} P_R(1) = P_{L1}(1) + P_{L2}(1) \\ P_R(2) = 1 - P_R(1) \\ P_{L1}(1) = P_{S1}(1) \cdot P_{CG}(1) = P_{S1}(1) - P_{S0}(1) \\ P_{L2}(1) = P_{S0}(1) \cdot P_{CBG}(1) \end{cases} \quad (8)$$

where $P_{S0}(1)$ and $P_{CBG}(1)$ are the success probability of S_0 and *CBG*, respectively.

2. Type 19 operator

- **Description:** It describes the unit turning unstable operation into normal operating. The input signals and output signals of Type 19 operator are denoted as *S* and *R*, respectively. *S* is a multistate signal flow, which contains an operating state, a faulting state, and *m* unstable operation states. And *m* unstable operation states are divided into two kinds, which are q unstable operation states turned into operation state by Type 19 operator, and *m-q* unstable operation states. *R* is also a two-state signal flow, which contains operating state and faulting state. *C* is the unit itself, i.e., Type 19 operator, which contains operating state and faulting state.

- **Operation rule table:**

V_S	V_C	V_R
0	0	0
0	*N*	*N*
1, …, *q*	0	0
1, …, *q*	*N*	*N*
q+1,…, *m*	0, *N*	*N*
N	0, *N*	*N*

- **GO operation formula:**

$$\begin{cases} P_R(0) = P_S(0) \cdot P_C(0) + \sum_{j=1}^{q} P_S(j) \cdot P_C(0) = \sum_{j=0}^{q} P_S(j) \cdot P_C(0) \\ P_R(N) = P_S(0) \cdot P_C(N) + \sum_{j=1}^{q} P_S(j) \cdot P_C(N) + \sum_{j=q+1}^{m} P_S(j) \cdot P_C(0) + \sum_{j=q+1}^{m} P_S(j) \cdot P_C(N) \\ \qquad + P_S(N) \cdot P_C(0) + P_S(N) \cdot P_C(N) \\ \qquad = \sum_{j=0}^{q} P_S(j) \cdot P_C(N) + \sum_{j=q+1}^{m} P_S(j) + P_S(N) \end{cases} \quad (9)$$

where $P_S(j)$ is state probability of unstable operation states for *S*, $j = 1, 2, \cdots, m$, $P_S(0)$ and $P_S(N)$ are state probability of operating state and faulting state for *S*, respectively,

$P_S(0) + P_S(N) + \sum_{j=1}^{q} P_S(j) + \sum_{j=q+1}^{m} P_S(j) = 1$; $P_C(0)$ and $P_C(N)$ are state probability of operating state and faulting state for C, $P_C(0) + P_C(N) = 1$.

3. Type 20 operator

- **Description:** It describes the signal flow of conditional operating mode, which is one of its input signal faults under the condition of another input signal success. The input signals and output signal of Type 20 operator are denoted as S_1, S_2, and R, respectively. S_1, S_2, and R have two states, which are state 1: success state and state 2: fault state. It is often used in standby structure.

- **Operation rule table:**

V_{S1}	V_{CG}	V_{S2}	V_R
1	1	1	2
1	2	2	1
2	1, 2	2	2

- **GO operation formula:**

$$\begin{cases} P_R(1) = P_{S1}(1) \cdot [1 - P_{CG}(1)] = P_{S1}(1) - P_{S2}(1) \\ \mathrm{P}_R(2) = 1 - P_R(1) \end{cases} \tag{10}$$

where $P_{S\alpha}(1)$, $P_{CG}(1)$, and $P_R(1)$ are the success probability of S_α, CG, and R, respectively, $\alpha = 1, 2$; $P_R(2)$ is the fault probability of R.

2.2. Signal flow

Signal flow represents specific fluid flow, such as oil, gas, electricity, and so on, or a logical process. It describes the relationships among the GO operator, its inputs, and outputs. And its attribution includes state value and state probability. The signal flow is used to connect the GO operator and as the direction of GO operation.

2.3. GO model

GO model is developed by using signal flows to connect GO operators according to the system principle diagram, engineering drawing, and function constitute directly. And it reflects the system characteristics visually. The proper GO model should be satisfied with:

- The operator in GO model must be labeled with type and number, especially the number which is unique. The first number in the GO operators represents the type of operator and the second number represents the numbering of the operators.

- There is at least one input GO operator (such as Type 4 or Type 5). Generally, the number of GO operator begins with the input GO operator.

- For each GO operator in GO model, its input signal flow must be the output signal flow of the other GO operators. Each signal flow must be labeled with the unique number, and the numbers on the signal line represent the signal flow numbering.
- The signal flow sequence must start with any input GO operator and end with the output signal flow of system. The GO model does not allow being a cyclic model. Generally, the number of signal flow should be labeled with the output signal flow of an input GO operator.

Example:

The structure diagram of pressurized water reactor (PWR) purification system and its GO model are, respectively, shown in **Figures 2** and **3**, and the GO operator type of component is presented in **Table 2**.

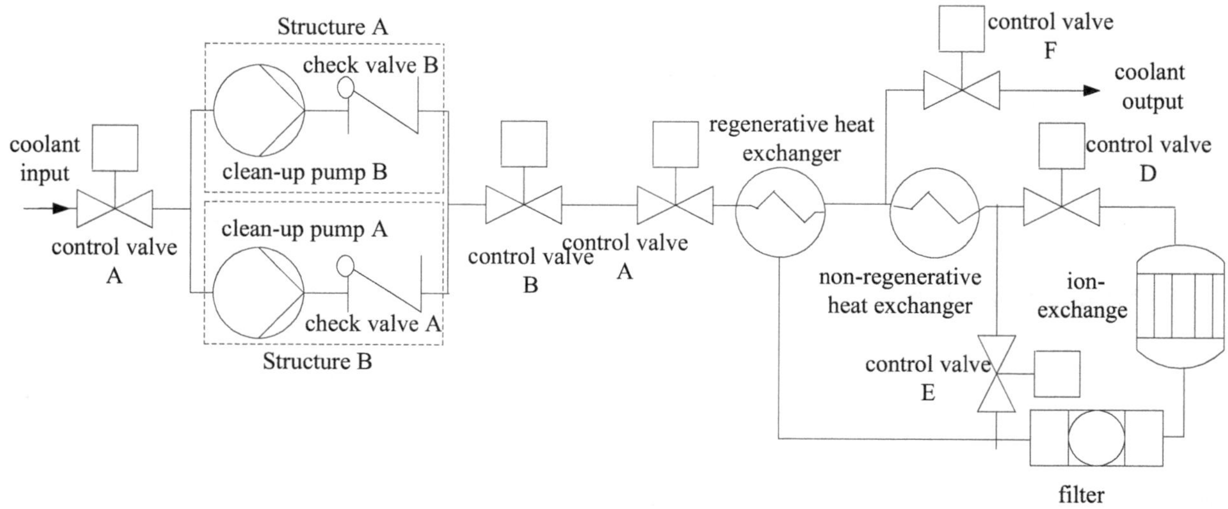

Figure 2. Structure diagram of PWR purification system.

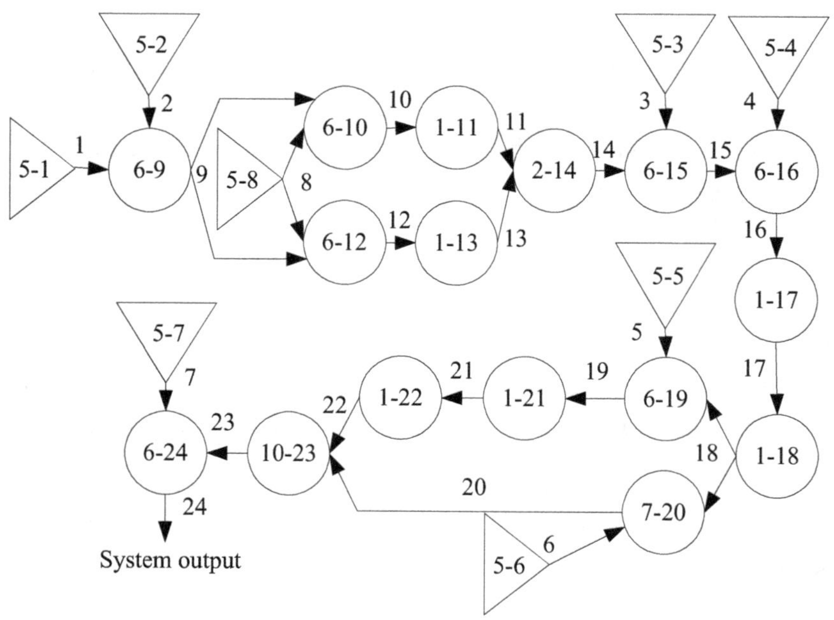

Figure 3. GO model of PWR purification system.

Operator number	Component	Operator type
1	Coolant input	5
2, 3, 4, 5, 6, 7	Control signal	5
8	Control power	5
9	Control valve A	6
10	Clean-up pump A	6
11	Check valve A	1
12	Clean-up pump B	6
13	Check valve B	1
15	Control valve B	6
16	Control valve C	6
17	Regenerative heat exchanger	1
18	Nonregenerative heat exchanger	1
19	Control valve D	6
20	Control valve E	7
21	Ion exchange	1
22	Filter	1
24	Control valve F	6

Table 2. GO operator type of component for **Figure 2**.

3. GO operation

GO operation is defined that it begins with output signal flow of input GO operators in GO model to calculate the state probability and state value of output signal flow for the next GO operator, and it will be finished until the system output signal flow is calculated after the sequence of signal flow. GO operation contains the quantitative and qualitative analyses. And the quantitative and qualitative analyses are conducted using the GO operation based on the GO algorithm, following the reliability analysis process of GO method. GO algorithm and GO analysis process are the key elements of GO operation.

3.1. GO algorithm

The operational efficiency and accuracy of analysis are thus affected by the GO algorithm. The GO algorithm is comprised of a state combination algorithm [4] and a probability formula algorithm [18–21]. The number of state combinations for a complex system is very large, and the probability of a combination state cannot be easily computed. The probability formula algorithm is faster and easier than the state combination algorithm, and it is the mainstream GO algorithm. Thus, this section illustrates two kinds of probability formula algorithms, which are direct GO algorithm and GO algorithm with shared signals.

3.1.1. Direct GO algorithm

Direct GO algorithm is based on the calculation of state probability for signal flow, and the proper direct GO algorithm should be satisfied with:

- The state probability of input GO operator is the state probability of its output signal flow. The output signal flow of input GO operator is the input signal flow of the next GO operator.
- The state probability of output signal flow for the next GO operator is calculated based on its GO operation formula and data, and this output signal flow will be the next GO operator's input signal flow.
- Based on the above rules, following the signal flow sequence, the output signal flow of every GO operator in GO model can be obtained, and the GO operation will be finished until the state probability of output signal flow represented system output is calculated.
- When GO operator is executed with quantitative analysis, it is not necessary to list the corresponding state combination.

3.1.2. GO algorithm with shared signals

In GO model, the output signal flow of a GO operator often connects multiple GO operators, and they are the input signal flows of more than one GO operators, such output signal flow is called shared signal. If the GO operation adopts the direct GO algorithm, the quantitative analysis results will have biases. Thus, GO algorithm with shared signals was proposed in order to obtain more accurate result.

1. Shared signal

There are two situations in GO model on shared signal, which are as follows:

- Completely contain. It means all items in formula of a signal flow's state probability have state probability of shared signal.
- Not completely contain. It means some items in formula of a signal flow's state probability have not state probability of shared signal.

The rules of the processing shared signals in GO algorithm with shared signals are as follows:

- Behind a shared signal, all of signal flows in sequence of GO model still have this shared signal, and there are two situations, which are completely contain and not completely contain.
- For multiple signal flows with the same shared signal, their joint state probability cannot be obtained by using the product of their state probabilities directly.
- For two signal flows completely contain the same one shared signal, their joint state probability can be obtained by using the product of their state probabilities to divide the state probability of shared signal.
- For multiple signal flows with the same one shared signal, their joint state probability formula can be obtained by turning a high stage term of shared signal in the formula into a one stage term of shared signal.

2. Probability formula

For a system with L shared signals, the probability formula of GO algorithm with shared signals is given by Eq. (10).

$$P_R = \sum_{K_1=0}^{1}\sum_{K_2=0}^{1}\cdots\sum_{K_l=0}^{1} P_{RK_1K_2\cdots K_l}\prod_{l=1}^{L}[(1-P_{Sl})(1-K_l)+P_{Sl}K_l] \tag{11}$$

where $P_{RK_1K_2\cdots K_L}$ is the cumulative probability of system output under a combination of all shared signals, and Sl is a shared signal in the system. $K_l = 0$ and $K_l = 1$ are the failure and success states of the shared signal l, respectively. P_{Sl} and P_R are success probability of shared signal l and system output, respectively. The item of $\prod_{l=1}^{L}[(1-P_{Sl})(1-K_l)+P_{Sl}K_l]$ is the state probability for each combination of shared signals.

3. Calculating form

It is difficult and complex to derive mathematical formulae for a complex system with a large number of shared signals. A new form of Eq. (10) involves probabilistic weighting of shared signals. The probabilistic weighting improves operation efficiency greatly and avoids the need for complex mathematical formulae. The calculation process is shown in **Table 3**. In **Table 3**, numbers 1 and 0 represent success state and failure state of a shared signal Sl, respectively.

The success probability of system can be obtained by Eq. (11).

$$P_R = \sum_{j=1}^{2^L} A_j B_j \tag{12}$$

where A_j is the state probability for each combination of shared signals, and B_j can be obtained by the GO operation of the system, which sets success and failure probability of a shared signal to 1 and 0 depending on the state of shared signal for each combination of shared signals.

3.2. Reliability analysis process of GO method

The reliability analysis process of GO method is the criterion and prerequisite for conducting quantitative analysis and qualitative analysis. Generally, the steps of GO analysis process are as follows:

State combination of shared signal				State probability of combination	Success probability of system
S_1	S_2	…	S_L		
0	0	0	0	A_1	B_1
0	0	0	1	A_2	B_2
⋮	⋮	⋮	⋮	⋮	⋮
1	1	1	1	$A_2{}^L$	$B_2{}^L$
Success probability of system				P_R	

Table 3. Calculation form of GO algorithm with shared signal.

Step 1. Conducting system analysis. The system analysis is the base of GO method, and it directly affects the developing GO model and conducting GO operation. First, to analyze system structure and system function constitutes according to the principle diagram, engineering drawing, or function flowchart of system. Second, to determine system characteristics, such as correlations, multistate, and so on. Then, to determine the interfaces, inputs, and output of system. Finally, to define the success rule of system according to system analysis result.

Step 2. Developing GO model. First, to select GO operator according to system analysis results, and then to establish GO model through the signal flow to connect GO operator.

Step 3. Processing data of GO operator. According to engineering practice, to obtain the state probabilities of GO operator.

Step 4. Operating quantitative analysis. If the GO model does not contain shared signal, the direct algorithm can be selected to conduct GO operation. If the GO model contains shared signal, the GO algorithm with shared signal should be selected to conduct GO operation. Otherwise, it will cause a big error.

Step 5. Operating qualitative analysis. Setting the reliability of a function GO operator in GO model is 0, and the reliabilities of other GO operators are kept constant; in this case, if system reliability is 0 by GO operation, this GO operator will be a one-order minimum cut set. Setting the reliabilities of two function GO operators in GO model is 0 except one-order minimum cut set, and the reliabilities of other GO operators are kept constant; in this case, if system reliability is 0 by GO operation, the two GO operators will be a two-order minimum cut set. In the same way, the higher-order minimum cut sets of system can be obtained.

Step 6. Evaluating system. The quantitative analysis result and qualitative analysis result can be used as a guidance and theoretical basis for improving system and fault diagnosis of system, and so on.

Above all, the reliability analysis of GO method is formulated, as shown in **Figure 4**.

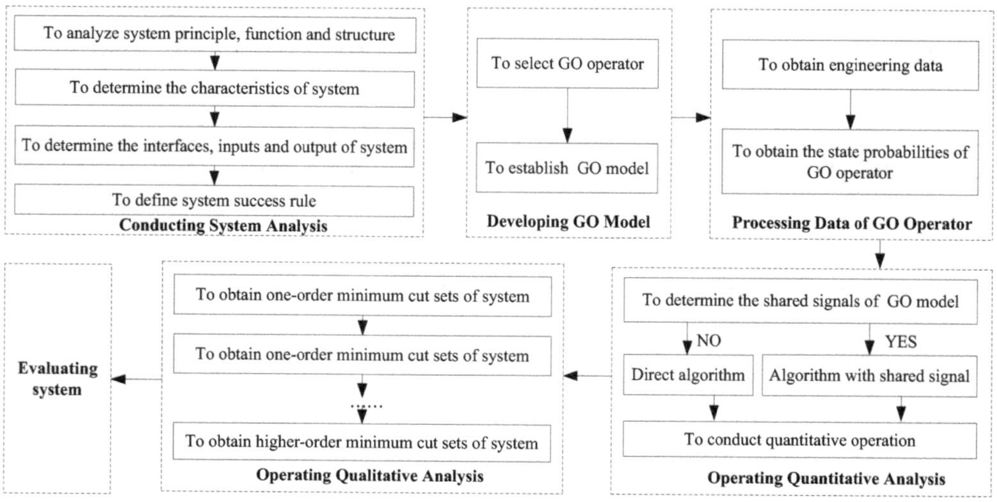

Figure 4. Reliability analysis process of GO method.

4. Example

In this section, taken a hydraulic oil supply system (HOSS) of an armored vehicle as a case, its reliability analysis is respectively conducted by GO method, FTA, and MCS in order to illustrate the usage and characteristics of GO method.

4.1. Reliability analysis of HOSS based on GO method, FTA, and MCS

4.1.1. Reliability analysis of HOSS based on GO method

1. Conducting system analysis

- **To analyze system principle, function, and structure:** HOSS supplies oil for an armored vehicle, pump-motor system, pump-motor control system, hydraulic torque converter, and lubrication system. HOSS consists of pressure oil tank, pump P1, P2, P3, and P4, oil filter LF1, LF2, and LF3, pressure relay, bypass valve LF2B and LF3B, one-way valve CV1 and CV2, constant pressure valve RV1, RV2, and RV3, hydraulic torque converter TC, radiator HE, and so on, as shown in **Figure 5**. Oil is extracted by P1 from oil pan via LF1, and then oil is injected into pressure oil tank via LF2 and case inner passage. When LF2

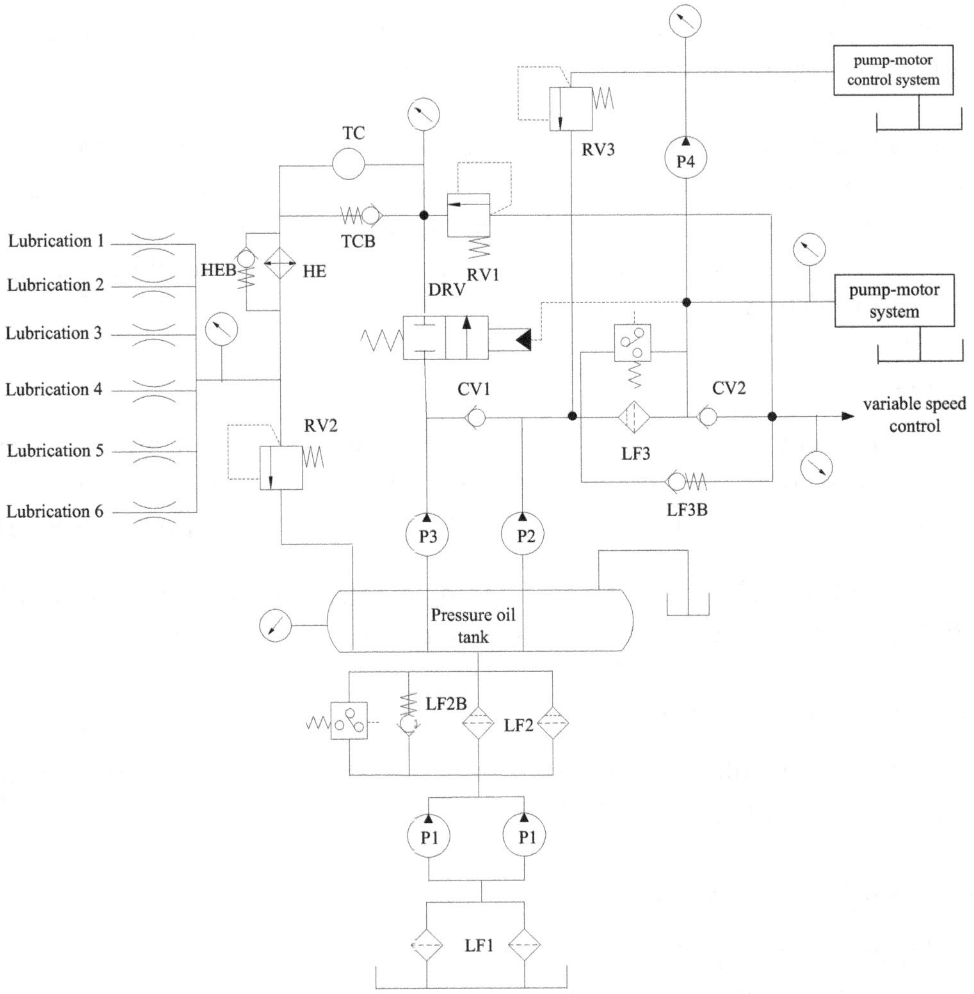

Figure 5. Diagram of HOSS.

group is obstructed and pressure between input and output becomes more than 0.5 mega Pascal, oil will be injected into pressure oil tank via LF2B. Oil is extracted by P2 from pressure oil tank; then oil is injected into CV2 via LF3 and then it is injected into hydraulic manifold block as the pressure oil provided for oil cylinder of variable speed control system and pump-motor control system by P4. LF3 and LF3B are another parallel structure, and they are same as LF2 group and LF2B. Because of the pressure of control oil which decreases a little at the situation of high speed, ingress oil of P2 can meet requirements of system. In addition, oil is extracted by P3 from pressure oil tank via DRV to TC. Then, ingress of oil is injected into lubrication system via HE. TC and TCB, and HE and HEB are same as LF2 group and LF2B. RV1, RV2, and RV3 are constant pressure valves of variable speed control and pump-motor system, lubrication system, and pump-motor control system, respectively.

- **To determine characteristics of system:** According to analysis of HOSS, the LF2 group and LF2B, LF3 and LF3B, TC and TCB, and HE and HEB are standby structures in HOSS. The standby equipments haven't the change-over switch.

- **To determine interfaces, input, and output of system:** According to analysis of HOSS, the oil from oil pan and pump group power are system inputs, and oil supply of variable speed control system, pump-motor system, pump-motor control system, hydraulic torque converter, and lubrication system are system outputs.

- **To define system success rule:** According to analysis of HOSS, success rule can be defined as a system that can provide oil to variable speed control system, pump-motor system, pump-motor control system, hydraulic torque converter, and lubrication system of an armored vehicle under high speed condition at steering situation without considering overload protection.

2. Developing GO model

- **To select GO operator:** According to the system analysis result and the types of GO operator, the functional GO operators and logical GO operators are selected to describe the units itself and logical relationships in HOSS, respectively, as presented in **Table 4**.

- **To establish GO model:** According to diagram of HOSS and analysis result of HOSS, the GO model of HOSS is developed from system input to system output, as shown in **Figure 6**.

3. Processing data of GO operator

According to statistical results of data from engineering, the success state probabilities of component in HOSS are presented in **Table 5**.

4. Operating quantitative analysis

It is shown in **Figure 6** that signal flow S1, S4, S7, S8, S15, S16, S17, and S22 are shared signals; the GO algorithm with shared signals should be adopted to conduct GO operation, and the calculation form is as presented in **Table 6**.

No.	Component	Operator type	No.	Component	Operator type
1	Oil pan	5	24	RV3	1
2, 3	LF1	1	25	P3	6
5, 6	P1	6	26	DRV	1
7	Pump group power	5	28	TC	1
9, 10	LF2	1	30	TCB	1
13	LF2B	1	32	HE	1
15	Pressure oil tank	1	34	HEB	1
16	P2	6	36	RV2	1
17	LF3	1	4, 8, 11, 27		2
18	CV2	1	37		10
20	LF3B	1	14, 21,31, 35		18A
22	RV1	1	12, 19, 29, 33		20
23	P4	6			

Table 4. GO operator in GO model of HOSS.

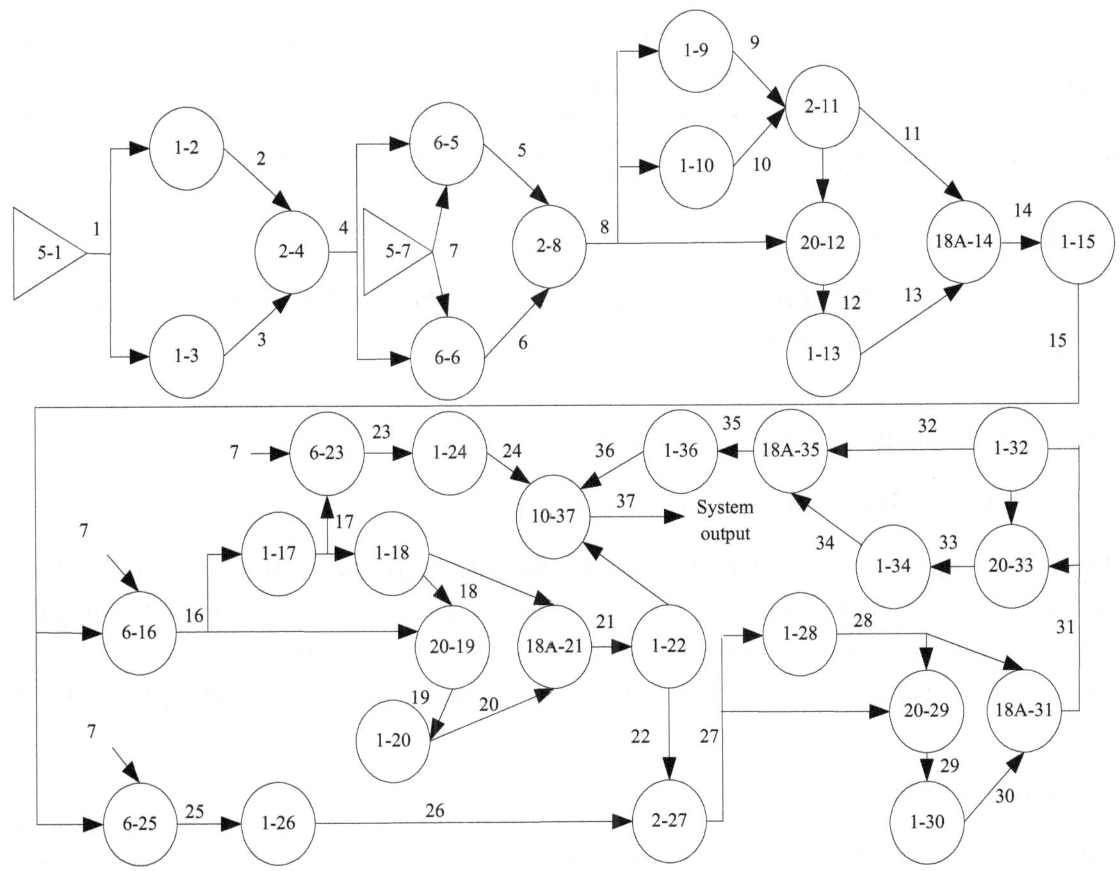

Figure 6. GO model of HOSS.

No.	Success state probability	No.	Success state probability	No.	Success state probability
1	0.99975006256	13	0.99867457268	24	0.99949819446
2	0.99413847887	15	0.99990821563	25	0.99867457268
3	0.99413847887	16	0.99950052524	26	0.99923302636
5	0.99950052524	17	0.99987532520	28	0.97865595028
6	0.99950052524	18	0.99891901058	30	0.99867457268
7	0.98676571976	20	0.99867457268	32	0.99548611722
9	0.99987532520	22	0.99949819446	34	0.99966858815
10	0.99987532520	23	0.99950052524	36	0.99949819446

Table 5. GO model of HOSS.

State of shared signal								State probability of combination	Success probability of system
S1	S4	S7	S8	S15	S16	S17	S22		
0	0	0	0	0	0	0	0	8.154e-32	0
0	0	0	0	0	0	0	1	1.620e-28	0
0	0	0	0	0	0	1	1	1.290e-24	0
⋮	⋮	⋮	⋮	⋮	⋮	⋮	⋮	⋮	0
1	1	1	1	1	1	1	1	0.98528272861	0.99846792597
Success probability of HOSS (P_{37})								0.9837732025	

Table 6. Quantitative analysis result by calculation form of GO algorithm with shared signal for **Figure 6**.

5. Operating qualitative analysis

According to the Step 5 in Section 3.2, all minimum cut sets of HOSS can be obtained by multiple GO operations, as presented in **Table 7**.

4.1.2. Reliability analysis of HOSS based on FTA and MCS

1. Reliability analysis of HOSS based on FTA

The reliability analysis process of FTA mainly contains developing fault tree model of system, obtaining all minimum cut sets of system by using Fussell-Vesely method, and obtaining the system success probability according to the minimum cut sets of system. In this case, the brief fault tree model of HOSS is shown in **Figure 7**, and the brief analysis processes are presented in **Table 8**.

2. Reliability analysis of HOSS based on MCS

The reliability analysis process of MCS mainly contains generating random numbers of success probability of GO operators, establishing simulation model, and obtaining success probability of system by operating a specified number of simulation times according to simulation model.

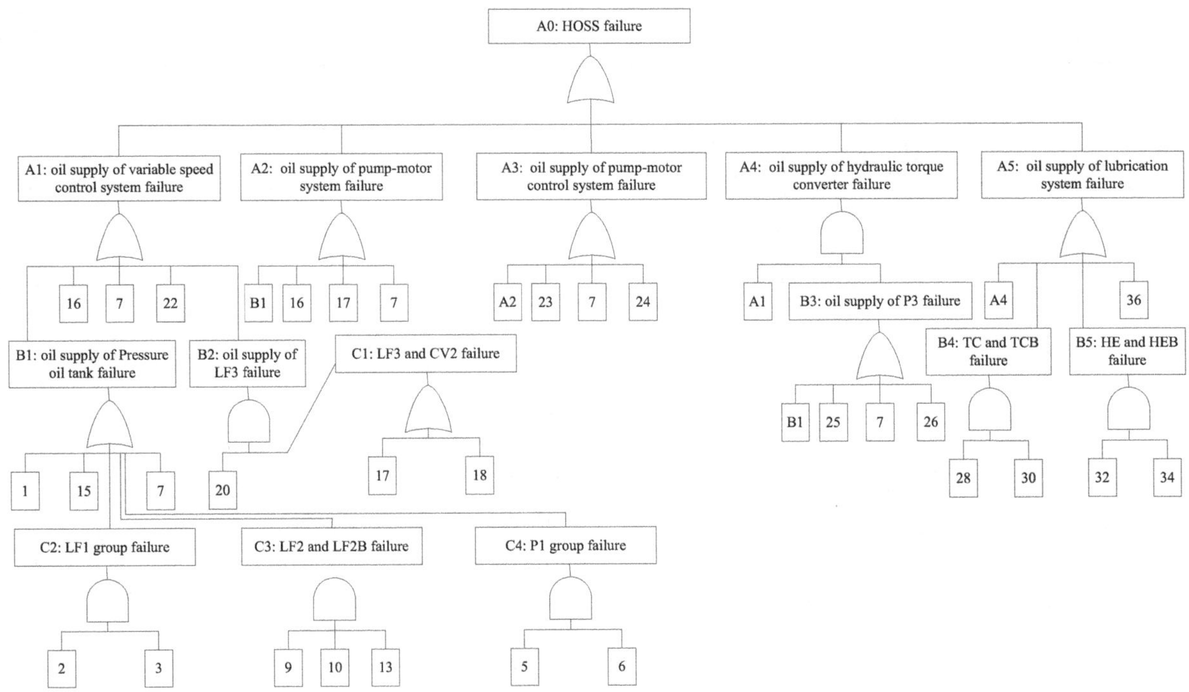

Figure 7. Brief fault tree model of HOSS.

Order	No.	Minimum cut sets	Order	No.	Minimum cut sets
1	1	Oil pan	2	2, 3	LF1 group
	7	Pump group power		5, 6	P1 group
	15	Pressure oil tank		18, 20	CV2, LF3B
	16	P2		28, 30	TC, TCB
	22	RV1		32, 34	HE, HEB
	23	P4	3	9, 10, 13	LF2 group, LF2B
	24	RV3			
	36	RV2			
	17	LF3			

Table 7. Qualitative analysis result by GO method for **Figure 6**.

In this case, the quantitative analysis result of HOSS by simulating 1 million times is 0.9838136000.

4.2. Comparison with FTA and MCS

According to reliability analysis process and analysis results of GO method, FTA, and MCS, the qualitative analysis results by the GO method are consistent with FTA, and the quantitative analysis result by the GO method is very close to the result by MCS. It illustrates that both the accurate quantitative analysis result and qualitative analysis result can be obtained by multiple GO operations based on GO method, and the comparisons of GO method, FTA, and MCS are presented in **Table 9**.

Brief analysis processes					Minimum cut sets	Order
A0	A1	B1	1	1	1	1
			C2	2, 3	2, 3	2
			C3	9, 10, 13	9, 10, 13	3
			15	15	15	1
			C4	5, 6	5, 6	2
			7	7	7	1
		16	16	16	16	1
		7	7	7	7	1
		B2	C1, 20	17, 20	-	-
				18, 20	18, 20	2
		22	22	22	22	1
	A2	B1	B1	B1	-	-
		16	16	16	16	1
		7	7	7	7	1
		17	17	17	17	1
	A3	A2	A2	**A2**	-	-
		23	23	23	23	1
		7	7	7	7	1
		24	24	24	24	1
	A4	A1, B3	A1, B1	A1, B1	-	-
			A1, 25	A1, 25	-	-
			A1, 7	A1, 7	-	-
			A1, 26	A1, 26	-	-
	A5	A4	A4	A4	-	-
		B4	28, 30	28, 30	28, 30	2
		B5	32, 34	32, 34	32, 34	2
		36	36	36	36	1

Table 8. Brief analysis processes of FTA for HOSS.

Feature	GO	FTA	MCS
Modeling oriented	Success	Failure	Logical
Modeling method	Decision-making tree	Fault tree	Logical language description
Model consistency	High	Poor	High
Model structure	Similar with schematic diagram	Hierarchy logical graph	Logical relationship
Model size	Small and compact	Hierarchy and large	Large

Feature	GO	FTA	MCS
Model element	Components and logical gate, characteristics	Failure event and logical gate	Components and logical language
Model description	Reflect system natural	Reflect cause and effect of failure	Reflect logical relation in system
Model sign	Various types and typical	Few types and not typical	Few types and not typical
Quantitative analysis	Accurate, stable	Approximative	Accurate, unstable
Qualitative analysis	Easy	Complex	-

Table 9. Comparisons of GO method, FTA, and MCS.

Author details

Yi Xiao-Jian[1,2]*, Shi Jian[3,4] and Hou Peng[2]

*Address all correspondence to: yixiaojianbit@sina.cn

1 Department of Overall Technology, China North Vehicle Research Institute, China

2 School of Mechatronical Engineering, Beijing Institute of Technology, China

3 Academy of Mathematics and Systems Science, Chinese Academy of Sciences, China

4 School of Mathematical Sciences, University of Chinese Academy of Sciences, China

References

[1] Williams RL, Gateley WY. Use of the GO methodology to directly generate minimal cut sets. In: Fussell IB, editor. Nuclear System Reliability Engineering and Risk Assessment. Pennsylvania: Society for Industrial and Applied Mathematics; 1977. pp. 825–849

[2] Yi X-J, Dhillon BS, Mu H-N. Reliability analysis using GO methodology: A review. In: 22nd ISSAT International Conference Reliability and Quality in Design (RQD 2016); 4–6 August 2016; Los Angeles, USA. 2016. pp. 16–53

[3] Yi X-J, Dong HP, Wang QF, Zhang Z. A new system reliability analysis method: The current development of GO methodology in China. WIT Transactions on Engineering Sciences. 2015;**109**:222–229

[4] Chu BB. GO Methodology: Overview Manual. EPRI NP-3123. Vol. 1. Washington: Electric Power Research Institute; 1983

[5] Chu BB. GO Methodology: Application and Comparison of the GO Methodology and Fault Tree Analysis. EPRI NP-3123. Vol. 2. Washington: Electric Power Research Institute; 1983

[6] Chu BB. GO Methodology: GO Modeling Manual. EPRI NP-3123. Vol. 3. Washington: Electric Power Research Institute; 1983

[7] Chu BB. GO Methodology: GO User's Manual. EPRI NP-3123. Vol. 4. Washington: Electric Power Research Institute; 1983

[8] Chu BB. GO Methodology: Program and User's Manual (IBM Version). EPRI NP-3123. Vol. 5. Washington: Electric Power Research Institute; 1983

[9] Yi X-J, Shi J, Dong HP, et al. Reliability analysis of repairable system with multiple fault modes based on GO methodology. ASCE-ASME Journal of Risk and Uncertainty in Engineering Systems, Part B: Mechanical Engineering. 2016. DOI: 10.1115/1.4030971

[10] Yi X-J, Shi J, Dong HP, et al. Reliability analysis of repairable system with multiple failure modes based on GO methodology. In: ASME 2014 International Mechanical Engineering Congress & Exposition; 14–20 November 2014; Montreal, Canada. 2014

[11] Yi X-J, Dhillon BS, Shi J, et al. Quantitative reliability analysis of repairable systems with closed-loop feedback based on GO methodology. Journal of the Brazilian Society of Mechanical Sciences and Engineering. 2016. DOI: 10.1007/s40430-016-0665-9

[12] Yi X-J, Shi J, Mu HN, et al. Reliability analysis on repairable system with dual input closed-loop link considering shutdown correlation based on goal oriented methodology. Journal of Donghua University (English edition). 2016;**33**(2):25–29

[13] Yi X-J, Shi J, Mu HN, et al. Reliability analysis of repairable system with multiple-input and multi-function component based on GO methodology. In: ASME 2015 International Mechanical Engineering Congress & Exposition. Paper No. IMECE2015-51289; 13–19 November 2015; Houston, Texas. 2015

[14] Yi X-J, Shi J, Mu HN, et al. Reliability analysis of repairable system with multiple-input and multi-function component based on GO methodology. ASCE-ASME Journal of Risk and Uncertainty in Engineering Systems, Part B: Mechanical Engineering. 2016. DOI: 10.1115/1.4034744

[15] Yi X-J, Dhillon BS, Mu HN, et al. Reliability analysis method for multi-state repairable systems based on goal oriented methodology. In: Proceedings of ASME 2016 International Mechanical Engineering Congress & Exposition. Paper No. IMECE2016-65380; 11–17 November 2016; Phoenix, USA. 2016

[16] Yi X-J, Dhillon BS, Shi J, et al. Reliability analysis method on repairable system with standby structure based on goal oriented methodology. Quality and Reliability Engineering International. 2015. DOI: 10.1002/qre.1953

[17] Mu HN, Liu JW, Yi X-J, et al. Reliability analysis method of phased-mission nuclear power equipment based on goal oriented methodology. In: 2016 IEEE International Conference on Industrial Engineering and Engineering Management (IEEM); 4–7 December 2016; Bali, Indonesia. 2016

[18] Shen ZP, Gao J, Huang XR. A new quantification algorithm for the GO methodology. Reliability Engineering and System Safety. 2000;**67**(3):241–247

[19] Shen ZP, Gao J. GO methodology and improved quantification of system reliability. Journal of Tsinghua University. 1999;**39**(6):15–19

[20] Shen ZP, Gao J, Huang XR. An exact algorithm dealing with shared signals in the GO methodology. Reliability Engineering and System Safety. 2001;**73**(2):177–181

[21] Yi X-J, Dhillon BS, Shi J, et al. A new reliability analysis method for heavy vehicle systems based on goal oriented methodology. Proceedings of the Institution of Mechanical Engineers, Part D: Journal of Automobile Engineering. 2016. DOI: 10.1177/0954407016671276

9

An Analysis of Software and Hardware Development in the PMU-Based Technology and Suggestions Regarding its Implementation in the Polish Power Grid

Michał Szewczyk

Abstract

The ongoing evolution of electric power systems (EPS), especially distribution systems within the EPS structure, is driven by the implementation of the smart grid framework. This requires new approaches and technologies to continue ensuring a reliable and secure supply to end users. Fluctuating output from solar photovoltaic and wind plants can cause voltage and power variations in the feeders. In the power grid framework, phasor measurement units (PMUs) are recognized to be an invaluable aid in ensuring the secure operation and stability of transmission systems. The synchrophasor technique requires a high-accuracy time stamping of all the measurements within the analyzed power system area. It must be emphasized that the harmonic injection from power electronic components such as fluorescent lighting, computers, and power inverters of motors and generators can increase total harmonic distortion (THD) levels on distribution feeders and modify the conventional patterns of voltage and current signals. Therefore, what is vital for the functional reliability of synchronous measurements is the implementation of measurement algorithms, which can realize high-accuracy measurements, both in quasi-static and dynamic EPS operating conditions. This article presents the results of software simulations and hardware tests of measurement algorithms that meet the requirements of the IEEE C37.118™-2011 Standard.

Keywords: PMU-based measurements technology, time synchronization, power system operation, reliability, power system adaptive automation

1. Introduction

The availability of high-precision timing sources, such as the Global Positioning System (GPS) and the IEEE 1588 compliant network clock sources [1–3] plus the networking capability [4–5] of protective relaying devices and systems, is fundamentally changing the way that many current and future protective relaying applications are or will be implemented. Synchrophasor measurements [6–10], i.e., phasor measurements with high-accuracy time stamping, are under consideration for many future protective relaying applications [11–20]. Synchrophasor measurements are also used in many other power system applications such as wide-area monitoring and situational awareness applications. This chapter focuses primarily on their proper physical realization resulting from the standard requirements. Fulfillment of these requirements can lead to the development of the relaying applications that can either be implemented or are considered for future implementations.

The use of high-accuracy measurements and reliable protective algorithms with adequately high-accuracy and prompt decision-making [21–28] should guarantee an effective operation and protection of the electric power systems (EPS) from the consequences of disturbances also in the power system area characterized by changeable operation frequency over a wide range. The adaptation of measurement and protective algorithms in cases under consideration has the frequency nature, i.e., it concentrates on such a change of the parameters of algorithm operation which guarantees the proper estimation of measurement and criterion values by changeable frequency of the input measurement signals: currents and voltages received from the primary circuit [22–23]. The measurement and protective algorithms used at present in the measurement-protection devices tend to be defined for the constant frequency 50/60 Hz of the input signal, and from the point of view of the synchronous measurement standard requirements, they are characterized by too big inaccuracies when the frequency of the input signals varies over a wide range [21–22]. Another problem for the formulation of the measurement algorithms is the determination of the level of the insensitivity of these algorithms to the existence of interfering components in the input signals ([22, 25, 27]). Several interfering components and factors can occur especially in the input measurement signals during the faults within the EPS ([14–15, 22, 25, 27]). This situation is accompanied by high penetration levels of intermittent dispersed generation (DG) [14–15], and it can significantly affect the EPS operation. The existence of thyristor frequency converter is the source of high harmonics, particularly the odd high 5th, 7th, 9th harmonics with the amplitude reaching up to 10% of the amplitude of the basic component [16]. Moreover, as penetration levels increase, concerns regarding dynamic interactions among DG units are becoming more important. Uncontrolled personal electric vehicle (PEV) charging can lead to the overloading of the distribution equipment and violations of low-voltage limits and their courses. Also, the DC component in the input current signals can reach significant values and has a considerable influence on the accuracy of the measuring algorithms [21–22]. In summary the measuring algorithms implemented in the digital measure-protective systems should be characterized by:

- High accuracy of operation over a wide range of frequency change
- Non-sensitivity or low-level sensitivity to the existence of high harmonics in the input measurement signals
- Low inaccuracy in the case of the existence of the DC component of the high level and longtime decay (in current signals)
- Maximum short time of transient states in the measurements or decision-making

2. Basic requirements resulting from the series of the standard C37.118™

The IEEE Std 1344™-1995 was the original synchrophasor standard. It was replaced in 2005 by IEEE Std C37.118-2005 [6]. The newest standard has been split into two standards: IEEE Std 37.118.1-2011 [7], covering measurement provisions, and IEEE Std 37.118.2™-2011 [8], covering data communication. In the IEEE standard from 2011, there is a significant change of the requirements for synchronous measurements and additional clarification for the phasor and synchronized phasor definitions. The concepts of total vector error (TVE) and compliance tests are retained and expanded, tests over temperature variation have been added, and dynamic performance tests have been introduced [7–9]. In addition, limits and characteristics of frequency measurement and rate of change of frequency (ROCOF) measurement have been developed.

Phasor representation of sinusoidal signals is commonly used in the AC power system analysis. The sinusoidal waveform defined in Eq. (7):

$$x(t) = X_{\mathrm{m}}(\omega t + \varphi) \tag{1}$$

is commonly represented as the phasor as shown in Eq. (2) [7]:

$$\underline{X} = X_{\mathrm{r}} + \mathrm{j}X_{\mathrm{i}} = \frac{X_{\mathrm{m}}}{\sqrt{2}} e^{\mathrm{j}\varphi} = \frac{X_{\mathrm{m}}}{\sqrt{2}}(\cos\varphi + \mathrm{j}\sin\varphi) \tag{2}$$

where the magnitude is the root-mean-square (RMS) value, $X_m/\sqrt{2}$, of the waveform and the subscripts r and i signify real and imaginary parts of a complex value in rectangular components. This phasor is defined for the angular frequency ω.

TVE is an expression of the difference between a "perfect" sample of a theoretical synchrophasor [7] and an estimate given by the unit under test at the same instant. The value is normalized and expressed as per unit of the theoretical phasor. TVE is defined in the following Eq. (7):

$$TVE(n) = \sqrt{\frac{\left(\hat{X}_{\mathrm{r}}(n) - X_{\mathrm{r}}(n)\right)^2 + \left(\hat{X}_{i}(n) - X_{\mathrm{i}}(n)\right)^2}{(X_r(n) + X_i(n))^2}} \tag{3}$$

where $\hat{X}_{\mathrm{r}}$ and $\hat{X}_{\mathrm{i}}$ are the sequences of estimates given by the unit under test and $X_r(n)$ and $X_i(n)$ are the sequences of theoretical values of the input signal at the times (n) assigned by the unit to those values. Synchrophasor measurements shall be evaluated using the TVE criterion of Eq. (3).

A phasor measurement unit (PMU) shall also calculate and be capable of reporting frequency and ROCOF [7]. For this measurement, the following standard definitions are used. If there is given a sinusoidal signal, as in the Eq. (7):

$$x(t) = X_{\mathrm{m}} \cos[\Psi(t)] \tag{4}$$

the frequency is defined as follows:

$$f(t) = \frac{1}{2\pi}\frac{\Psi(t)}{dt} \tag{5}$$

Then, the ROCOF (rate of change of frequency) is defined:

$$\text{ROCOF}(t) = \frac{df(t)}{dt} \tag{6}$$

As mentioned above, the series of the Standard C37.118™ was developed for synchronized phasor measurement systems in power systems. They define a synchronized phasor (synchrophasor), frequency, and rate of change of frequency (ROCOF) measurements. They also describe time tag and synchronization requirements for measurements of all three of these quantities. Next, C37.118™ specifies methods for evaluating these measurements and requirements for compliance with the standard under both static and dynamic conditions. The "old" 2005 version of the standard, commonly followed by equipment manufacturers and system integrators, specifies the performance of phasor measurements only under steady-state conditions. The latest revision of the standard (the 2011 update) extends the synchrophasor definition. It also specifies measurement requirements and test conditions. Steady-state requirements of TVE and ROCOF (resulting from the IEEE C37.118.1–2011 Standard) are included in [7, 9].

The standard C37.118.1–2011 defines the measurement response time and measurement delay time [7]. The measurement response time is the time to transition between two steady-state measurements before and after a step change is applied to the input. The PMU shall support data reporting (by recording of output). This reporting shall be done at submultiples (F_s) of the nominal power-line (system) frequency (**Table 1**) [7].

According to the standard, the support for other reporting rates is optional and includes higher rates like 100/s or 120/s or rates lower than 10/s such as 1/s. The rates lower than 10/s are not subject to dynamic requirements. In this case no filtering is required, and lower-rate data (<10/s) can be provided directly by selecting every nth sample from a higher-rate stream. The reporting rate and performance class are often the largest factors influencing the accuracy of the measurements. These determine the measurement window to be used, filtering, and the length of the interval over which an event will be reported.

To comply with the standard, a PMU shall provide synchrophasor, frequency, and ROCOF measurements that meet the requirements in a given class [7]. These requirements shall be met at all times and under all configurations irrespective of whether the PMU function is a standalone physical unit or included as part of a multifunction unit. So, all compliance tests are to be performed with all parameters set to standard reference conditions, except those being varied

System frequency	50 Hz			60 Hz					
Reporting rates (F_s, frames per second)	10	25	50	10	12	15	20	30	60

Table 1. Required PMU reporting rates [7].

as specified for the test. The reference condition specified for each test is the value of the quantity being tested when others are unvarying. Only the parameters specified for each requirement shall be varied as the effects shall be considered independently. Reference conditions for all tests are defined in [7].

3. General characteristics of filtration

The main objective of filtration in the measuring and automation devices within the EPS is to pass signal components lying in a selected frequency band and to suppress signal components of the other frequencies. It is because of the existence of the interfering components in the input signals which can occur especially during faults within the EPS, high penetration levels of intermittent DG, and today's widely used thyristor frequency converters.

Generally, a linear filter can be described by the equation:

$$y(t) = \int_{-\infty}^{\infty} x(t-\tau)\ h(\tau)d\tau \tag{7}$$

where y and x are the output and input signals and h is the response function of the filter.

After the Fourier transform of the Eq. (7) is obtained:

$$Y(j\omega) = X(j\omega)\ \ H(j\omega) \tag{8}$$

Eqs. (7) and (8), respectively, characterize the filter completely in the time and frequency domain. The function $h(\tau)$ is important for the characteristics of the filter like the function $H(j\omega)$ which represents its frequency spectrum.

In principle, the shorter the interval of the transient state of the filter, the worse the filtration properties. Thus, better filtration properties lead to the extension of time to the stable output response. In practice, it is a trade-off between the short time of the transient state and high quality of filtration, depending on the application requirements. In particular the short transient states of the filtration are required in the power system protection devices. Short stabilization time should be kept in order to take fast and proper decisions during the disturbance, the change of network configuration, the discharge of power, etc. Therefore, it is recommended that filters should be used with fast-decay w-functions or w-functions described in the period T_w and outside this range: taking a value 0. Concluding, filtering of input measurement signals should provide:

- Fast stabilization of output signals after a step change of an input signal (ensuring short time of the measurement)
- Effective elimination or dumping of the interfering (unusable) components
- Possible low computational effort (quasi real-time measurement)

Digital filtration is based on the discrete values of an input signal. Using this technique, a discrete output signal can be obtained and is dependent on the input signal and filter characteristics. The characteristics of the filter can be described using the frequency and time characteristics.

Considering the frequency characteristics of filters, both analog and digital, the following are groups of filters:

- Low-pass filters
- Band-pass filters
- High-pass filters
- Stopband filters

From this group, low-pass and band-pass filters are especially widely used in the power system protection automation.

Analyzing the time characteristics, two groups of digital filters can be distinguished:

- Non-recursive filters of finite impulse response (FIR)
- Recursive filters of infinite impulse response (IIR)

A digital non-recursive filter can be described by Eq. (9):

$$y(n) = \sum_{k=0}^{p-1} a_{(k)} x(n-k) \tag{9}$$

where x and y are the input and output samples, $a_{(k)}$ is the filter coefficients, and p is the number of samples in the window.

Eq. (9) of a non-recursive filter is a discrete convolution of filter coefficients and an input signal, limited to p samples of the input signal.

A digital non-recursive filter can be described by Eq. (10):

$$y(n) = \sum_{k=0}^{p-1} a_{(k)} x(n-k) + \sum_{k=1}^{w} b_{(k)} y(n-k) \tag{10}$$

where $a_{(k)}$ and $b_{(k)}$ are the filter coefficients and w is number of coefficients $b_{(k)}$, $w < (p - 1)$.

Eq. (10) shows that actual response of that filter is not only a function of an input signal and filter coefficients but also a function of w-numbers of the previous values of the output signal. After using the Z-transform, we can obtain:

- For a non-recursive filter:

$$H(z) = \frac{Y(z)}{X(z)} = \sum_{k=0}^{p-1} a_{(k)} z^{-k} \tag{11}$$

- For a recursive filter:

$$H(z) = \frac{Y(z)}{X(z)} = \frac{\sum_{k=0}^{p-1} a_{(k)} z^{-k}}{1 - \sum_{k=1}^{w} b_{(k)} z^{-k}} \tag{12}$$

To obtain the discrete spectra of the filter, the operator z shall be substituted by the equivalent operator $e^{-j\omega\ T_i}$:

- For a non-recursive filter:

$$H^*(j\omega) = \sum_{k=0}^{p-1} a_{(k)} e^{-jk\omega\ T_i} \quad \text{for } \omega < \frac{\pi}{T_i} \tag{13}$$

- For a recursive filter:

$$H(j\omega) = \frac{\sum_{k=0}^{p-1} a_{(k)} e^{-jk\omega\ T_i}}{1 - \sum_{k=1}^{w} b_{(k)} e^{-jk\omega\ T_i}} \quad \text{for } \omega < \frac{\pi}{T_i} \tag{14}$$

where T_i is the sampling period.

Recursive filters are rarely used in power system automation. The cause is a long stabilization time of an output signal after a step change of an input signal and a relatively high sensitivity to the slight changes of values of the coefficients $a_{(k)}$ and $b_{(k)}$.

The possibilities of optimizing non-recursive filters are limited. The optimization due to the required frequency response, the required high dynamics, and the coefficients of the filter and quick transition from band-pass to stopband frequently leads to high-order filters that are difficult to implement in real time. Therefore, optimization of non-recursive filtering should be pursued by selecting the permissible length of the measurement window and possibly a simple form of the coefficients function (coefficients $a_{(k)}$) that define the required spectral characteristics.

4. Experimental setup

In this study, MATLAB with Simulink and Signal Processing Toolbox was used for software simulations. MATLAB is a popular programming language developed by MathWorks [29]. MATLAB can analyze data, develop algorithms, and create models and applications. The Simulink toolbox is a block diagram environment for multidomain simulation and model-based design. It supports simulation, automatic code generation, and continuous test and verification of embedded systems. Simulink provides a graphical editor, customizable block libraries, and solvers for modeling and simulating dynamic systems. It is integrated with MATLAB. It makes it possible to incorporate MATLAB algorithms into models and export simulation results to MATLAB for further analysis. The Signal Processing Toolbox provides industry-standard algorithms and applications for analog and digital signal processing (DSP). Among other things, it makes it possible to visualize signals in time and frequency domains, compute FFTs for spectral analysis, and design FIR and IIR filters. Algorithms in the toolbox can be used as a basis for developing custom algorithms.

The second stage of this study was performed using microprocessor-based Automatic Relay Test System ARTES 440 II manufactured by KoCoS Company (**Figure 1a**), and a digital fault recorder RZ-40 (**Figure 2**), manufactured by Energotest, with the implemented functionality of a PMU unit (both algorithms and other special functions). The ARTES 440 II [30] is used for carrying out general operating tests and for testing the configured excitation and tripping characteristics of various protection devices, such as distance protection relays, overcurrent

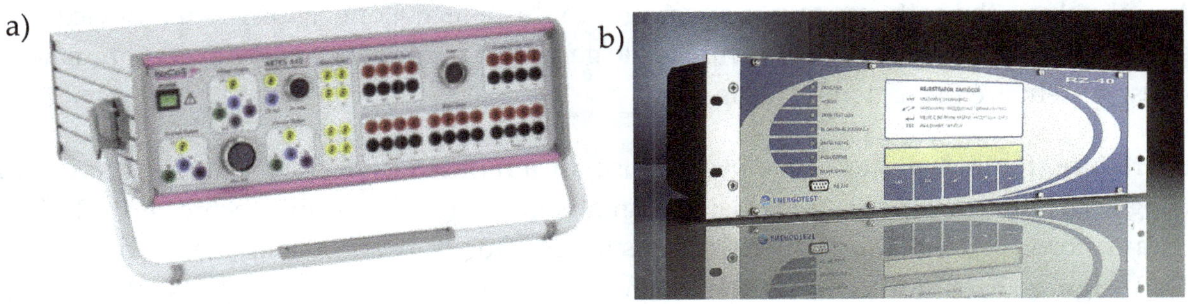

Figure 1. Real photos of (a) ARTES 440 II [30] and (b) RZ-40 fault recorder [31].

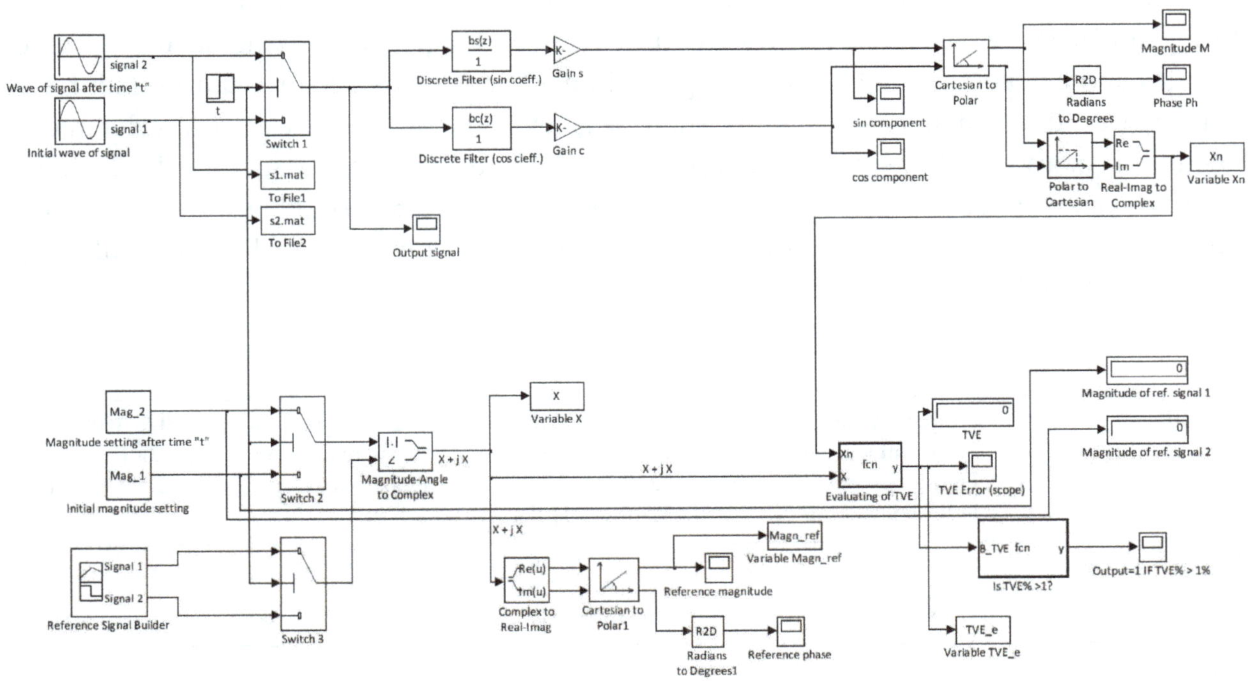

Figure 2. MATLAB/Simulink model used in the software simulations.

relays, and voltage and frequency relays. In addition, the test instrument can also be used as a three-phase function generator which is freely configurable with regard to amplitude, frequency, and phase relation. As mentioned earlier, ARTES 440 II has four voltage and six current amplifiers whose output signals can be set independently from one another as regards amplitude, frequency, and phase angle. The test quantities are calculated from the parameters entered via the software and are supplied to the device under test by means of digital-analog converters and amplifiers. Because the test quantities are generated synthetically, they are unaffected by disturbances in the incoming supply.

The currents generated by the test system are available via the sockets of the current output group. The six current channels are freely configurable as regards phase, amplitude, and frequency. The current amplifiers of the ARTES 440 II provide a maximum test current of 25 A per channel. If higher test currents are required, the amplifier outputs can be operated in parallel. The output values of the current and voltage amplifiers are monitored by the system during tests (internal feedback measurement). If the output values do not agree with the setpoint values, the software issues a warning to this effect. Connections to the device under test are made via safety sockets or a multipole generator output socket. Detailed parameters of the test system are presented in **Table 2**.

Disturbance (fault) recorders have been in use for a number of years and have evolved from analog recording devices to units using digital signal processing and recording techniques. Digital records can be easily collected, transmitted, stored, printed, and analyzed [31]. Also, RZ-40 (**Figure 1b**), a fault recorder selected for this study, typically contains directly measured analog channels, as well as event or binary channels. This allows the recorder to capture the time sequence of analog power system quantities, along with breaker contacts, logic-state changes, event contacts, etc. State-of-the-art recorders typically include calculated analog quantities and logic functions to ensure that pertinent power system information is captured during an event. RZ-40 fault recorder is designed for logging of instantaneous values of voltages and currents as well as binary signals in electric power structures during failures or disturbances. It has many additional functions, e.g., event recorder and electrical value meter. RZ-40 has many functions and wide-open hardware and software structure. Each of the functions can be used independently without lowering parameters of other functions. That is why it was possible to implement additional software and functions of PMUs to fulfill the requirements of the IEEE the C37.118™ series of the Standard.

General	**Frequency range** **Transient signals** **Frequency resolution** **Frequency accuracy**	**DC to 3 kHz** **DC to 4 kHz** **0.001 Hz** **Error < ±0.01%**	**Phase angle** **Phase resolution** **Phase accuracy**	**0 to 360°** **0.001°** **Error < ±0.1°[1]**
Voltage amplifiers	Resolution THD	13 mV < 0.05%[1]	Accuracy	Error < ±0.05%[1]
Current amplifiers	Resolution THD	1 mA < 0.05%[1]	Accuracy	Error < ±0.05%[1]

[1]For the frequency range of 10 to 200 Hz.

Table 2. Detailed accuracy and resolution parameters of the analog outputs of ARTES II [30].

The other benefit of using this digital fault recorder to realize the functions of PMUs is the fact that many of them are already working (especially in the Polish Power Grid infrastructure). Thus, the implementation of synchronous measurements using these devices will have low hardware costs of the initial wide-area measurements infrastructure. The only thing which needs to be done is to update the software of the functioning devices to the newest one which has the optimized and modified algorithms realizing the PMU functionality. Additionally, existing and future infrastructure facilitates making long-term pilot programs. These programs can lead to the development of new and long-awaited functions to ensure high-power grid availability, stability, and reliability, especially in the power system automation scope. Therefore, the main objective of this study is to initially check the PMU properties of this device according to the IEEE the C37.118™ series of the Standard.

4.1. Software simulations in MATLAB

The first stage of the study was to work out a model in MATLAB/Simulink environment (**Figure 3**). This model was used to simulate various tests according to the requirements of C37.118™ series of the Standard, especially C37.118.1–2011 [7–9]. It has been used exclusively for research and teaching purposes to demonstrate the properties of orthogonal filters.

On the basis of several other considerations and analyses, e.g., [21–22], full-cycle filters with sine and cosine widows were implemented. Because of the necessity to evaluate TVE (Eq. (3)), there is a reference signal builder in the simulator. The magnitude, phase, and frequency of input signals are freely configurable. The step output "t" is used to switch from signal 1 to signal 2 after a configured time *t*. Signals 1 and 2 are taken out to Files 1 and 2, a full signal sequence is visible on the scope "Output signal." A generated signal sequence is delivered to the full-cycle filters block with sine and cosine windows. The required coefficients are loaded from MATLAB workspace according to the frequency of the time window. Then, the orthogonal components are used to compute magnitude (M) and phase (Ph). Calculated values of the magnitude and phase (in complex variable X_n) are delivered to the function block "evaluating of TVE." The slope of TVE can be viewed on the scope and is exported to variable TVE_e. Additionally, the B_TVE function block is used for evaluation purposes if TVE is below its reference value of 1%. The model presented in **Figure 2** does not have the ROCOF

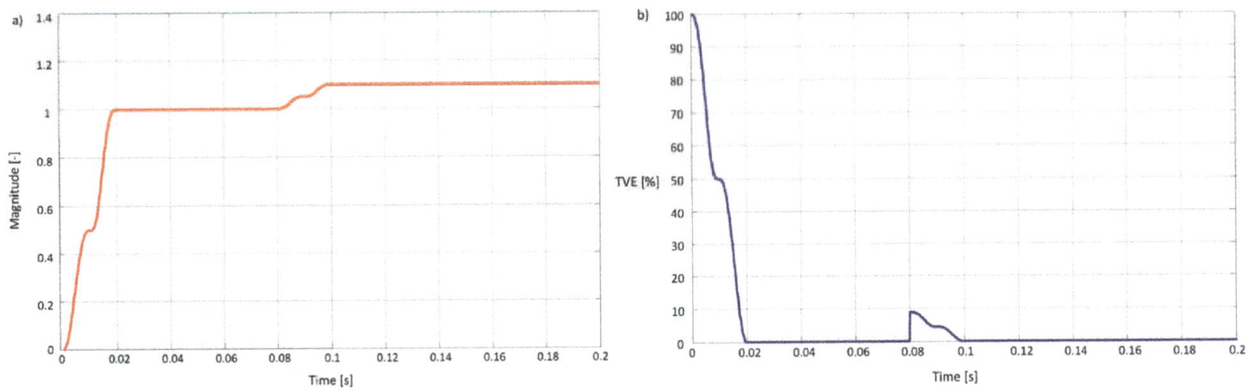

Figure 3. Magnitude and TVE course after a step change of the magnitude in t = 80 ms (a) magnitude course and (b) TVE course.

measurement module. Although the implementation of ROCOF measurements poses no problems (and indeed this was done), the main tests in the dynamic conditions were carried out on a real hardware device, and only a simplified model is presented in **Figure 2**.

Several testing conditions have been simulated using the developed model. Sample results of these tests are presented in **Figures 3** to **6**. All simulations were realized for the sampling frequency f_s = 4 kHz. **Figure 3** presents magnitude and the evaluated TVE course after a step change of input signal magnitude of +10% (in relative values), for the nominal frequency of input signal (f_1 = 50 Hz) and the window frequency of filter (f_w) equal to the nominal frequency. This step change was performed in 80 ms from the beginning of simulation.

Figure 4 shows the courses of TVE for the same f_1 and f_w parameters as before: there is no change of the magnitude, but there is a step change of phase of π/6 (**Figure 4a**) and 5π/6 (**Figure 4b**) radians. Next, again f_i and f_w are equal to 50 Hz: there is no change of the magnitude and phase, but there is a step change of frequency of −0.5 Hz (**Figure 4c**) and −1 Hz (**Figure 4d**). Finally, **Figure 5** shows the situation, when the frequency of the filter window (f_w = 48.78 H) is almost fully correlated with the frequency of the input signal (f_1 = f_w = 48.78 Hz (**Figure 5a**)) and closely correlated (f_1 = 48.5 Hz, f_w = 48.78 (**Figure 5b**)). Of course for the last test, the coefficients of the filter are different from the first three tests and evaluated and loaded from the MATLAB workspace. In the example, for the selected frequency f_w = 48.78 Hz, there are 82 samples in the filter window, and the filter window is almost

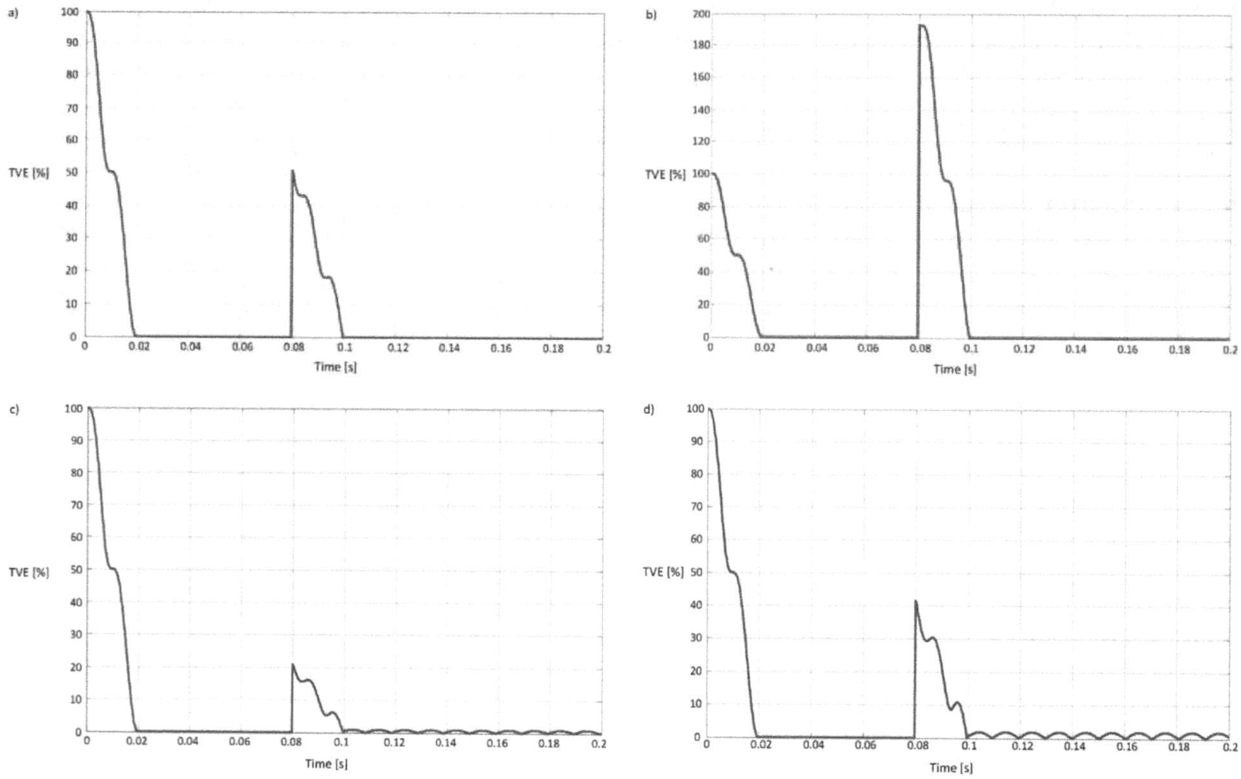

Figure 4. TVE courses after a step change in t = 80 ms of the input signal: (a) phase of π/6 radians, (b) phase of 5π/6 radians, (c) frequency of 0.5 Hz, and (d) frequency of 1 Hz.

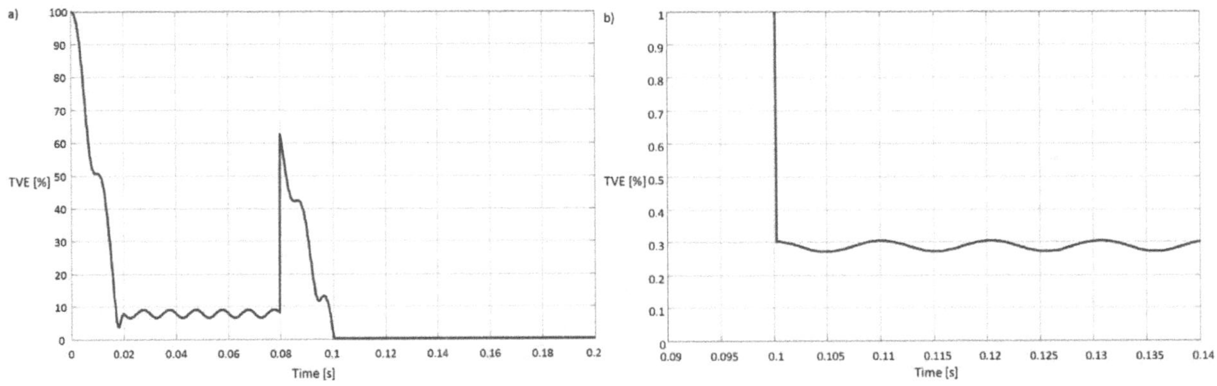

Figure 5. TVE courses for filter window frequency f_w = 48.78 Hz and the input signal of frequency: (a) f_1 = 48.78 and (b) f_1 = 48.5 (zoomed).

fully matched to the frequency of the input signal also equal to 48.78 Hz. The number of samples is a natural number.

Therefore, taking the sampling frequency of 4 kHz, the other two nearest frequencies of the fundamental component of an input signal, for which the window filter can be almost fully correlated with the input signal, are 49.28 (81 samples in the filter window) and 48.19 (83 samples in the filter window). It should be emphasized that this phenomenon is dependent on the sampling frequency. When the sampling frequency is 2 kHz, there are 40 samples in the window (for the frequency of the fundamental component equal to 50 Hz). The nearest two natural numbers of samples are 39 and 41. Evaluating 2000/39 gives the frequency of the window equal to 51.28 Hz and 2000/41 gives f_w = 48.78 Hz. Comparing these frequencies with the ones for f_s = 4 kHz (relatively 50.63 Hz and 49.28 Hz), it can be noticed that lowering the sampling frequency leads to wider frequency ranges in which the algorithm works with the nonmatched frequencies of windows and generates inaccuracies exceeding the limits defined by the standard. On the other hand, raising the sampling frequency leads to the increase in the computational effort and time.

Analyzing software simulations, it can be seen that for the assumed sampling frequency, TVE exceeds the limit of 1% for the frequencies differing from the nominal more than about 0.5 Hz. Therefore, the adaptive technique using switching between filter windows frequencies can be considered to comply with the standard limitations as well as with other solutions described in [21–22].

4.2. Testing and validation of the RZ-40 device by ARTES II

Figures 6–8 show sample hardware tests carried out with the ARTES II and RZ-40 devices. These tests were realized for various conditions and cases described in the IEEE C37.118.1–2011 Standard.

Each of the tests is realized in 1 second. In the half of this period (500 ms from the beginning of test), there is a step change of the given quantity. **Figure 6** shows the courses of the magnitude

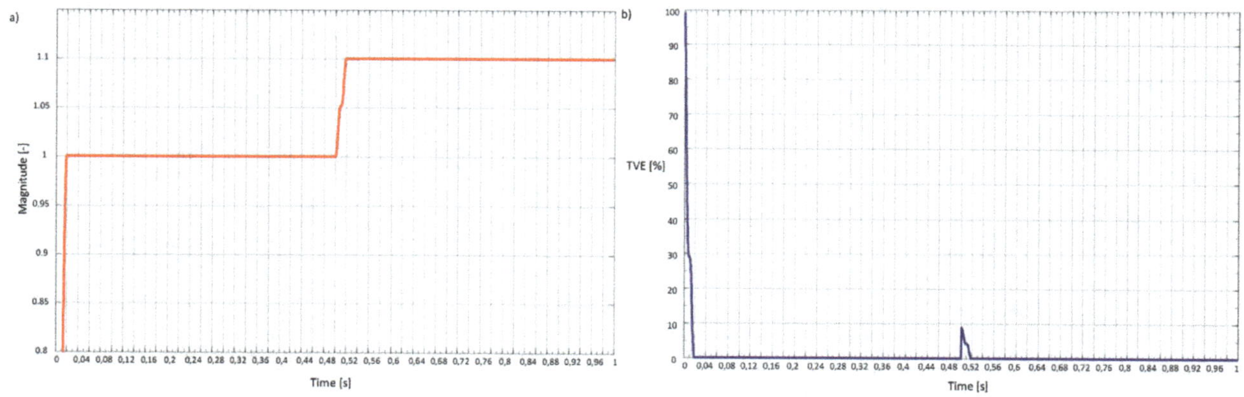

Figure 6. Magnitude and TVE course after a step change of the magnitude in t = 500 ms: (a) magnitude recorded in RZ-40 device and (b) TVE course (evaluated).

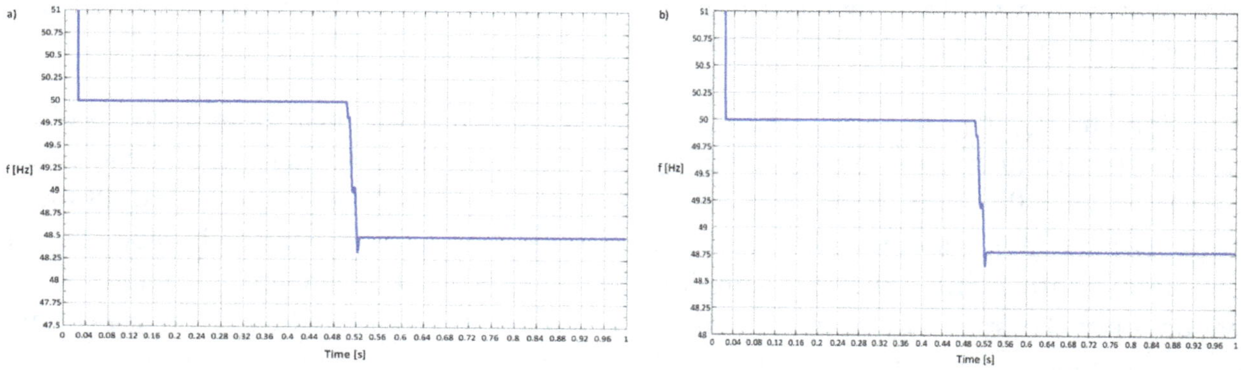

Figure 7. Frequency courses evaluated in RZ-40 device after a step change of the input signal frequency in t = 500 ms: (a) f_1 = 45.5 Hz and (b) f_1 = 45.78 Hz.

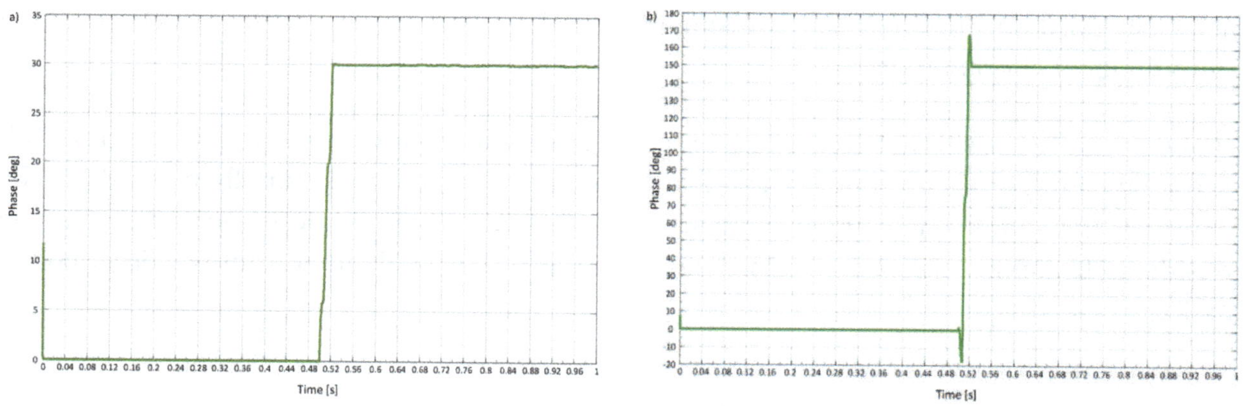

Figure 8. Phase courses recorded in RZ-40 device after a step change of the input signal phase in t = 500 ms: (a) from 0 to 30° and (b) from 0 to 150°.

(**Figure 6a**) and TVE (**Figure 6b**) during the step change of that first on the level + 10% (relative values). The first course is recorded using the disturbance recorder ability implemented in the ET Manager software supplied by the manufacturer. The TVE is evaluated using the recorded values. **Figure 7** presents the evaluated frequency for 48.5 Hz (**Figure 7a**) and 48.78 (**Figure 7b**). Lastly, **Figure 8** shows the phase courses for the step change of phase from 0 to 30° (**Figure 8a**) and from 0 to 150° (**Figure 8b**).

As it can be observed, there are full-cycle window filters implemented in the RZ-40 devices. It can be concluded from the time period between the occurrence of a step change (0.50 s) and the moment of the stable response to this step (0.52 s). Any fluctuations in the initial period of tests should not be considered as they are the results of transient states of starting simulations (MATLAB/Simulink) or transients after injecting voltages on the RZ-40 inputs when the previous input magnitude was 0.

As mentioned previously, the RZ-40 has been tested in accordance with the C37.118.2011™ Standard and the "2014" amendment. In particular, the implemented measurement algorithms focused on providing high accuracy and stability of frequency determination across a wide range of frequency changes in measurement signals. **Tables 3** and **4** show the results of frequency estimation for a monoharmonic signal with set fundamental frequencies (45 to 55 Hz in 1 Hz steps) for successive 50 samples. Similar measurements were made for further 100 and 200 samples. The results of the measurements are shown in **Figures 9** and **10**. Based on the obtained results, it can be stated that the algorithm is characterized by high-frequency estimation accuracy (average error of measurement below 2 mHz) and high stability (maximal standard deviation at 1 mHz for the extreme frequency of the tested range). The dynamic tests

	45 Hz	**46 Hz**	**47 Hz**	**48 Hz**	**49 Hz**	**50 Hz**
Frequency (mean) [Hz]	44,998,237	45,998,779	46,999,410	47,998,901	48,999,969	49,999,111
Standard deviation [mHz]	0,894	<0,001	0,840	<0,001	<0,001	0,415
Mean freq. Error [mHz]	−1763	−1221	−0,590	−1099	−0,031	−0,889
Average percentage error [−]	−0,004	−0,003	−0,001	−0,002	<0,001	−0,002

Table 3. Results of frequency estimation in the RZ-40 for the 50 successive samples (1).

	51 Hz	**52 Hz**	**53 Hz**	**54 Hz**	**55 Hz**
Frequency (mean) [Hz]	50,998,157	51,999,145	52,998,279	53,999,208	54,998,621
Standard deviation [mHz]	0,395	0,000	0,395	0,344	0,720
Mean error [mHz]	−1843	−0,855	−1721	−0,792	−1379
Average percentage error[−]	−0,004	−0,002	−0,003	−0,001	−0,003

Table 4. Results of frequency estimation in the RZ-40 for the 50 successive samples (2).

performed indicated excellent properties of the algorithm for potential applications in the power system. The results obtained have also been confirmed by independent tests carried out at the Fraunhofer Institute in Magdeburg using a reference device[1].

Synchrophasor measurements should be synchronized with the UTC time with the accuracy sufficient to meet the accuracy requirements of the C37.118 Standard [3]. Time-stamp error should be of just a few (single) μs. In order to fulfill this requirement, the accuracy of the time source should be about ten times higher than the level of the expected time error. To synchronize the RZ-40 unit, the GPS receiver EM-406A and then EM-506 was taken as the source of the time signal. The EM-406A is a 20-channel GPS receiver, and the EM-506 one is a 48-channel receiver. High accuracy of measurements and high precision of time stamps make it possible to develop many new or improved applications. One of them is the SmartLoad system [13, 31].

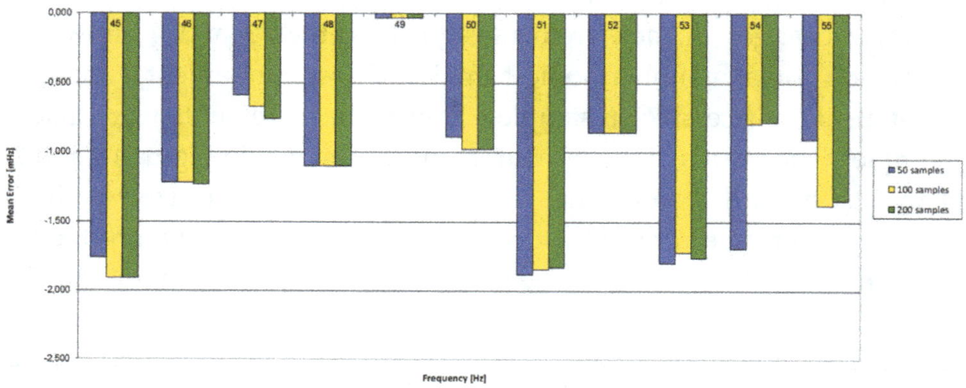

Figure 9. Mean error of frequency estimation.

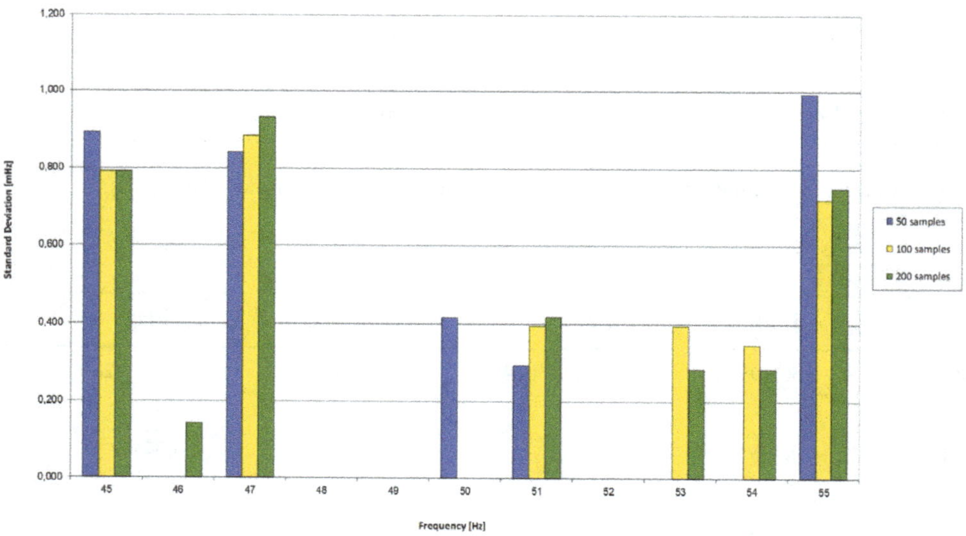

Figure 10. Standard deviation of frequency estimation.

[1] Tests commissioned by Energotest Sp. z o.o.

This system is an advanced system designed to make a quasi real-time balance of power and adaptive load-off in the case of a possible active power deficit in the supervised area of a power network. This system has already been implemented. Some new implementations are under development and will be the subject of subsequent publications. It seems that they can significantly improve the reliability of the power system's operation with the benefits of a well-implemented PMU-based measurements technology.

5. Discussion and conclusions

The IEEE C37.118.1.2011 Standard defines a phasor measurement unit (PMU), which can be a stand-alone physical unit or a functional module within another physical unit. The Standard does not specify any hardware, software, or a method for computing phasors, frequency, or ROCOF. In all likelihood the best solution, especially in the initial period of implementation, is to incorporate the PMU functionality into the existing hardware devices commonly used in the power grid. This minimizes costs and facilitates realizing potential pilot programs. The pilot programs seem to be the essential part of the PMU and synchronous measurement studies. Considering the RZ-40 unit is by no means accidental. This unit has a modular and scalar hardware architecture. It is very easy to change the input analog cards, DSP module, and communication module or to update the firmware of the unit including the measurement algorithms and other functionalities [28]. Last but not least, it is also a very popular unit in the Polish Power Grid.

Analyzing the requirements for the synchronous measurements resulting from the C37.118™ series of the Standard, significant changes can be noticed between the 2005 and 2011 versions. The requirements in the version of the Standard dated 2005 were defined only for static conditions and were very imprecise. This leads to their free interpretation by manufactures because several of them were not directly defined or defined on the levels which give many different solutions allowing compliance with the Standard. Although many algorithms which are used may comply with the 2005 Standard, they may have different properties in the dynamic stages, especially during the fast transient states. That is why the Standard from 2011 emerges. It is much more precise, and it defines the requirements both in the static and dynamic conditions. It also defines many important time limits for the measurement algorithms realizing synchronous measurements. It was concluded that the compliance problems with the Standard in the case of wide-area measurements are dependent on their target applications. It was almost impossible to build the wide-area protection systems based on the Standard of 2005. As a matter of fact, it defined some practical requirements in steady conditions, but it is the requirements in dynamic states that are most vital for the protections. In addition these "dynamic" requirements should be comparable in different devices realizing PMU functions. All this makes synchronous measurements very hard to realize, mainly as regards factors related to the PMU-expected higher functionalities. Simulations realized in this study indicate the possibility of developing fast and reliable adaptive measurement algorithms to comply with the C37.118.1–2011 requirements. Several tests confirm that for the RZ-40 device. Detailed analyses of the 2011 Standard point that some of the requirements are very hard or even impossible to meet. Some of the difficulties in meeting the requirements probably result from simple typo-errors they contain. Other requirements are feasible but hard to implement. They cause computational problems resulting from increasing the response time and error on the output. This seems to be confirmed by Amendment 1 to the Standard released at the end of March 2014 [9] entitled "Modification of Selected Performance Requirements," in which many of the required parameters, mainly applying to the dynamic requirements, are revised. It should be emphasized that most of these parameters were met by the RZ-40 during the tests discussed in this study for limits defined in 2011. Summing up, on the basis of the tests defined in the Standard and conducted in this study, it is possible to meet the Standard requirements by the devices like RZ-40. It should be observed that these are "synthetic tests" and probably many other devices can comply with the Standard. On the other hand, most of these "others" were only tested and confirmed by the manufacturers to comply with the standard of 2005. Therefore, long-term pilot tests are essential to compare the properties of different devices with the PMU functionality. These tests should be conducted using different PMU devices working in the same locations of the power grid. The initial forecasting time of such tests is from about 6 months to a year. Further research is needed to analyze the results. Some of the most noticeable changes to the power distribution landscape are being driven by the proliferation of intermittent DG, PEVs, microgrids, and power electronic components. The application of PMUs in distribution systems is still under-explored or is at the implementation stage. This is mostly due to the recognized technical problems (explained and revised in this study) and economic factors. However, since the drivers of change and needs for implementation are rapidly increasing, there is a growing interest in applying PMUs in the EPS, which will consequently engender a new body of research.

Acknowledgements

The author would like to thank Energotest Sp. z o.o. for providing opportunity to use the RZ-40 device in his study and contributing in the software and hardware development, both Energotest Sp. z o.o. and KoCoS Company for the invaluable hardware and software support as well as hardware and software components.

Author details

Michał Szewczyk

Address all correspondence to: michal.szewczyk@polsl.pl

Silesian University of Technology, Gliwice, Poland

References

[1] IEEE Std 1588™-2008. IEEE Standard for a Precision Clock Synchronization Protocol for Networked Measurement and Control Systems

[2] PC37.238. IEEE Draft Standard Profile for Use of IEEE 1588 Precision Time Protocol in Power System Applications

[3] Szewczyk M. Time synchronization for synchronous measurements in electric power systems with reference to the IEEE C37.118TM standard - selected tests and recommendations. Przegląd Elektrotechniczny. 2015;**4**:144-148. DOI: 10.15199/48.2015.04.32

[4] Szewczyk M. Selected analyses of teletransmission and teleinformatic structures in electrical power. Przegląd Elektrotechniczny. 2014;**3**:1-5. DOI: 10.12915/pe.2014.03.1

[5] Szewczyk M. Analysis of the requirements of reliability and quality for systems and data transmission equipment in modern power systems. Przegląd Elektrotechniczny. 2014;**3**:84-89. DOI: 10.12915/pe.2014.03.1

[6] C37.118 rev. IEEE Standard for Synchrophasor Measurements for Power Systems. 2005

[7] C37.118.1 rev. IEEE Standard for Synchrophasor Measurements for Power Systems. 2011

[8] C37.118.2 rev. IEEE Standard for Synchrophasor Measurements for Power Systems. 2011

[9] IEEE Standard for Synchrophasor Measurements for Power Systems, Amendment 1. Modification of Selected Performance Requirements. 27 March, 2014

[10] Szewczyk M. Standard requirements for systems realizing synchronous measurements in the power system infrastructure. Przegląd Elektrotechniczny. 2014;**3**:80-83. DOI: 10.12915/pe.2014.03.16

[11] Wache M, Murray DC. Application of Synchrophasor measurements for distribution networks. IEEE Power and Energy Society General Meeting. Jul. 2011:1972-1975

[12] Sanchez-Ayala G, Aguerc JR, Elizondo D, Lelic M. Current trends on applications of PMUs in distribution systems, innovative smart grid technologies (ISGT). 2013. IEEE. Feb. 2013;**24-27**:1-6

[13] Szewczyk M. Conditions for the improvement and proper functioning of power system automation equipment in the present and the expected future structure of the electric power sector. Przegląd Elektrotechniczny. 2015;**5**:179-186. DOI: 10.15199/48.2015.05.40

[14] Halinka A, Szewczyk M. Distance protections in the power system lines with connected Wind Farms. Gesche Krause, editor. From Turbine to Wind Farms - Technical Requirements and Spin-Off Products. InTech; April 2011:135-160. DOI: 10.5772/15955

[15] Halinka A, Rzepka P, Szablicki M, et al. Impact of proper functioning of power system automation for the safety of power system in the consideration of the new technical and economical solutions in polish power grid. Przegląd Elektrotechniczny. 2011;**2**:140-143

[16] Borghetti A, Nucci CA, Paolone M, Ciappi G, Solari A. Synchronized phasors monitoring during the islanding maneuver of an active distribution network. IEEE Transactions on Smart Grid. March 2011;**2**(1). Article number 5680630:70-79

[17] Meng W, Wang X, Wang Z, Kamwa I. Impact of causality on performance of Phasor measurement unit algorithms. IEEE Transactions on Power Systems. 2017:1-11. DOI: 10.1109/TPWRS.2017.2734662

[18] Roscoe AJ, Burt GM, McDonald JR. Frequency and fundamental signal measurement algorithms for distributed control and protection applications. IET Generation, Transmission and Distribution. 2009;**3**(5):485-495

[19] Kundu P, Pradhan AK. Wide area measurement based protection support during power swing. International Journal of Electrical Power & Energy Systems. December 2014;**63**:546-554. DOI: 10.1016/j.ijepes.2014.06.009

[20] Shima KS, Kimb ST, Leec JH, et al. Detection of low-frequency oscillation using synchrophasor in wide-area rolling blackouts. International Journal of Electrical Power & Energy Systems. December 2014;**63**:1015-1022. DOI: 10.1016/j.ijepes.2014.06.069

[21] Rebizant W, Szafran J, Wiszniewski A. Digital Signal Processing in Power System Protection and Control. Springer-Verlag London; 2013. DOI: 10.1007/978-0-85729-802-7

[22] Halinka A., Sowa P, Szewczyk M. Measurement algorithms of selected electric parameters in wide range of frequency change. Proc. of SIP2001. Honolulu. USA: 155–159

[23] Kamwa I, Samantaray SR, Joos SR. Wide frequency range adaptive phasor and frequency PMU algorithms. IEEE Transactions on Smart Grid. March 2014;**5**(2). Article number 6575186:569-579

[24] Chakir M, Kamwa I, Le H. Extended C37.118.1 PMU algorithms for joint tracking of fundamental and harmonic phasors in stressed power systems and microgrids. IEEE Transactions on Power Delivery. June 2014;**29**(3). Article number 6810880:1465-1480

[25] Phadke AG, Kasztenny B. Synchronized phasor and frequency measurement under transient conditions. IEEE Transactions on Power Delivery. 2009;**24**(1):89-95

[26] Roscoe AJ, Burt GM, Rietveld G. Improving frequency and ROCOF accuracy during faults, for P class Phasor measurement units. 2013 IEEE International Workshop on Applied Measurements for Power Systems, AMPS 2013 – Proceedings 2013. Article number 6656233. 97–102

[27] Premerlani W, Kasztenny B, Adamiak M. Development and implementation of a synchrophasor estimator capable of measurements under dynamic conditions. IEEE Transactions on Power Delivery. January 2008;**23**(1):109-123

[28] A Study of the Measurement-Path Accuracy of Digital Protection Devices in a Wide Range of the Frequency Change. Unpublished report of the research project sponsored by the Polish National Research Committee. Research manager of the project: Michał Szewczyk

[29] www.mathworks.com

[30] www.kocos.de

[31] www.energotest.com.pl

10

Reliability Design of Mechanical System-Like Water-Dispensing System in Refrigerator Subjected to Repetitive Impact Loading

Seong-woo Woo

Abstract

Based on field data and parametric accelerated life tests (ALT), the mechanical system-like water-dispensing system in a bottom-mounted freezer (BMF) was redesigned to find out the missing design parameter in the design phase. To carry out parametric ALTs using a force/momentum balance, the simple mechanical loads of the water-dispensing system were analyzed. At the first ALT, the hinge and front corner of dispenser lever were fractured. The failure shapes found experimentally were similar to those of the failed samples in field. Dispenser lever in water-dispensing system was modified as having its fillets and ribs. At the second ALT, the modified dispenser lever also was fractured because of not having enough strength for impact loading at its front corner. The missing design parameters of the dispenser lever were not enough to have corner fillet rounding and rib thickness. After parameter ALTs with corrective action plans, reliability of the newly designed water-dispensing system is assured to have B1 life 10 years with a failure rate of 0.1%/year.

Keywords: reliability design, water-dispensing system, parametric accelerated life testing, missing design parameter

1. Introduction

As customers want to have water-dispensing function, **Figure 1** shows the bottom-mounted freezer (BMF) refrigerator with the newly designed water-dispensing system. As shown in **Figure 1(b)**, it consists of the dispenser cover, spring, and dispenser lever. To dispense water for a product lifetime, the dispenser system needs to be designed to withstand the operating conditions subjected to it by the consumers who use the BMF refrigerator. Dispensing water in the BMF refrigerator has several operating steps: (1) press the lever and (2) dispense water.

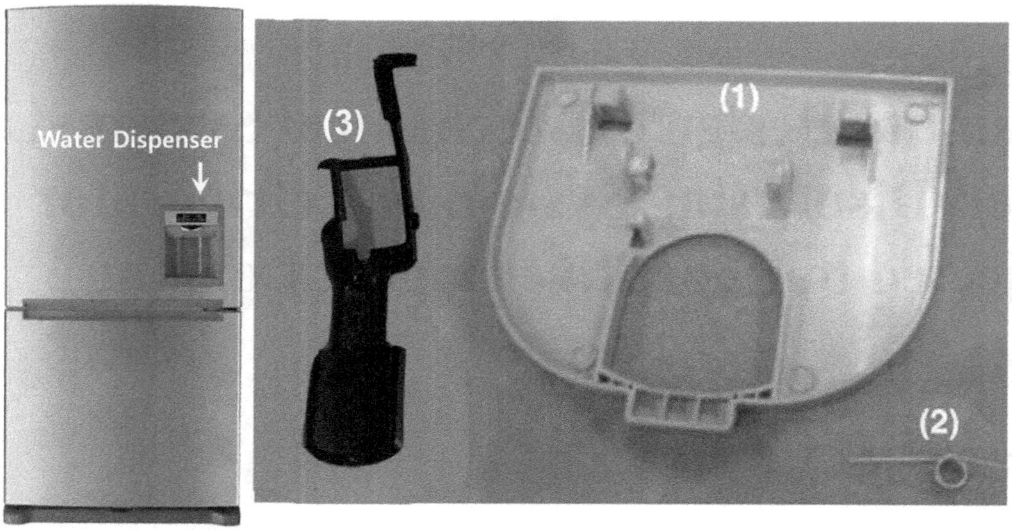

Figure 1. BMF refrigerator with the newly designed water-dispensing system. (a) BMF refrigerator. (b) mechanical parts of dispenser: (1) dispenser cover, (2) spring, and (3) dispenser lever.

He will push the dispenser lever if customer want to drink the cooling water. Consequently, the water-dispenser system will have a variety of repetitive mechanical loads when the consumer uses it, though depending on the consumer usage conditions.

The water-dispensing system in field had been fracturing, causing end-users to replace their product. When subjected to repetitive stresses in operating refrigerator, the failed water-dispensing system came from the design flaws. Market data also showed that the returned products had critical design flaws in structure, including stress risers—sharp corner angles and thin ribs—in the water-dispensing system. The design flaws that could not withstand the repetitive impact loads on the water-dispensing system could cause a crack to occur. Thus, the reliability design of the new water-dispensing system is required to robustly withstand repetitive loads under customer usage conditions (**Figure 2**).

A typical pattern of repeated load or overloading may cause structural failure in product lifetime under the customer usage conditions. Many engineers think such possibility can be assessed: (1) mathematical modeling like Newtonian method, (2) the time response of system simulation for (random) dynamic loads, (3) the rain-flow counting method, and (4) miner's rule that the system damage can be estimated [1–3]. However, because there are a lot of assumptions, this analytic methodology is exact but complex to reproduce the product failures due to the design flaws in field.

Robust design skills like Taguchi methods and statistical design experiments (SDE) [4–9] were studied by engineers and statisticians many years ago. Especially, Taguchi's robust design method utilizes parameter design to place it in a location where random noise factors do not cause effects and determines the optimal design parameters and their levels. As utilizing interactions between control factors and noise factors, the purpose of parameter design is to find the proper control factors that make the system's performance robust in relation to changes in the noise factors. In the robust design process, the noise factors are assigned to an outer array, and the control factors are assigned to an inner array in a complex matrix (**Figure 3**).

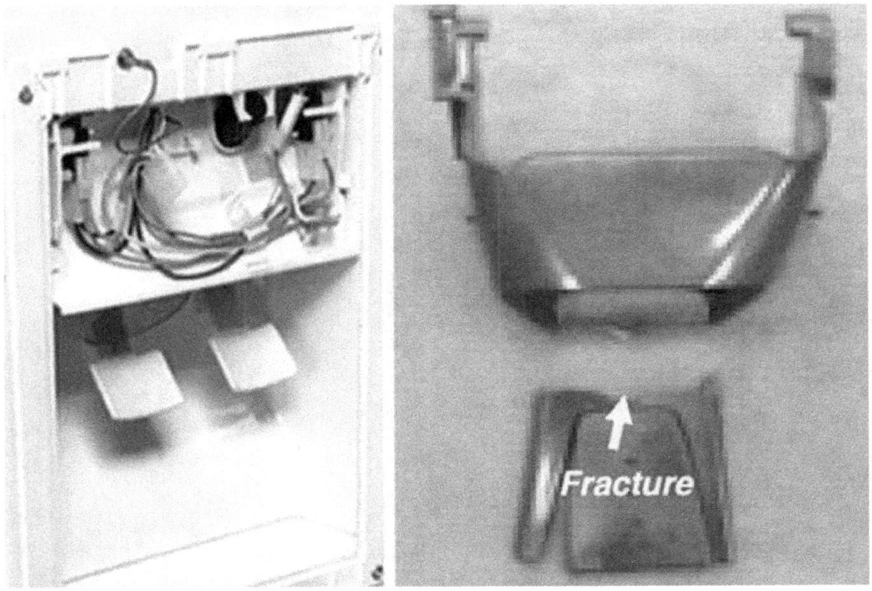

Figure 2. A damaged product after use.

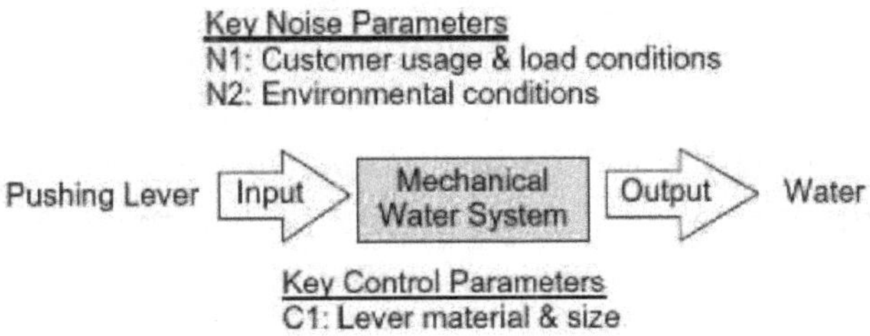

Figure 3. Parameter diagram of water-dispensing system.

However, as the noise array is calculated repeatedly for every row in the control array repetitive, experimental iterations in the Taguchi product array would require a lot of computing time. Because a mechanical structure has complex shape, Taguchi method would require infinite iterations. It is not easy to search out solutions in the robust design process, though a mechanical structure is simple. Newly designed refrigerators with the missing design parameters may result in reliability disasters and pay the huge quality costs.

In this study, we suggest a new parametric accelerated life testing (ALT) method that can enhance the reliability design of newly designed water-dispensing system. To confirm the design parameters, parametric ALT will utilize load analysis, new sample size equations with accelerated factor, and the modifications of dispenser system. It will benefit to confirm the final design of new product. Another case study of this new reliability testing methodology [10–23] will be suggested.

2. Load analysis

As seen in **Figure 4**, the mechanical dispensing system works like the functional design concept. To properly operate the water-dispensing system, its mechanical lever assembly consists of many mechanical structural parts. When a cup touches on the lever to release the

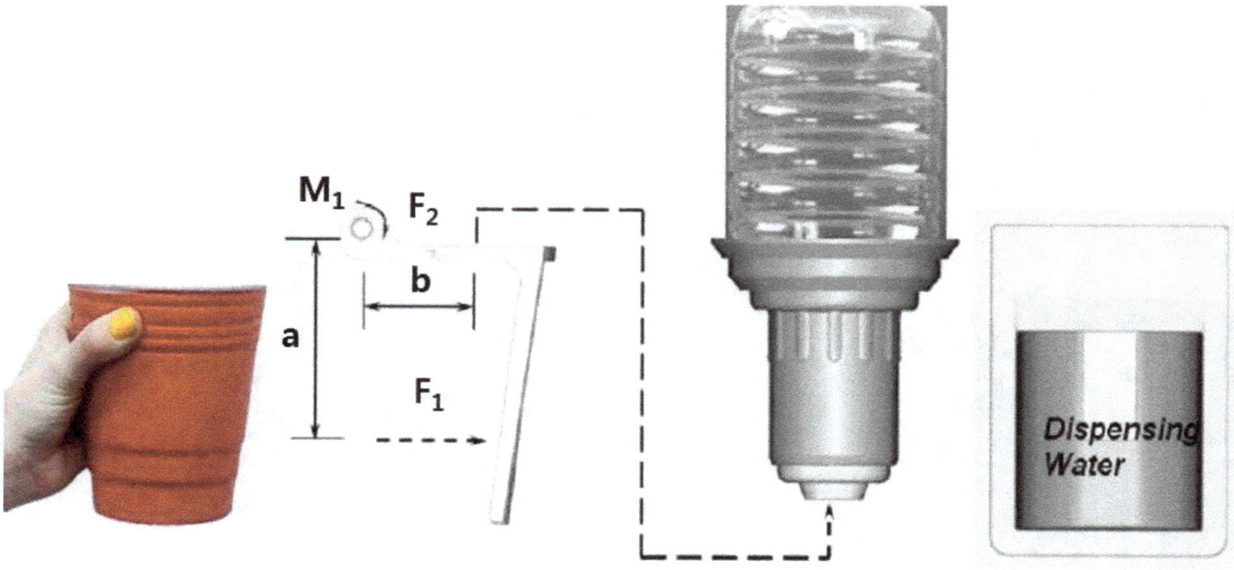

Figure 4. Functional design concept of the mechanical dispensing system.

water, water will dispense. Depending on the end-user usage conditions, the lever assembly goes through repetitive impact loads in the water-dispensing process. The concentrated stress of the pressing cup reveals sharp corner angles. As a result, the impact force applies on the hinge of lever. When designing the mechanical dispensing system, it is a critical design step to withstand these repetitive stresses. In the United States, the typical consumer claims the refrigerator to release water from 4 to 20 times a day.

From the free-body diagram of the simple lever system, the force and momentum at hinge can be represented as

$$F = F_1 \tag{1}$$

$$M = aF_1. \tag{2}$$

Because the stress of the dispenser lever depends on the applied force of cup, the life-stress model (LS model) [24] can be represented as

$$TF = A(S)^{-n} = A(F)^{-\lambda}. \tag{3}$$

The acceleration factor (AF) can be derived as

$$AF = \left(\frac{S_1}{S_0}\right)^n = \left(\frac{F_1}{F_0}\right)^\lambda. \tag{4}$$

3. Parametric accelerated life testing (ALT) in water-dispensing system

To derive the sample size equation and carry out parametric ALT, probability concepts in reliability engineering should be made out. First of all, the characteristic life η_{MLE} from the maximum likelihood estimation (MLE) can be derived as:

$$\eta_{MLE}^{\beta} = \sum_{i=1}^{n} \frac{t_i^{\beta}}{r}. \tag{5}$$

If the confidence level is $100(1-\alpha)$ and the number of failure is $r \geq 1$, the characteristic life, η_{α}, would be estimated from Eq. (4)

$$\eta_{\alpha}^{\beta} = \frac{2r}{\chi_{\alpha}^{2}(2r+2)} \cdot \eta_{MLE}^{\beta} = \frac{2}{\chi_{\alpha}^{2}(2r+2)} \cdot \sum_{i=1}^{n} t_i^{\beta}, \quad \text{for } r \geq 1. \tag{6}$$

If there are no failures, p-value is α and In $(1/\alpha)$ is mathematically equivalent to chi-squared value, $\frac{\chi_{\alpha}^{2}(2)}{2}$. The characteristic life, η_{α}, would be expressed as:

$$\eta_{\alpha}^{\beta} == \frac{2}{\chi_{\alpha}^{2}(2)} \cdot \sum_{i=1}^{n} t_i^{\beta} = \frac{1}{\ln\frac{1}{\alpha}} \cdot \sum_{i=1}^{n} t_i^{\beta}, \quad \text{for } r = 0. \tag{7}$$

Because Eq. (6) is established for all cases $r \geq 0$, it can be redefined as:

$$\eta_{\alpha}^{\beta} = \frac{2}{\chi_{\alpha}^{2}(2r+2)} \cdot \sum_{i=1}^{n} t_i^{\beta}, \quad \text{for } r \geq 0. \tag{8}$$

From the Weibull reliability function, we can find out that the characteristic life can be converted into L_B life as:

$$R(t) = e^{-\left(\frac{L_{BX}}{\eta}\right)^{\beta}} = 1 - x. \tag{9}$$

After logarithmic transformation, Eq. (9) can be expressed as:

$$L_{BX}^{\beta} = \left(\ln\frac{1}{1-x}\right) \cdot \eta^{\beta}. \tag{10}$$

If the estimated characteristic life of p-value α, η_{α}, in Eq. (8), is substituted into Eq. (10), we obtain the BX life equation:

$$L_{BX}^{\beta} = \frac{2}{\chi_{\alpha}^{2}(2r+2)} \cdot \left(\ln\frac{1}{1-x}\right) \cdot \sum_{i=1}^{n} t_i^{\beta}. \tag{11}$$

For a 60% confidence level, the first term $\frac{\chi_{\alpha}^{2}(2r+2)}{2}$ in Eq. (11) can be approximated to $(r+1)$ [25]. By Taylor expansion, if the cumulative failure rate, x, is below 20%, the second term $\ln\frac{1}{1-x}$ approaches to x, and it can be represented as:

$$L_{BX}^{\beta} = \frac{x}{r+1} \cdot \sum_{i=1}^{n} t_i^{\beta}. \tag{12}$$

Most lifetime testing has insufficient samples. The allowed number of failures would not have as much as that of the sample size.

$$\sum_{i=1}^{n} t_i^{\beta} = \sum_{i=1}^{r} t_i^{\beta} + (n-r)h^{\beta} \geq (n-r)h^{\beta}. \tag{13}$$

If Eq. (13) is substituted into Eq. (12), the *BX* life equation can be modified as follows:

$$L_{BX}^{\beta} \geq \frac{x}{r+1} \cdot \cdot (n-r)\, h^{\beta} \geq L_{BX}^{*\beta}. \tag{14}$$

Then, sample size equation with the number of failures can also be modified as:

$$n \geq (r+1) \cdot \frac{1}{x} \cdot \left(\frac{L_{BX}^{*}}{h} \right)^{\beta} + r. \tag{15}$$

From the sample size Eq. (15), we can go on parametric ALT testing under any failure condition ($r \geq 0$). Consequently, it also confirms whether the failure mechanism and the test method are proper.

If the acceleration factors in Eq. (4) are replaced with the planned testing time h, Eq. (15) will be modified as:

$$n \geq (r+1) \cdot \frac{1}{x} \cdot \left(\frac{L_{BX}^{*}}{AF \cdot h_a} \right)^{\beta} + r. \tag{16}$$

Reliability target of the new water-dispensing system is over B1 life 10 years. The operating conditions and cycles of the dispenser system were examined, based on the customer usage conditions. Under B1 life 10 years, if the objective number of life cycles L_{BX} and AF are given, the actual required test cycles h_a can be obtained from Eq. (16). ALT equipment can then be conducted in accordance with the operation procedure of the dispenser system. In parameter ALT, we can obtain the missing design parameters.

4. Laboratory experiments

For the water-dispensing system in BMF refrigerator, the working conditions of customer are about 0–43°C with a relative humidity ranging from 0 to 95% and 0.2–0.24 g's of acceleration. The water dispensing happens approximately 4–20 times per day. With a product life cycle design point for 10 years, water-dispensing system approximately happens for 73,000 usage cycles.

The maximum force expected by the consumer in dispensing water was 19.6 N. For accelerated testing, the applied force makes from double to 39.2 N. With a quotient, λ, of 2, the total *AF* was approximately 4.0 using Eq. (4). To find out missing design parameter of the newly designed water-dispensing system, reliability target can be set to more than the B1 life 10 years. Presumed the shape parameter β was 2.0 and x was 0.01, the actual test cycles calculated in Eq. (16) were 65,000 cycles for sample size 8 units. If parametric ALT for water-dispensing system fails less than once during 65,000 cycles, it will be guaranteed to have a B1 life 10 years with about a 60% level of confidence (**Figures 5** and **6**).

Figure 7(a) and **(b)** shows the failed product from the field and the first accelerated life testing, respectively. The failure sites in the field and the first ALT occurred at the hinge and front corner of dispenser levers as a result of high-impact stress. **Figure 8** represents a graphical analysis if

Figure 5. Equipment used in accelerated life testing.

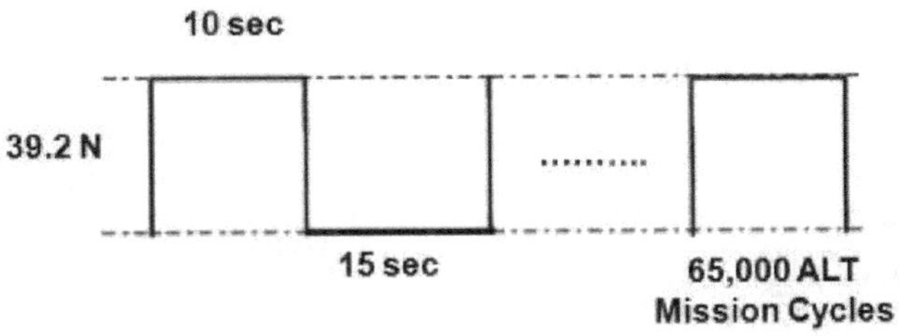

Figure 6. Duty cycles of repetitive impact load F on the water-dispensing system.

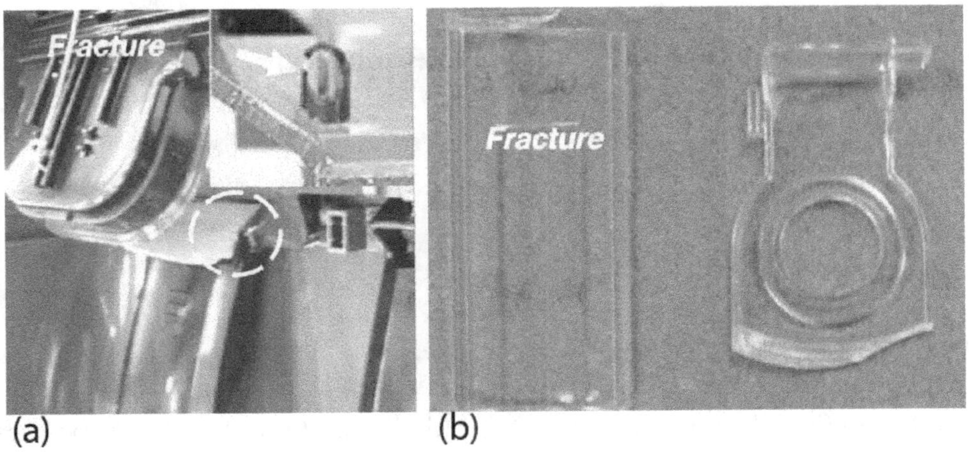

Figure 7. Failed products in field and crack after first ALT. (a) Failed products in field. (b) Crack after first ALT.

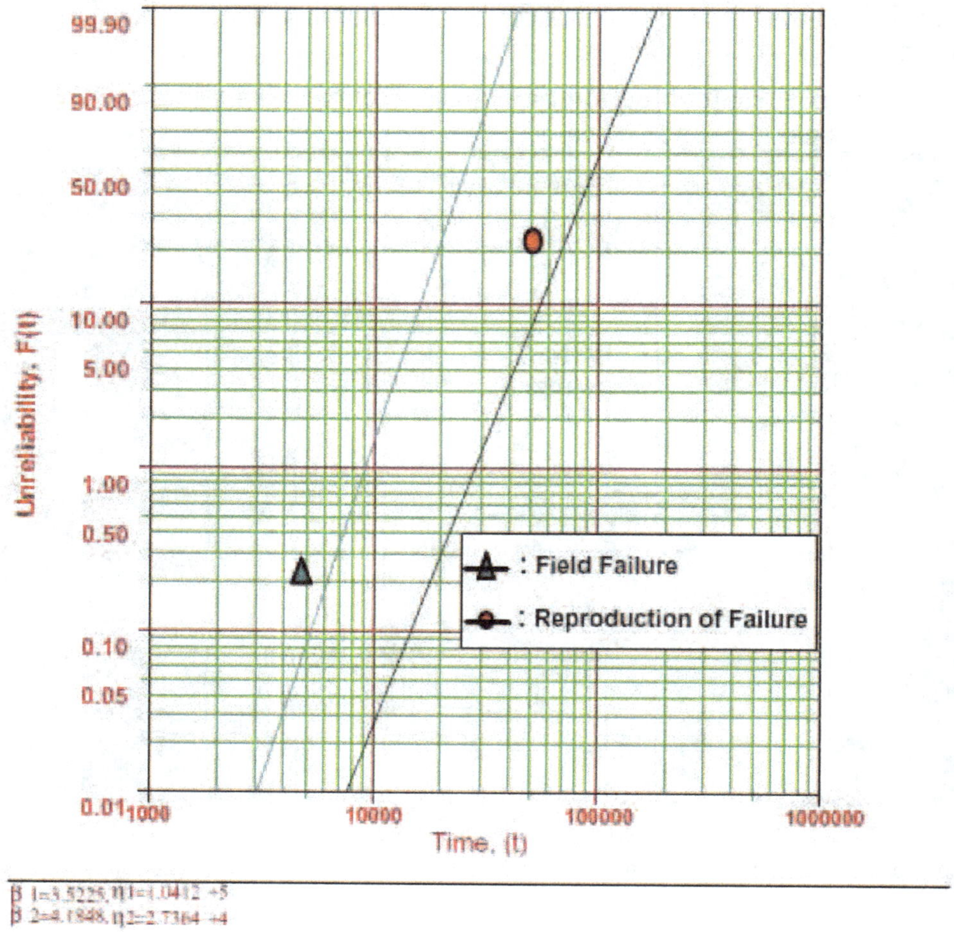

Figure 8. Field data and first ALT on Weibull chart.

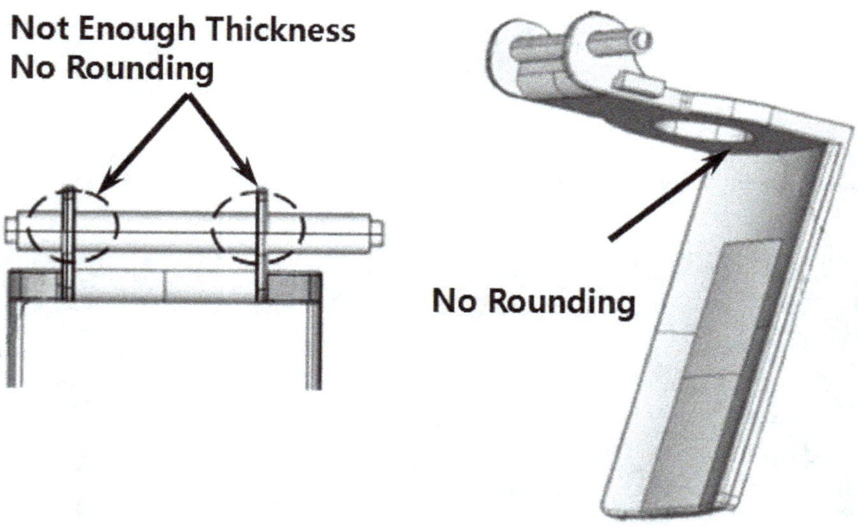

Figure 9. Structure of failing dispenser lever in field.

the ALT results and field data are plotted on Weibull distributions. For the shape parameter, the estimated value on the chart was 2.0. For the final design, the shape parameter was determined to be 3.5. The reduction factor R was 0.01 from the experiment data—product lifetime L_B, acceleration factor *AF*, actual mission cycles h_a, and shape parameter β. It means that R is β power of product lifetime versus testing cycle. Consequently, we know that this parameter ALT is effective to decrease the sample size because reduction factor is less than 0.1 from Eq. (16).

As seen in **Figures 7(b)** and **9**, the fracture of the dispenser lever in the first and second ALTs occurred in its hinge and front corner. As shown in **Table 1**, the missing design parameters of the dispenser lever can be listed. If dispenser lever is subjected to the repetitive impact load, we can conclude that its design flaws can result in a fracture.

To withstand the fracture of dispenser lever due to the repetitive food stresses, the dispenser lever was redesigned as follows: (1) increasing the rib rounding of hinge, C1, from R0 to R2.0 mm; (2) increasing the front corner rounding, C2, from R0 to R1.5 mm; (3) increasing the rib

CTQ	Parameters			Unit
Crack	KNP	N1	Impact loading	N
	KCP	C1	Hinge rib rounding, fillet1	mm
		C2	Front corner rounding, fillet2	mm
		C3	Hinge rib thickness, rib1	mm
		C4	Front side rounding, fillet3	mm
		C5	Front lever thickness, rib2	mm

Table 1. Vital parameters based on ALTs.

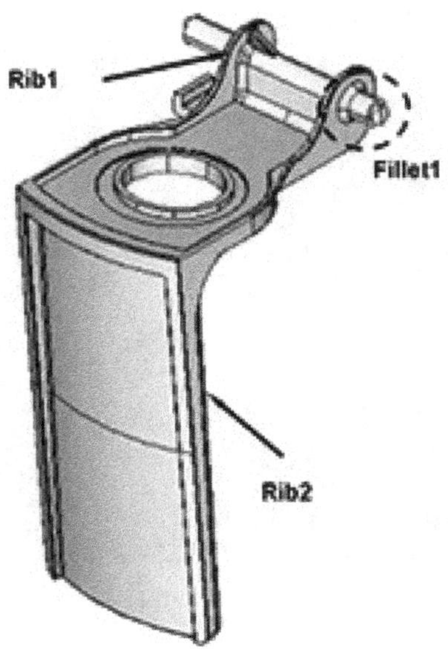

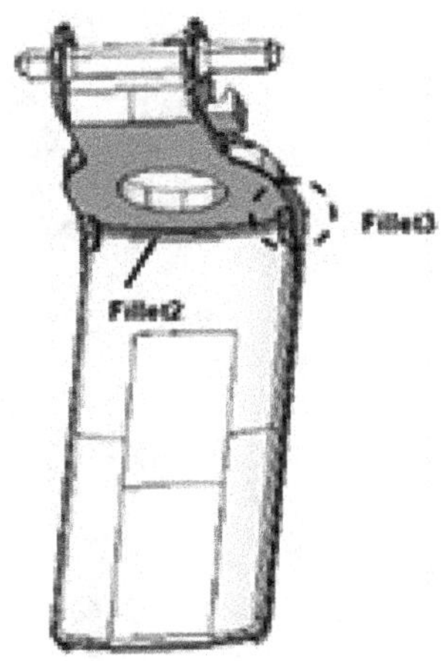

C1: Fillet1 R0 → R1.5 (1st ALT) → R2.0 (2nd ALT)

C2: Fillet2 R0 → R1.5 (1st ALT)

C3: Rib1 T1 → T1.8 (1st ALT)

C4: Fillet3 R0 → R8 (1st ALT) → R11 (2nd ALT)

C5: Rib2 T3 → T4 (3rd ALT)

Table 2. Redesigned dispenser lever.

thickness of hinge, C3, from T1 to T1.8; (4) increasing the front side rounding, C4, from R0 to R11 mm; and (5) thickening the front lever, C5, from T3 to T4 mm (**Table 2**).

As the design flaws make better, the parameter design criterion of the newly designed samples was secured to have more than the reliability target—B1 life 10 years. The confirmed value, β, on the Weibull chart was 3.5. For the second ALT with sample size 8 units, the actual test cycles in Eq. (16) were 38,000. In the third ALT, there were no design problems of the water-dispensing system until the test was carried out to 68,000 cycles. We therefore concluded that the modified design parameters found from first and second ALT were effective.

Table 3 summarizes the ALT results. **Figure 10** also shows the results of the first ALT and third ALT plotted in the Weibull distribution. With the modified design parameters, final samples of the water-dispensing system were guaranteed to reliability target—B1 life 10 years.

	1st ALT	**2nd ALT**	**3rd ALT**
	Initial design	**Second design**	**Final design**
In 38,000 cycles, lever has no crack	25,000 cycles: 2/8 fracture	32,000 cycles: 1/8 fracture	38,000 cycles: 8/8 OK 56,000 cycles: 8/8 OK 68,000 cycles: 1/8 fracture
Freezer Drawer Structure		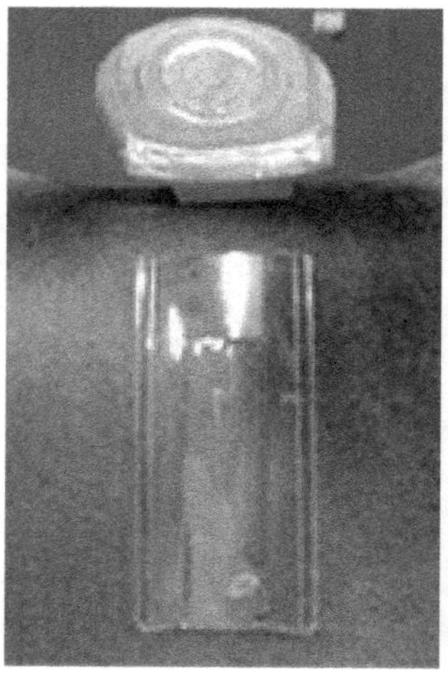	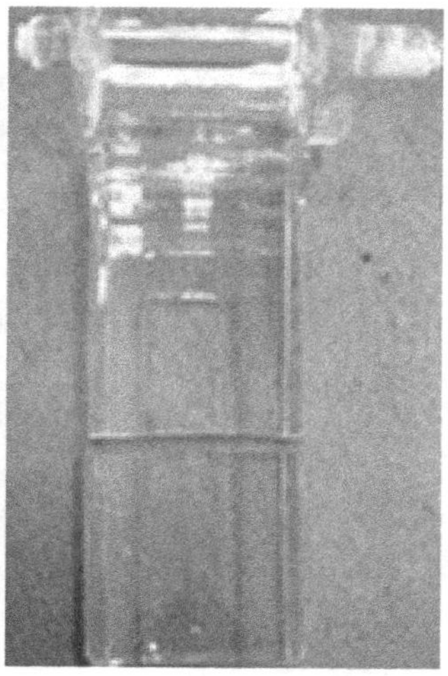
Material and specification	C1: Fillet 1 R0 → R1.5 C2: Fillet2 R0 → R1.5 C3: Rib1 T1 → T1.8 C4: Fillet3 R0 → R8	C1: Fillet 1 R1.5 → R2.0 C2: Fillet 2 R8.0 → R11.0	C3: Rib2 T3.0 → T4.0

Table 3. Results of ALTs.

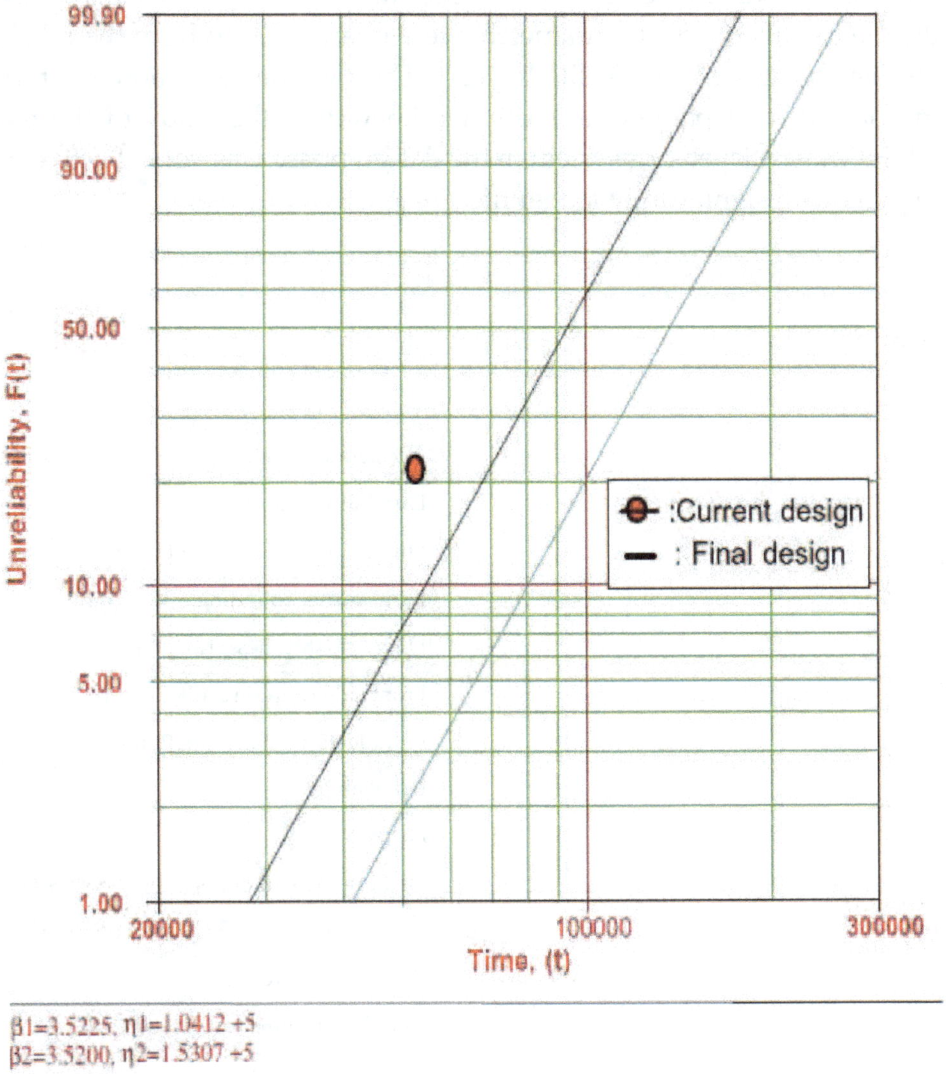

Figure 10. Results of ALT plotted in Weibull chart.

5. Conclusions

We suggested a new reliability design method for newly designed water-dispensing system in BMF refrigerators. The missing design parameters for the water-dispensing system were identified through parameter ALTs. At that time, the reliability target was B1 life 10 years.

As previously seen in the first ALT, the hinge and front corner of dispenser lever were fracturing because of design flaws due to repetitive impacting stress. As increasing the fillets and ribs of the dispenser lever, the water-dispensing lever system was corrected. During the second ALT, the front corner of dispenser lever also fractured because they did not have enough strength to withstand the repetitive impact loads. As additional reinforced ribs of dispenser lever were provided, we knew that the design of water-dispensing system was improved robustly.

As a result of these modified design parameters, there were no problems in the third ALT. Consequently, the modified design parameters were guaranteed to have the reliability requirement of water-dispensing system—B1 life 10 years. Through the inspection of returned products in field, load analysis, and parametric ALTs, we knew that the study of the missing design parameters of the water-dispensing system in the design phase was very effective in redesigning more reliable parts with significantly longer life.

Nomenclature

AF	Acceleration factor
BX	Durability index
C1	Hinge rib rounding, fillet1, mm
C2	Front corner rounding, fillet2, mm
C3	Hinge rib thickness, rib1, mm
C4	Front side rounding, fillet3, mm
C5	Front lever thickness, rib2, mm
$F(t)$	Unreliability
F	Force, *N*
F1	Pushing force under accelerated stress conditions, *N*
F0	Pushing force under normal conditions, *N*
h	Testing cycles (or cycles)
h^*	Non-dimensional testing cycles, $h^* = h/L_B \geq 1$
KCP	Key control parameter
KNP	Key noise parameter
L_B	Target B_X life and $x = 0.01X$, on the condition that $x \leq 0.2$
n	Number of test samples
N1	Consumer pushing force, *N*
R	Reduction factor, $R = (L_B/AF \cdot h)^\beta \geq 1$
r	Failed numbers
S	Stress
S_1	Mechanical stress under accelerated stress conditions
S_0	Mechanical stress under normal conditions
t_i	Test time for each sample
TF	Time to failure
X	Accumulated failure rate, %
x	$x = 0.01 \cdot X$, on condition that $x \leq 0.2$
Greek symbols	
η	Characteristic life

λ	Cumulative damage exponent
μ	Friction coefficient
Superscripts	
β	Shape parameter in a Weibull distribution
n	Stress dependence, $n = -\left[\frac{\partial \ln(T_f)}{\partial \ln(S)}\right]_T$
Subscripts	
0	Normal stress conditions
1	Accelerated stress conditions

Author details

Seong-woo Woo

Address all correspondence to: twinwoo@yahoo.com

Reliability Association of Korea, Seoul, Korea

References

[1] Matsuishi M, Endo T. Fatigue of metals subjected to varying stress. The Japan Society of Mechanical Engineers. Fukuoka, Japan, March; 1968

[2] Mott RL. Machine Elements in Mechanical Design. 4th ed. Upper Saddle River: Pearson Prentice Hall; 2004, pp. 190–192

[3] Palmgren AG. Die Lebensdauer von Kugellagern. Zeitschrift des Vereines Deutscher Ingenieure. 1924;**68**(14):339–341

[4] Taguchi G. Off-line and on-line quality control systems. Proceedings of the International Conference on Quality Control. Tokyo, Japan; 1978

[5] Taguchi G, Shih-Chung T. Introduction to Quality Engineering: Bringing Quality Engineering Upstream. New York: American Society of Mechanical Engineering; 1992

[6] Ashley S. Applying Taguchi's quality engineering to technology development. Mechanical Engineering. July; 1992

[7] Wilkins J. Putting Taguchi methods to work to solve design flaws. Quality Progress. 2000;**33**(5):55–59

[8] Phadke M. Quality Engineering Using Robust Design. Englewood Cliffs: Prentice Hall; 1989

[9] Byrne D, Taguchi S. The Taguchi approach to parameter design. Quality Progress. 1987;**20**(12): 19–26

[10] Woo S, Pecht M. Failure analysis and redesign of a helix upper dispenser. Engineering Failure Analysis. 2008;**15**(4):642–653

[11] Woo S, O'Neal D, Pecht M. Improving the reliability of a water dispenser lever in a refrigerator subjected to repetitive stresses. Engineering Failure Analysis. 2009;**16**(5): 1597–1606

[12] Woo S, O'Neal D, Pecht M. Design of a hinge kit system in a kimchi refrigerator receiving repetitive stresses. Engineering Failure Analysis. 2009;**16**(5):1655–1665

[13] Woo S, Ryu D, Pecht M. Design evaluation of a French refrigerator drawer system subjected to repeated food storage loads. Engineering Failure Analysis. 2009;**16**(7):2224–2234

[14] Woo S, O'Neal D, Pecht M. Failure analysis and redesign of the evaporator tubing in a kimchi refrigerator. Engineering Failure Analysis. 2010;**17**(2):369–379

[15] Woo S, O'Neal D, Pecht M. Reliability design of a reciprocating compressor suction reed valve in a common refrigerator subjected to repetitive pressure loads. Engineering Failure Analysis. 2010;**17**(4):979–991

[16] Woo S, Pecht M, O'Neal D. Reliability design and case study of a refrigerator compressor subjected to repetitive loads. International Journal of Refrigeration. 2009;**32**(3): 478–486

[17] Woo S, O'Neal D, Pecht M. Reliability design of residential sized refrigerators subjected to repetitive random vibration loads during rail transport. Engineering Failure Analysis. 2011;**18**(5):1322–1332

[18] Woo S. Chapter 11: The reliability design of mechanical system and its Parametric ALT. In: Handbook of Materials Failure Analysis with Case Studies from the Chemicals, Concrete and Power Industries. Butterworth Heinemann: Elsevier; 2015. pp. 259–276

[19] Woo S, O'Neal D. Reliability design of mechanical systems subject to repetitive stresses. Recent Patents on Mechanical Engineering. 2015;**8**(4):222–234

[20] Woo S, O'Neal D. Improving the reliability of a domestic refrigerator compressor subjected to repetitive loading. Engineering. 2016;**8**(3):99–115

[21] Woo S, O'Neal D. Design of the hinge kit system subjected to repetitive loading in a commercial refrigerator. Challenge Journal of Structural Mechanics. 2016;**2**(2):75–84

[22] Woo S. Reliability design of ice-maker system subjected to repetitive loading. Engineering. 2016;**8**(9):618–632

[23] Woo S. Reliability Design of Mechanical Systems: A Guide for Mechanical and Civil Engineers. 1st ed. Switzerland: Springer; 2017

[24] McPherson J. Accelerated testing, packaging, electronic materials handbook. ASM International. 1989;**1**:887–894

[25] Ryu D, Chang S. Novel concept for reliability technology. Microelectronics Reliability. 2005;**45**(3):611–622

11

Resource Planning to Service Restoration in Power Distribution Systems

Magdiel Schmitz,
Maria Clara Ferreira Almeida da Silva,
Vinícius Jacques Garcia, Daniel Bernardon,
Lynceo Favigna Braghirolli and Júlio Fonini

Abstract

Whenever there are extreme weather events, electric power distribution systems are generally affected largely because they are highly subject by their constructive nature: overhead networks. In this context, the management of maintenance actions is generally referred to as emergency service order, usually associated with a lack of supply and requiring human intervention. The key issue for the resource planning refers to an estimation of service time that allows for more assertive planning possible. This chapter proposes a predictive modelling of emergency services for resource planning when considering the geographic dispersion of such services and also the time windows that comprise the amount of service time demanded. After presenting the methodological procedures, a case study depicts the application of the proposed method in order to support proactive service routing.

Keywords: power distribution systems, decision-making, predictive modelling, emergency services, service restoration

1. Introduction

Certain regions are subject to frequent adverse weather events of sufficient intensity to cause damage to the electrical networks, requiring human intervention to verify and eliminate the causes of such disruption and thus leading to the creation of emergency service orders. These orders, with a high degree of uncertainty and urgency, along with commercial orders, with a previously defined route, become a task of great relevance for the network operation center of the electric distribution systems and qualifying even more the highly dynamic environment involved. The key issue for the resource planning, with regard to the service crews, refers to an estimation of service time that allows for more assertive planning possible, improving the crew throughput.

Considering the need to define a typical behaviour to those orders, this chapter proposes a predictive modelling of emergency services for resource planning when considering the geographic dispersion of such services and also the time windows that comprise the amount of service time demanded. The methodology to resource planning presented in this chapter is based on the predictive modelling of emergency service by defining random variables to denote how much service time is demanded in each portion of the whole geographic area considered and also how important such demands may be in the sense of reliability indexes, as number of customers and the amount of power not supplied.

Following this definition of random variables, one may obtain the service time to attend emergency services stratified by each region and also restricted to a certain time window, that is, 1-h interval. With this information, a decision-making process may be conducted to define the number of hours demanded by maintenance crews that will be needed to attend all the services in the 1-h interval. The whole picture when assuming 1 day demand will be possible by collecting all these 1-h intervals, furnishing the geographic location and the on-site service time to a possible proactive-routing approach.

In order to show how the methodology may be applied, a case study is developed considering a horizon of emergency service occurrences, from a given Brazilian Power Utility. The purpose of this case study is not only about the past but also about how to update the case study of actual occurrences in such a way that they may be considered for future estimations.

The process of conducting statistical analyses to construct the time series for each given 1-h interval is depicted in a graphic user interface also allowing the timeline description of the estimation of service time dispersed over the geographic area assumed, followed by the decision-making process of selecting the most relevant random variables to denote the service time demanded.

2. Service restoration by repair crews

Although the advent of smart grids has conducted to increasing the level of automation [1], the need for human intervention is particularly necessary in cases of extreme weather events or even in the occurrence of collisions in overhead networks, which become exposed by itself and may cause the lack of power supply.

This interruption refers to the switching from normal to the emergency operating condition [2], in all of the networks or even in a restricted part of it, conducting customers to the lack of power supply and affecting continuity indexes. In addition, companies still have the billing process affected by not addressing the load demand for a certain period.

The use of repair crews becomes imperative precisely to overcome the limitations that interfere with the remote control equipment, since the insulation of the defect can even be performed automatically but its correction requires human interference.

From this crew management comes the well-known emergency dispatching problem to repair crews already scheduled with pre-programmed services, which leads to a change of course and describes the problem of dynamic routing of vehicles [3]. Exactly in the context of service operations in electric power distribution utilities, the problem of this work is defined in [4].

The nature of the optimization problem has its roots on minimizing the waiting time between the detection of the defect and the arrival of crews at the service location, because the less this time the less the time without supply, thus contributing to the reliability of the electric power distribution system.

Another aspect worth mentioning is the random nature of events: the occurrence of emergency orders. In case of a reactive system, each new pending emergency order, or a set of them, gives rise to the problem of dynamic routing [5]. In addition, the routing problem associated with the context of this work also presents an important particularity: the service time in each service location, which is exactly the estimated service time for resolution. It follows that the dispatch problem is closed to the problem of minimum latency [6].

From the conclusions of [7], it can be observed that a decision that is completely reactive, responding to the disturbances generated by the creation of emergency orders, can lead to a sequence of mistaken decisions due to the loss of the 'global vision' related to the whole problem depicted after analysing all the past events over the day.

This chapter aims to anticipate the occurrence of emergency services in order to provide a proactive approach to dispatch, in which emergency occurrences can be assumed to be probable and therefore likely to be included in routing solutions of programmed orders.

3. Demand forecasting

The routing problem is usually addressed by assuming that attributes (customer demands, travel and service times, orders criticality) are previously known. However, most of the real cases present some level of uncertainty [8]. Anticipating these uncertainties already in planning phase is important because it allows a more precisely routing plan and favouring the crew management, operational cost reduction and also avoiding penalties from regulatory agencies through improvement of reliability indicators and the revenue from the energy sale, by reducing supply disruptions.

According to [9], the data of vehicle-routing problems can be 'Static and deterministic', 'Static and stochastic', 'Dynamic and deterministic' and 'Dynamic and stochastic'. Exactly, this last class is that related to this work, referred many times as partially dynamic, this approach assumes that some of the unknown data are in stochastic format. A proactive planning can be realized by transforming the stochastic knowledge in dummy consumers with expected service time and temporal and spatial position.

With such definitions, the next section describes how to address the uncertainty by performing demand-forecasting techniques.

3.1. Demand forecasting

Considering the external service orders of electric power utilities, we treat two different types of situations, the commercial and the emergency orders. The commercial ones have a typical input behaviour, known deadline and, as a consequence of these characteristics, can be planned in advance, without technical or legal problems involving power failures. By contrast, the emergency orders are randomly generated, with the corresponding dynamic character [10].

As to the attendance capacity, all the available crews have commercial and emergency demand during the workday, overlapping one each other and respecting the attendance priority. This behaviour can be seen in **Figure 1**, where there is a considerable difference between the planned route and the route performed mainly due to the occurrence of emergency orders: they have higher priority with respect to commercial orders and cause modification on the route every time they are considered.

These constant changings on planned route by knowing customers and their attributes progressively classify the problem addressed as dynamic and stochastic [7]. One possible alternative that is adopted in this work is associated with an attempt to predict the emergency occurrences and answering when and where they may occur [11].

Indispensable in the planning and in the strategies for decision-making, demand forecasting oriented towards a future period should rely on precise techniques [11]. Of the commonly used techniques to perform the forecasting, the exponential-smoothing approach can be applied to predict continuous and discrete variables.

The exponential smoothing provides the service time to the analysis horizon, using Eq. (1), according to [12]

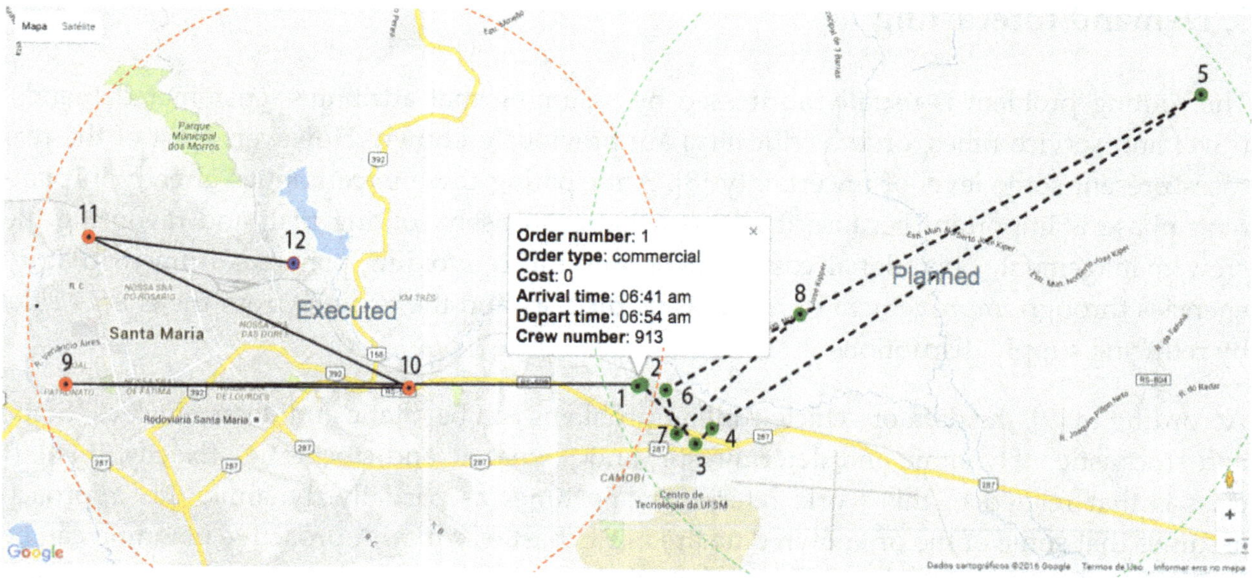

Figure 1. Routing of crews with emergency-order fulfilment.

$$F_{t+1} = \alpha x_t + (\alpha)F_t \quad (1)$$

where t = actual period; F_{t+1} = future forecasting; x_t = demand in the period t; F_t = forecasting to t; α ($0 < \alpha < 1$) is a smoothing constant.

With these definitions, what one has is the theoretical basis to realize the forecasts of time of service from the available time series.

However, it is necessary to consider that only the service time forecasting is not enough to answer the questions pointed out by this work, which are the service time related to a certain part of the geographic area considered and also related to a time interval defined [13, 14]. Exactly these challenges will be illustrated in the next section, which describes the methodological procedures that are employed.

4. Proposed methodology

Following the contribution by Ferrucci et al. [7], inspired by the paper of Ichoua et al. [15], the approach proposed in this work also considers the classification of orders information that has already occurred, based at service time of each request, as pointed out in **Figures 2** and **3**.

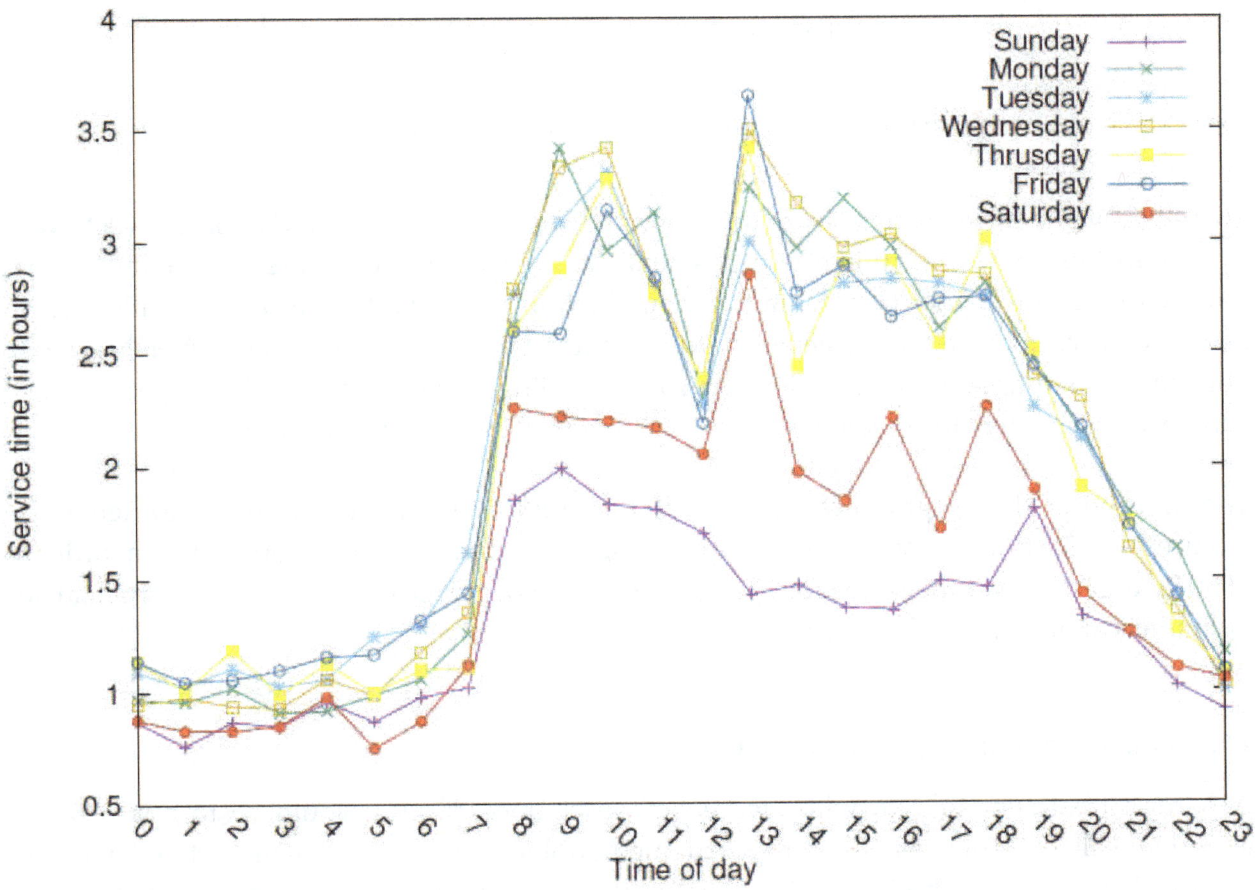

Figure 2. Variation analysis from the historical information. Day of the week and time of the day.

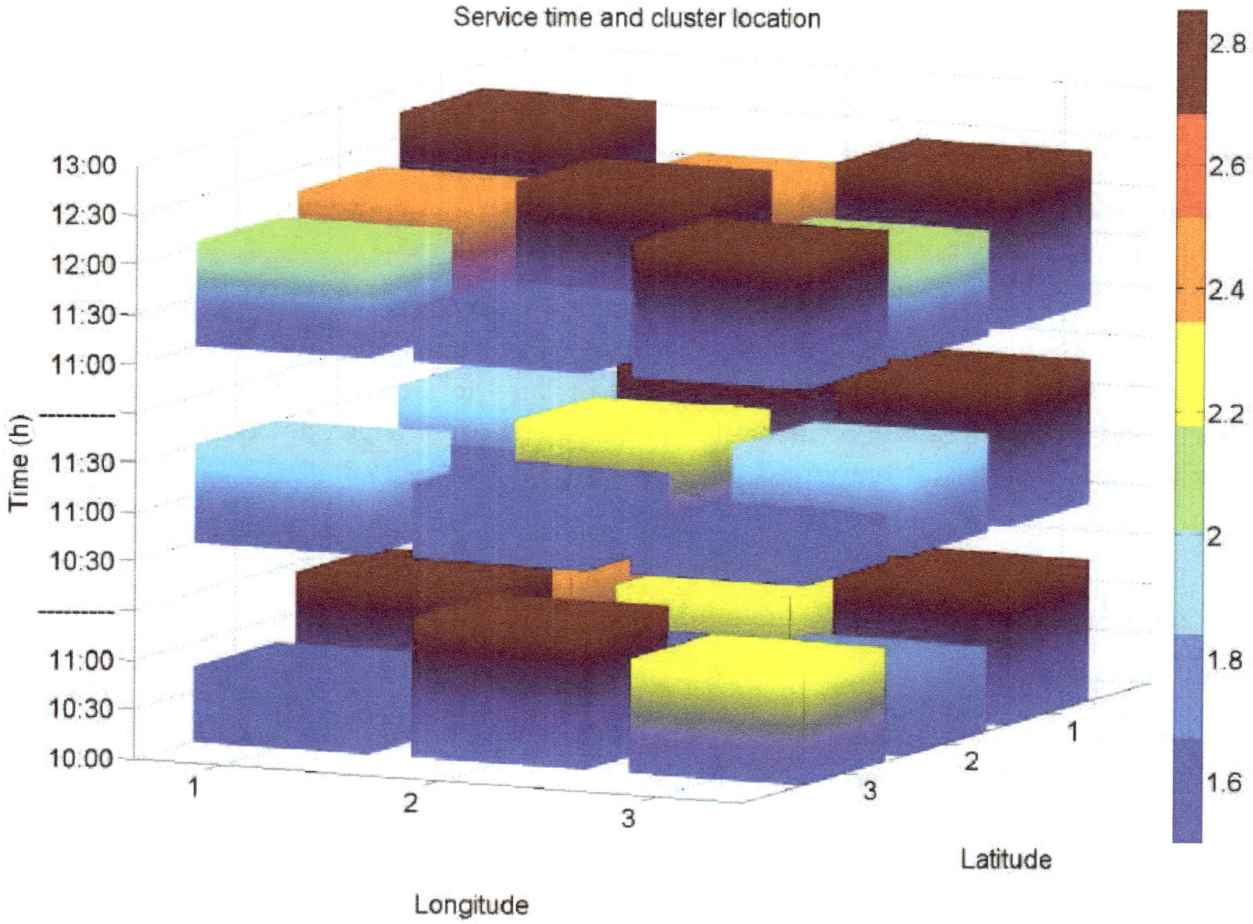

Figure 3. Variation analysis from the historical information. Geographical and time interval.

There is a well-defined pattern from the requested service time for each day of the week, very similar to business days unlike non-business days. In addition, **Figure 2** also shows that request demand is higher from 8 am to 8 pm. The last feature with regard to the influence on future requests is related to the geographical location: **Figure 3** depicts several layers, each one referring to one time interval, with their corresponding surface composed by cuboids; each cuboid's height refers to a level of service time required in the respective area.

The division of the geographical area and the identification of time windows to define the required service time are performed by stratifying and separating the historical data in order to identify a particular time series for each part of the geographical area and also particular for each interval of 1 day of the week.

4.1. Demand-forecasting methodology

A natural treatment to deal with emergency-order uncertainties would be to optimize the expected values. However, the proposed methodology approaches a restrictive model of chances [16], where each viable solution must be at an acceptable and predetermined level of failure. In the presented model (see Algorithm 1), this constraint is represented by a coefficient, corresponding to the maximum value of variation allowed. This artifice prevents exceptional disturbances from leading to distortions in the statistic analysis process.

The relationship between the scale of the geographical area considered and the coefficient of variation is performed both temporally and spatially. In order to increase the forecasting accuracy, historical data are stratified to each quadrant of the area considered for each day of the week and for each hour of the day. The methodology to proceed with this stratification is detailed in the algorithm of **Table 1**.

Beta (β) corresponds to the maximum value of variation allowed, normally 0.5, without considering those regions without emergency-order occurrences. Therefore, the referred coefficient of variation has the following limits:

$$0 < \text{coefficient of variation} \leq \beta \qquad (2)$$

Instead of manually testing multiple values for the coefficient of variation, the algorithm minimizes a nonlinear function obtained from an interpolation in order to determine the best value for that coefficient. This procedure aims to minimize the size of the quadrants obtained taking into account the constraint related to the coefficient of variation.

The main contribution of this work is the stratification of the data in order to have a more accurate forecast, since the coefficient of variation is calculated for each defined quadrant.

5. Case study

This section describes the computational study, performed to evaluate the proposed model. The model presented in algorithm of **Table 1** was implemented in the computational tool *GNU Octave*, chosen for its efficiency features in numerical processing and matrix computation, as well as being open-source software.

5.1. Instance

The model was tested using an actual historical data set, provided by a power distribution utility from southern Brazil. Constituted of 9408 emergency orders in a period of 12 months, the sample is endowed with several attributes, as latitude, longitude, year, month, day, weekday, hour of day and service time, used for the analysis performed.

5.2. Parameters setting

Size: Vector representing the values referring to the side of the area analysed in the calculation. This measure is in kilometres. The higher the number of analysed values, the greater is the accuracy of the function obtained by the interpolation.

Another point to be considered is that the lower the value reported on size vector: the minimum value for the area side size may vary, depending on the study region, that is, urban or

Algorithm 1—Geographic area optimization

Input: Database reading containing the following data for each order: latitude, longitude, year, month, day, day of week, time and service time.

Function: Temporal clustering of orders per hour and day of the week.

Input: size vector = [0.4:0.2:1]; #Vector with area side size values

```
for size = 0.4 to 1 do
    Construction of the grid with the definition of the geographical positions for the size considered.
    Order assignment to the respective areas based on the geographical position.
    for area = 1 to n do #Where n is the total of areas
        Select a specific area of the grid.
        for day of the week = 1 to 7 do
            for hour = 1 to 24 do
                Calculation of the coefficient Of Variation for the selected area.
            end-for
        end-for
    end-for
end-for
#Thus, we have a matrix with the coefficients of variation associated with the respective area side size, for each area of the grid, day of the week and time of day.
#Elimination of data with high randomness
for day of the week = 1 até 7 do
    for hour = 1 to 24 do
        for size = 0.4 to 1 do
            for 0 < coefficient Of Variation <= beta do
                Calculation of the mean Coefficient Of Variation for all areas
            end-for
        end-for
        Determine the characteristic function by the Lagrange interpolation procedure.
        Find the minimum of the function size (mean Coefficient Of Variation) in the range: 0 <= mean Coefficient Of Variation<=beta
    end-for
end-for
```

Table 1. Geographic area optimization algorithm.

rural area. Therefore, the following vector sizes were chosen, measured in kilometres: sizeVector = [0.4 0.6 0.8 1].

Beta (β): As previously informed, it corresponds to the maximum value to the coefficient of variation allowed to ensure the lowest acceptable variance. This value was set to 0.5.

Interpolation: In order to obtain a polynomial function that best represents the analysed data set, the Lagrange interpolation is used. It is important to highlight that the interpolation can be made with any other function which best describes the data set. The Lagrange interpolation method can construct a polynomial of n degree, which coincides with a given function in n + 1 points, according to the description of the method next presented by [17].

Lagrange interpolation theorem: Given n+1 distinct points, $z_0, z_1, \ldots, z_n$ and n+1 values, $w_0, w_1, \ldots, w_n$, there exists a unique polynomial $p_n(z) \in P_n$ for which

$$p_n(z_i) = w_i, i = 0, 1, \ldots n \tag{3}$$

Let us introduce the following polynomials of degree n:

$$l_n(z) = \frac{(z - z_0)(z - z_1)\ldots(z - z_{k-1})(z - z_{k+1})\ldots(z - z_n)}{(z_k - z_0)(z_k - z_1)\ldots(z_k - z_{k-1})(z_k - z_{k+1})\ldots(z_k - z_n)} \tag{4}$$

where k = 0,1,2,…,n.

Now let

$$G_n = (z - z_0)(z - z_1)\ldots(z - z_n) \tag{5}$$

Then

$$G_k' = (z_k - z_0)(z_k - z_1)\ldots(z_k - z_{k-1})(z_k - z_{k+1})\ldots(z_k - z_n) \tag{6}$$

and hence it follows from Eq. (4) that

$$l_k(z) = \frac{G_n(z)}{(z - z_k)G'_n(z_k)} \tag{7}$$

Eq. (7) becomes

$$p_n(z) = \sum_{k=0}^{n} w_k \frac{G_n(z)}{(z - Z_k)G_n'(zk)} \tag{8}$$

Eq. (8) is known as the *Lagrange Interpolation Formula*, and the polynomials $l_k(z)$ are called the *interpolating* or *sampling* of the Lagrange interpolation. In this problem, the z_n points are the *size* sample, and the w_n values the *coefficients of variation*.

5.3. Results

Table 2 presents the optimization result, with the cell values corresponding to the coefficient of variation and the area side size.

Hour of the day	Day of the week						
	1	2	3	4	5	6	7
1	(NaN,NaN)	(NaN,NaN)	(NaN,NaN)	(NaN,NaN)	(NaN,NaN)	(NaN,NaN)	(NaN,NaN)
2	(0, 0.6)	(NaN,NaN)	(NaN,NaN)	(NaN,NaN)	(0, 0.6)	(NaN,NaN)	(NaN,NaN)
3	(NaN,NaN)	(NaN,NaN)	(NaN,NaN)	(0, 0.6)	(NaN,NaN)	(NaN,NaN)	(NaN,NaN)
4	(NaN,NaN)	(NaN,NaN)	(NaN,NaN)	(NaN,NaN)	(NaN,NaN)	(NaN,NaN)	(NaN,NaN)
5	(NaN,NaN)	(NaN,NaN)	(NaN,NaN)	(NaN,NaN)	(NaN,NaN)	(NaN,NaN)	(NaN,NaN)
6	(NaN,NaN)	(NaN,NaN)	(NaN,NaN)	(NaN,NaN)	(0.5, 0.44207)	(NaN,NaN)	(NaN,NaN)
7	(NaN,NaN)	(0.5, 0.4)	(0, 0.4)	(0, 0.5)	(0, 0.66667)	(0, 0.6)	(NaN,NaN)
8	(0, 0.5)	(0, 0.6)	(0, 0.4)	(0, 0.4)	(0, 0.4)	(0.5, 0.4)	(0, 0.4)
9	(0.5, 0.4)	(0.5, 0.4)	(0, 0.4)	(0,5, 0.4)	(0, 0.4)	(0.5, 0.4)	(0, 0.4)
10	(0.5, 0.4)	(0.5, 0.4)	(0,24203, 0.4)	(0.28101, 0.4)	(0.5, 0.4)	(0, 0.4)	(0.27432, 0.4)
11	(0.5, 0.4)	(0.5, 0.4)	(0,5, 0.4)	(0, 0.4)	(0.5, 0.4)	(0.5, 0.4)	(0, 0.4)
12	(0, 0.4)	(0, 0.4)	(0.5, 0.4)	(0, 0.4)	(0, 0.4)	(0.5, 0.4)	(0, 0.4)
13	(0, 0.4)	(0.42894, 0.4)	(0.5, 0.4)	(0.42321, 0.4)	(0, 0.4)	(0, 0.4)	(0, 0.4)
14	(0.5, 0.4)	(0.5, 0.4)	(0, 0.4)	(0, 0.4)	(0, 0.4)	(0.5, 0.4)	(0.5, 0.4)
15	(0, 0.4)	(0.5, 0.4)	(0, 0.4)	(0, 0.4)	(0, 0.4)	(0, 0.4)	(0.5, 0.4)
16	(0.22726, 0.47263)	(0, 0.4)	(0.13927, 0.4)	(0, 0.4)	(0, 0.4)	(0.4, 0.4)	(0.5, 0.4)
17	(0, 0.4)	(0.10581, 0.48655)	(0, 0.4)	(0.5, 0.4)	(0.5, 0.4)	(0.5, 0.4)	(0, 0.4)
18	(0.13379, 0.4)	(0, 0.4)	(0, 0.4)	(0, 0.4)	(0.5, 0.4)	(0, 0.4)	(0.11458, 0.4)
19	(0.5, 0.4)	(0, 0.4)	(0,33325, 0.4)	(0.5, 0.4)	(0.5, 0.4)	(0, 0.4)	(0.5, 0.4)
20	(0.5, 0.4)	(0.5, 0.4)	(0,5, 0.4)	(0.5, 0.4)	(0. 0.4)	(0, 0.4)	(0, 0.4)
21	(0, 0.4)	(0, 0.4)	(0.30439, 0.4)	(0,5, 0.4)	(0.28037, 0.4)	(0.5, 0.4)	(0, 0.4)
22	(0, 0.4)	(0.5, 0.4)	(0.5, 0.4)	(0.23448, 0.4)	(0.5, 0.4)	(0, 0.4)	(0.5, 0.4)
23	(0.5, 0.4)	(0, 0.66667)	(0.41580, 0.4)	(0, 0.4)	(0, 0.4)	(0.5, 0.4)	(0, 0.4)
24	(0, 0.54451)	(0, 0.66667)	(0, 0.4)	(0, 0.4)	(0.5, 0.4)	(0.5, 0.4)	(0, 0.54451)

Note that the NaN indicates that there was no coefficient of variation within the established limits.

Table 2. Optimization result (coefficient of variation, area side size).

The results presented in **Table 2** show that not all hours of the day respect the coefficient of variation constraint: one can note that periods with more incidences of emergency orders (e.g. from 19 to 23 h) presented more precise data, analytically we can see that in the period with less emergency-orders occurrence (e.g. from 1 to 6 h), the data did not present the necessary reliability.

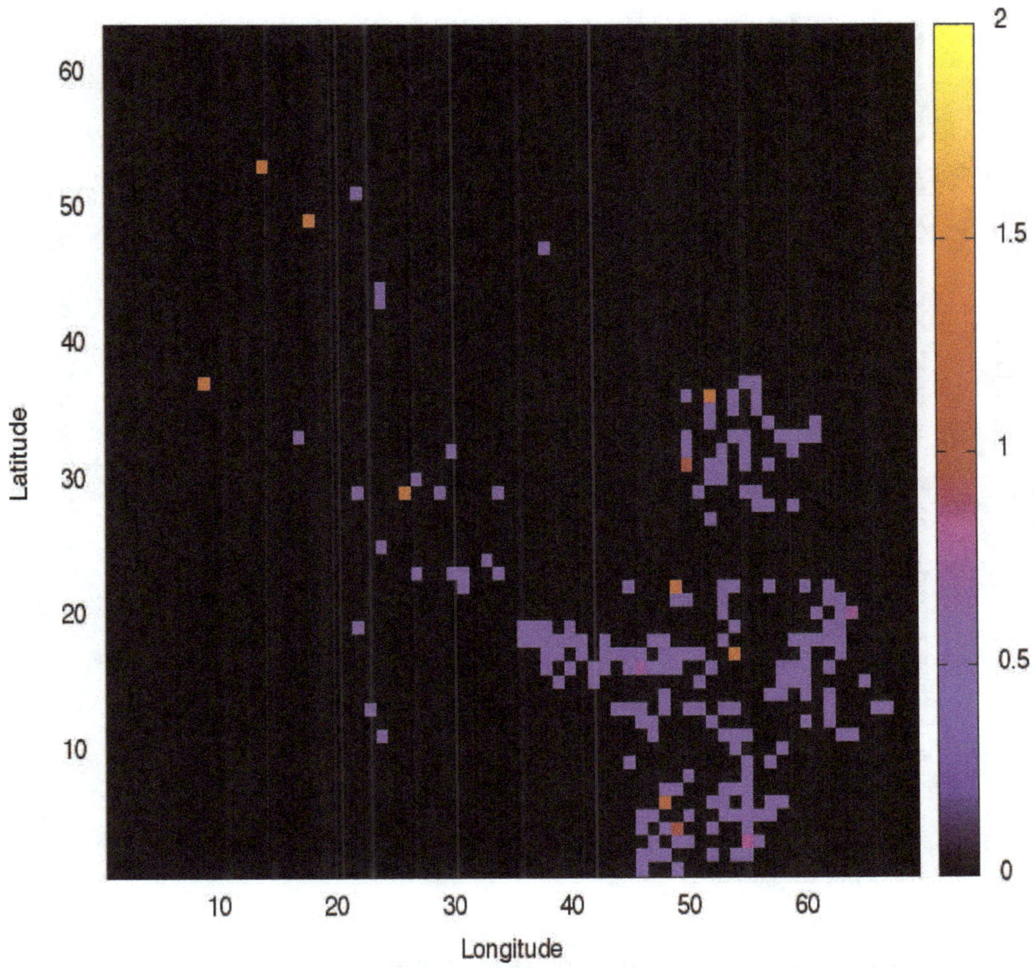

Figure 4. Service time history at Tuesday/19 h.

Figure 4 presents one solution, with the service time required for emergency-order attendance in the corresponding geographical area and restrict to a given hour of the day, in this case, it is a Tuesday evening, 19 h. From **Table 2**, one notes that the side area size of each quadrant is 0.4 km, with a coefficient of variation equal to 0.33325. It is worth noting that the dispersion depicted in **Figure 4** is for only those quadrants that respect the coefficient of variation constraint.

From the results of service times of **Figures 4–8**, one may note that the more precise the stratification, and the consequent reduction in the size of the side of the area (0.4 km), the greater the dispersion of the areas with the greatest demand. Such behaviour makes it more precise to define the area with the highest demand.

Figures 9–12 depict the number of selected areas to forecast service time, considering a Tuesday and varying the area side size from 0.4 to 1 km. An important relationship between area side size and the number of selected areas: as area side size increases, the number of areas decreases and also the accuracy to define the dummy nodes to be further used in a routing solution approach.

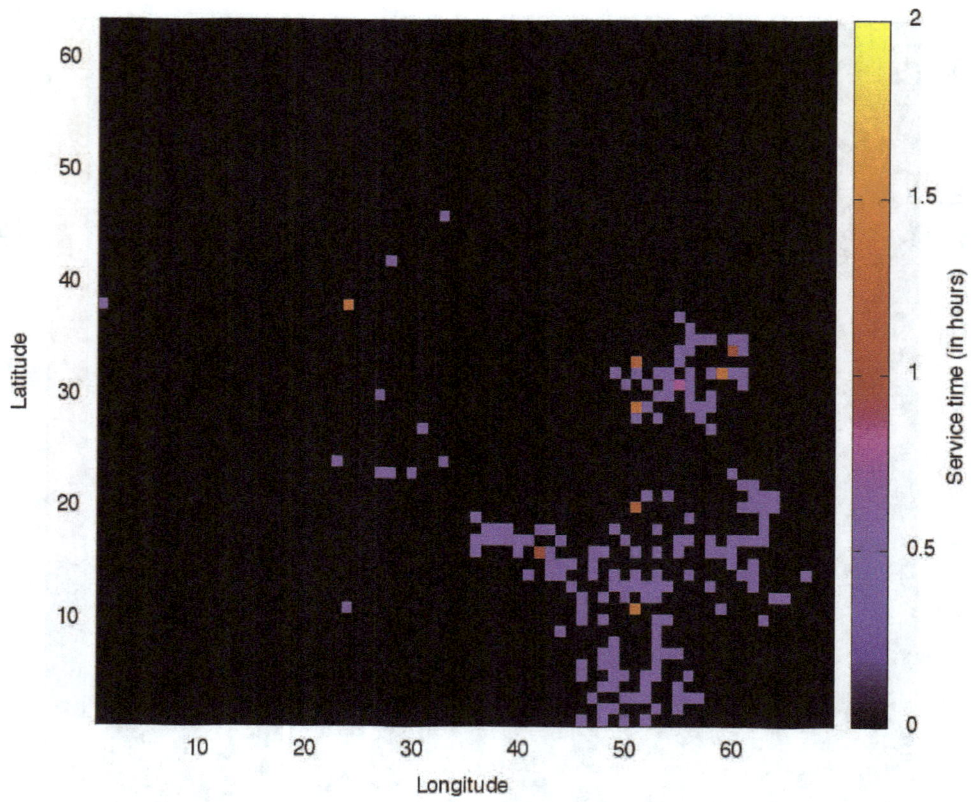

Figure 5. Service time history at Tuesday/20 h.

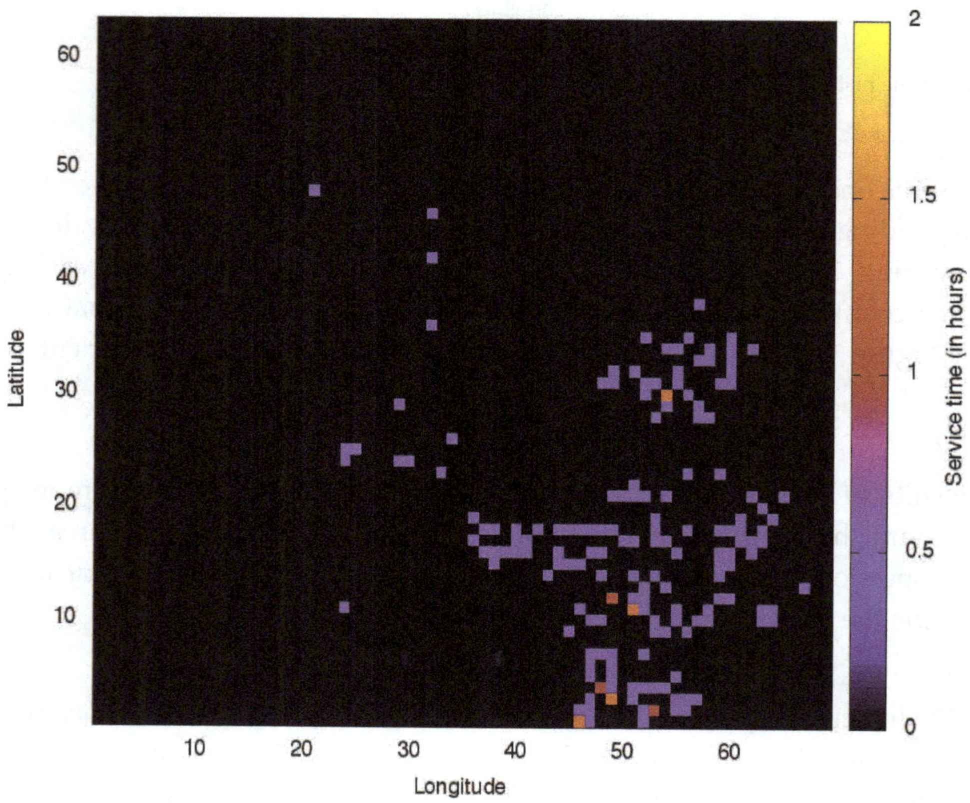

Figure 6. Service time history at Tuesday/21 h.

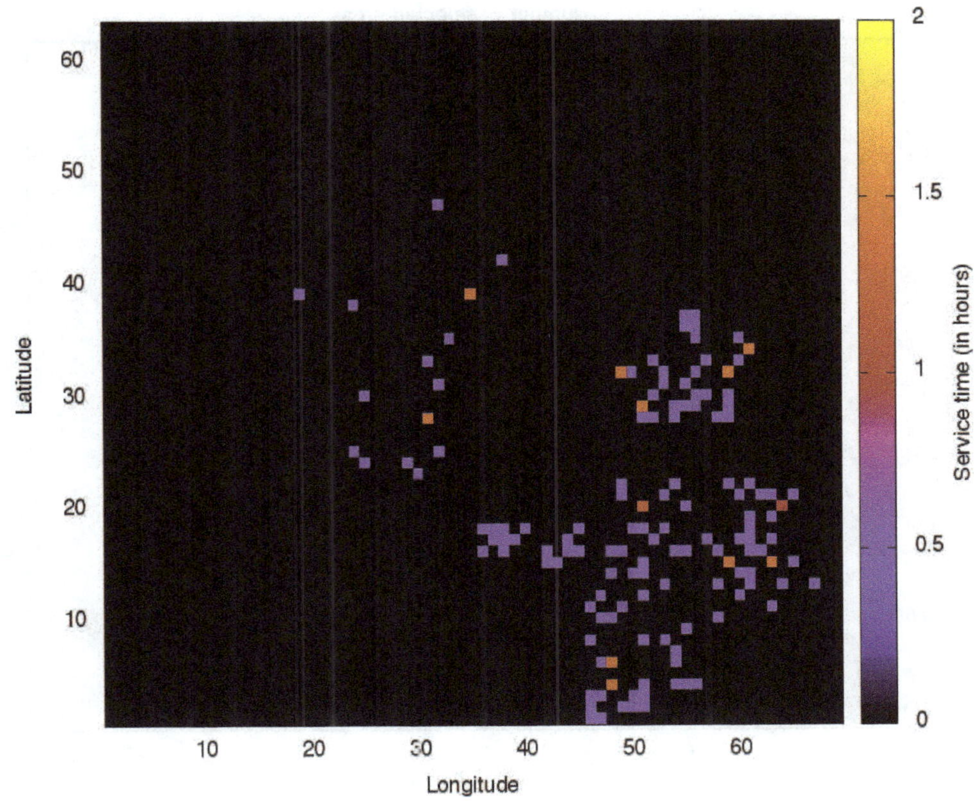

Figure 7. Service time history at Tuesday/22 h.

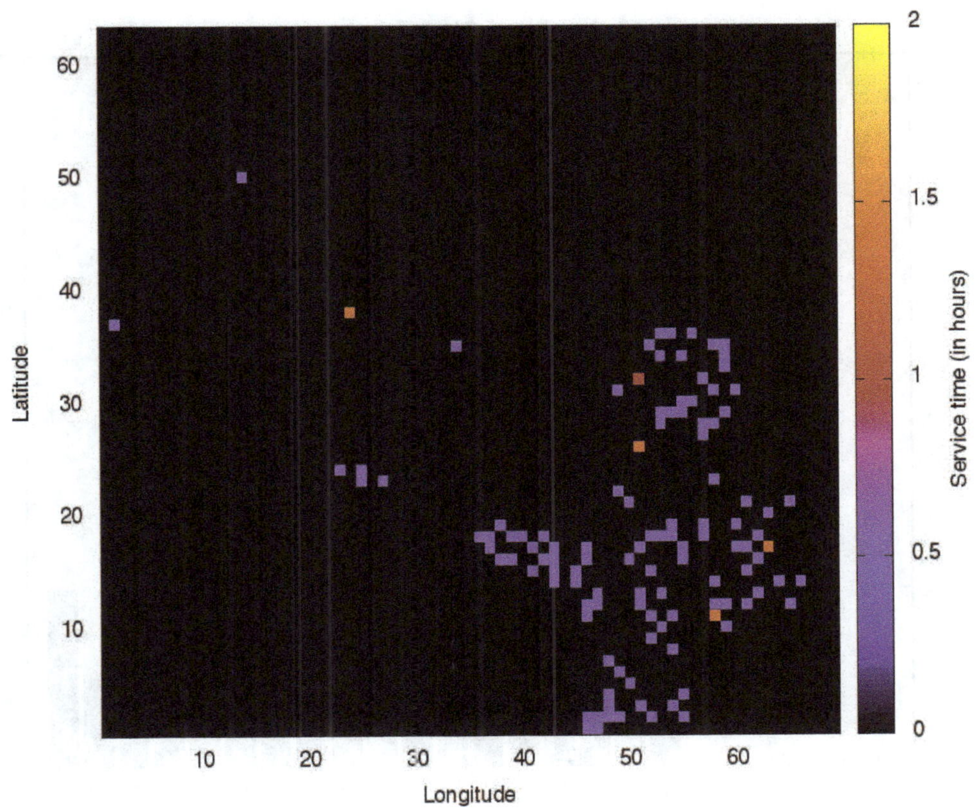

Figure 8. Service time history at Tuesday/23 h.

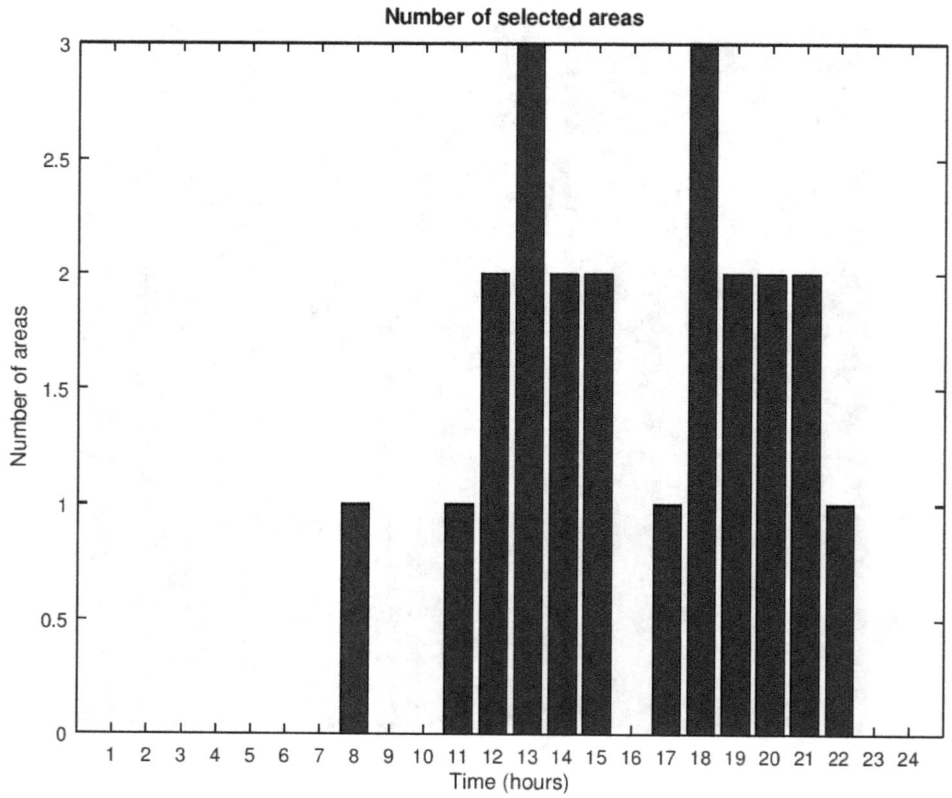

Figure 9. Number of selected areas at Tuesday for area side size of 0.4 km.

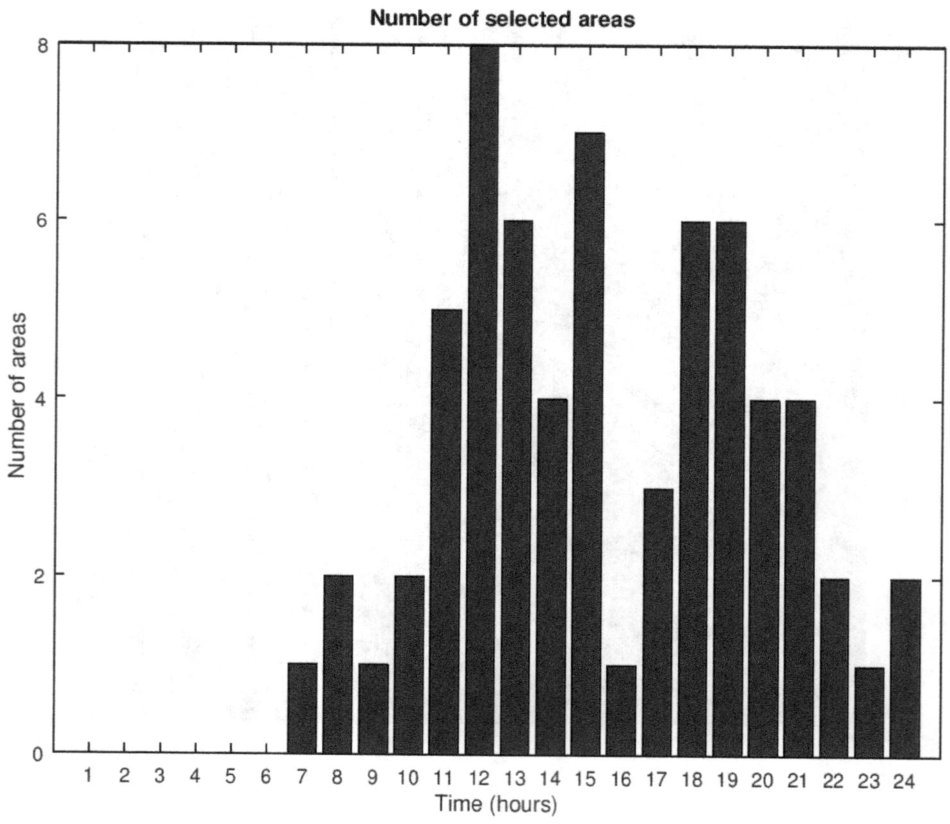

Figure 10. Number of selected areas at Tuesday for area side size of 0.6 km.

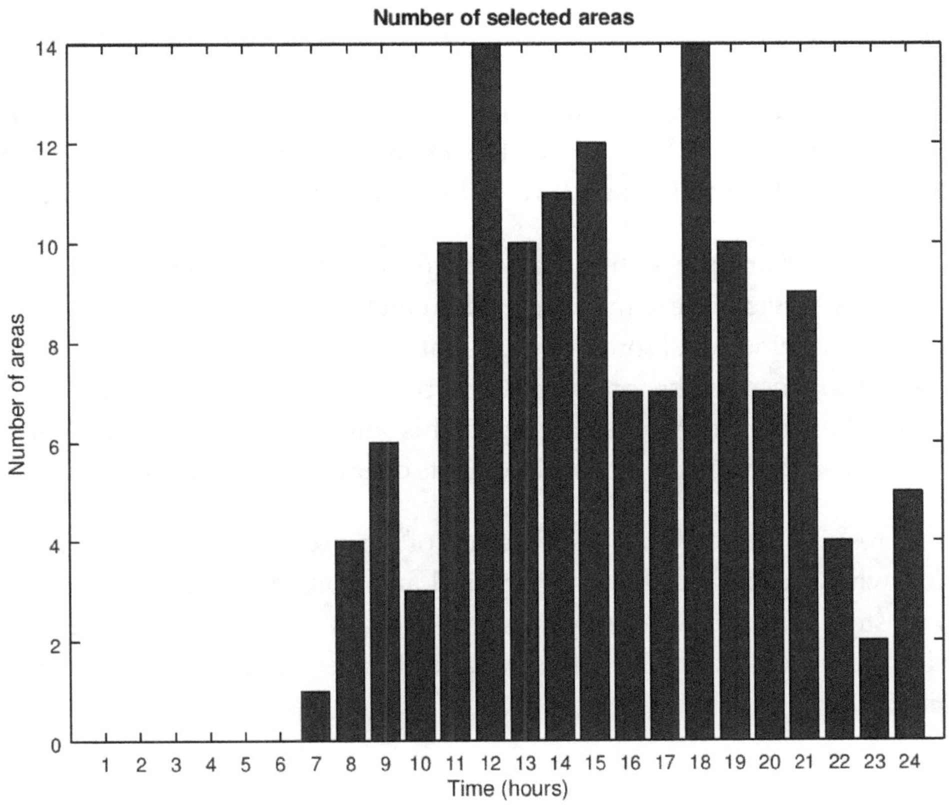

Figure 11. Number of selected areas at Tuesday for area side size of 0.8 km.

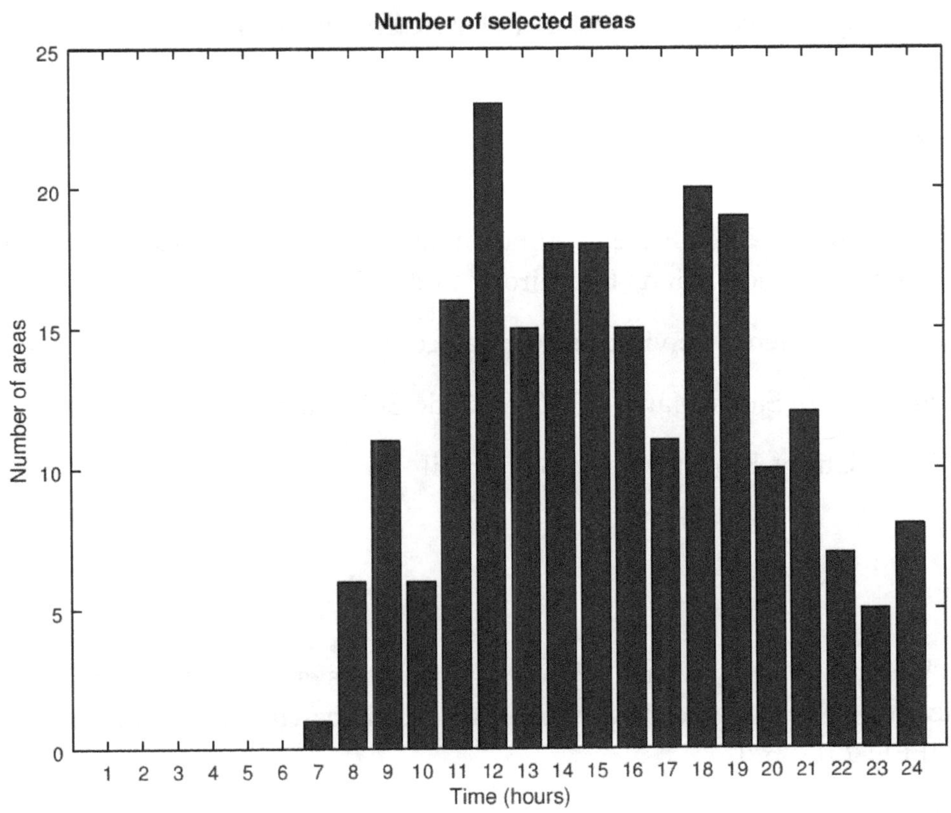

Figure 12. Number of selected areas at Tuesday for area side size of 1 km.

6. Conclusion

An inherent issue of the vehicle-routing problem is due to uncertainty of real-time planning attached to the emergency orders, where the latency of an order is minimized generally conflicting with the goal of minimizing travel time, since the pre-established route is changed.

This work has presented an approach to support proactive real-time vehicle routing, by using an algorithm to estimate service time demand related to emergency orders and considering geographical attributes and time windows. From the results obtained by this approach, one may successfully use dummy nodes obtained by the forecast method proposed to be included in a further real-time routing solution. The main purpose is to support this solution in order to minimize the waiting time and the latency, by decreasing the displacements over the route directions over the day.

Computational results indicate that the forecast of service demand can be performed with great precision when optimizing the geographical area considered, showing that the integration of derived stochastic knowledge results in a more accurate dimensioning of the service teams, and significant reductions in the discrepancy between the planned and executed routes may be obtained.

Acknowledgements

The authors of this chapter would like to thank the Power Distribution Company RGE Sul by the technical and financial support on the Research and Development project called 'Dynamic operations planning', linked to the National Electric Energy Agency(ANEEL).

Author details

Magdiel Schmitz[1], Maria Clara Ferreira Almeida da Silva[1], Vinícius Jacques Garcia[1]*, Daniel Bernardon[1], Lynceo Favigna Braghirolli[1] and Júlio Fonini[2]

*Address all correspondence to: viniciusjg@gmail.com

1 Federal University of Santa Maria, Rio Grande do Sul, Brazil

2 RGE Sul Power Utility, Rio Grande do Sul, Brazil

References

[1] Arif A, Wang Z, Wang J, Chen C. Power distribution system outage management with co-optimization of repairs, reconfiguration, and DG dispatch. In: Transaction on Smart Grid, editor. IEEE; 2017. p. 1

[2] Murphy L, Wu F. Comprehensive analysis of distribution automation systems. Technical Report M90/72. Berkeley: University of California; 1990

[3] Psaraftis HN. In: Golden BL and Assad AA. (Eds). Vehicle routing: Methods and studies. Dynamic Vehicle Routing Problems. North Holland, Amsterdam, The Netherlands: Dalctraf. 1988. pp. 223–248

[4] Garcia VJ, Bernardon DP, Abaide AR, Bassi OA, Dhein G. Reliability assessment by coordinating maintenance vehicles in electric power distribution systems. Procedia-Social and Behavioral Sciences. 2014;**111**:1045–1053

[5] Psaraftis HN. Dynamic vehicle routing: Status and prospects. Annals of Operations Research. 1995;**61**:143–164

[6] Angel-Bello F, Alvarez A, García I. Two improved formulations for the minimum latency problem. Applied Mathematical Modelling. 2013;**37**(4):2257–2266

[7] Ferrucci F, Bock S, Gendreau M. A pro-active real-time control approach for dynamic vehicle routing problems dealing with the delivery of urgent goods. European Journal of Operational Research. 2013;**225**(1):130–141

[8] Gounaris C, et al. An adaptive memory programming framework for the robust capacitated vehicle routing problem. Transportation Science. 2014

[9] Toth P, Vigo D. Vehicle Routing Problems, Methods, and Applications. DEI, University of Bologna. Bologna Italy. 2nd ed. SIAM; 2014

[10] Taha HA. Operations Research: An Introduction. Pearson/Prentice Hall; New Jersey. 2007. p. 557

[11] Makridakis S. METAFORECASTING: Ways of improving forecasting accuracy and usefulness. International Journal of Forecasting. 1988;**4**:467–491

[12] Hillier FS, Lieberman GJ. Introduction to Operations Research: Cases Developed. New York, NY: McGrawHill; 2001. p. 6

[13] Guimarães I, Garcia VJ, Bernardon DP, Foninni J. Emergency prediction in electric utilities: A case study from South Brazil. In: European Conference on Modelling and Simulation; Regensburg, Germany; 2016. p. 30

[14] Garcia VJ, et al. Emergency work orders in electric distribution utilities: From business process definition to quantitative improvements. In: 47th International Universities' Power Engineering Conference; 2012

[15] Ichoua S, Gendreau M, Potvin J. Exploiting knowledge about future demands for real-time vehicle dispatching. Transportation Science. 2006;**40**(2):211–225

[16] Braaten S, et al. Heuristics for the robust vehicle routing problem with time windows. Expert Systems with Applications. 2017;**77**:136–147

[17] Zayed AI, Butzer PL. Chapter 3—Lagrange interpolation and sampling theorems. In: Marvasti FA, editor. Nonuniform Sampling: Theory and Practice. Springer Science + Business Media, LCC; Kluwer Academic/Plenum Publishers, New York. 2001. pp. 123–127

12

Optimum Maintenance Policy for Equipment over Changing of the Operation Environment

Ibrahima dit Bouran Sidibe and Imene Djelloul

Abstract

This chapter investigates optimization of maintenance policy of a repairable equipment whose lifetime distribution depends on the operating environment severity. The considered equipment is undergone to a maintenance policy which consists of repairing minimally at failure and maintaining after operating periods. The periodic maintenance is preventive maintenance (PM) and allows reducing consequently the equipment age but with higher cost than minimal repair. In addition, the equipment has to operate at least in two operating environments with different severity. Therefore, in this analysis, the equipment lifetime distribution function depends on the operating severity. Under these hypotheses, a mathematical modeling of the maintenance cost per unit of time is proposed and discussed. This cost is mathematically analyzed in order to derive optimal periods between preventive maintenance (PM) and the optimal condition under which these exist.

Keywords: minimal repair, preventive repair, repairable equipment, several operating environments

1. Introduction

To reduce the failure risk of production equipments, preventive maintenance or replacement activities should be performed in appropriate schedules. The search of these appropriate schedules has led to the development and implementation of maintenance optimization policies for stochastic degrading production equipments. Indeed, the literature on this matter is already extensive, growing rapidly and also very heterogeneous. Accordingly, this chapter focuses only to some relevant and fundamental works on the maintenance theory. Early in [1, 2], several models appeared on the optimization of replacement or maintenance policies on infinite time horizon. In these works, the authors mainly discussed about the optimality conditions of theses maintenance models. Subsequently to these works, many extensions of the previous models were proposed on finite time span [3, 4] and also on infinite time horizon in the literature. For survey, the reader may refer, for example, to [5–8] and the references therein. We note that in most cited works, the authors assumed that the equipment lifetime distribution is parametrically characterized and well known. However, Coolen and his coauthors [9, 10] showed that this assumption impacts clearly the optimal replacement age and its cost per unit of time when the equipments undergo an age replacement policy (ARP). Recently in [11], de Jonge et al. pointed out also the weakness of the assumption on knowing of the equipment lifetime distribution and proposed a parametric modeling of ARP for new equipment with an uncertainty on the parameters of the equipment lifetime distribution. In this work, de Jonge and his coauthors used Bayesian approach to model the uncertainty on the parameters and figured out that this uncertainty has effects on the optimal policy (age and cost) under ARP.

Another way, most existing models merely rely only on a classical assumption which states that the operating environment is steady and has not any effect on the equipment characteristic and its lifetime distribution. Roughly speaking, they assume that the degradation process is the same during the equipment's life cycle. This is a restrictive assumption in many industrial areas where production equipment may have experiences under different operating environments with their own severity degree that impacts the equipment performance. For example, the degradation process of the mining machinery is impacted by the severity level of the environment where the machinery is being exploited. Another example may be the engines used for oil extraction. The degradation process of such equipment depends on whether they are operated onshore or offshore. In some other industries, production equipments are first operated in a given environment and then moved to another location where this latter might be more or less severe than the first. In the same way, many companies operate their equipments at home for several years before shipping them to their subsidiaries in other countries where they would be subjected to more severe operating conditions. Therefore, suitable maintenance strategies, integrating the heterogeneous operating conditions, should be developed to assess the degradation of such equipments.

In this chapter, a preventive maintenance is investigated for such equipment subject to random failures. The equipments are assumed to have an experience under two operating environments. In fact, each operating environment is characterized by its own degree of severity, which impacts the equipment lifetime distribution. Therefore, the equipment lifetime distributions follow then a different distribution depending on the operating environments. To reduce the failure occurrence risk during operating under both operating environments, the equipment undergoes to an periodic preventive maintenance (PM). However, the equipment is subjected to minimal repair at failure. The objective consists then on evaluating the optimal age to perform periodic preventive repair in order to minimize the expected maintenance cost per unit of time. This expected cost is induced by the costs of minimal and preventive repairs. This policy was already discussed by Nakagawa in [12], in which Nakawaga considered that the equipment lifetime remains the same during the operation. Nakagawa analyzed mathematically the periodic and sequential maintenance policies. Therefore, our chapter can be considered as an extension of Nakawaga work.

The remainder parts of the chapter are organized as follows. The analyzed problem is briefly introduced in Section 2. This section proposes a mathematical formulation of the total maintenance cost. Section 3 focuses on the maintenance cost analysis in order to derive the optimal conditions which ensure minimal total cost per unit of time. In this same section, an heuristic is proposed to find the optimal number and period between preventive actions on both environments. Numerical experiments are conducted to illustrate the proposed approach on the one hand, and on the other hand, the accuracy and robustness of model are demonstrated through the simulation in Section 4. At the end, a conclusion and future works are drawn in the last section.

2. Mathematical formulation of the maintenance cost

In this section, modeling of the maintenance policy is going to be proposed. This modeling takes into account different hypotheses of our analysis. In fact, our equipment has to be used under two operating environments with different severities denoted by $j = 1$ and $j = 2$ which stand, respectively, for the first and second environments. Therefore, the equipment spends T1 and T2 respectively in operating environment 1 and 2. Therefore, the operation duration is the combination of both durations $T_1 + T_2$. The equipment operates successively on both environments in order to perform its missions. During this operation, the equipment undergoes by two types of maintenance actions. Roughly speaking, the equipment is going to be repaired minimally at failure and preventively after some x_j operating periods. The minimal repair costs c_{mj} and allows that the equipment reaches the same reliability just before its failure. However, the preventive repair costs C_{pj} such as $C_p >> c_{mj}$. Therefore, this preventive repair impacts the equipment according to its age and its hazard function. First, the preventive action reduces the equipment age to zero. Second, the preventive action modifies the hazard function such as the hazard function after repair becomes higher than its hazard before. That involves that the wear-out process of the equipment degrades more after the preventive action than before **Figure 1**.

2.1. Preventive maintenance cost

During operation, the equipment undergoes by preventive action after each x_1 and x_2 unit of time, respectively, on the first and second environments. Each of these preventives actions costs C_{p1} on the first and C_{p2} on the second environment. In addition, the number of preventive actions is n_1 and n_2, respectively, on the first and second environments. Therefore, the total preventive repair costs

$$C_{TP} = n_1{}^* c_{p1} + n_2{}^* c_{p2}, \tag{1}$$

during the length of operation $(T_1 + T_2) = (n_1{}^* x_1 + n_2{}^* x_2)$.

2.2. Minimal repair cost

The minimal repair is performed regardless of the preventive actions. The minimal repair is performed at failures in order that equipment reaches the same reliability just before failing. Each minimal repair costs c_{m1} and c_{m2}, respectively, on the first and second environments.

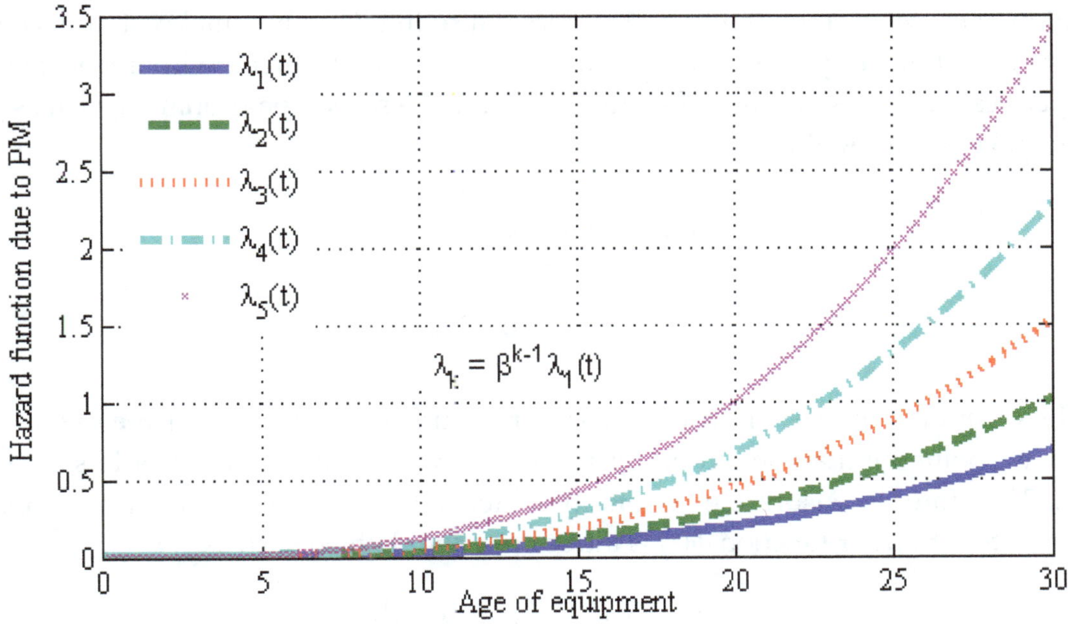

Figure 1. Evolution of hazard function due to preventive maintenance.

Therefore, the cost of minimal repair, on the kth interval with a duration x_j, is product of expected number of failure by the cost of a minimal repair c_{mj}. From Thompson analysis [13], the expected number of renewal on the interval $[0; x_j]$ coincides with the integration of hazard function on $[0; x_j]$. Then, the minimal repair costs during the kth interval are given by

$$C_{mj} = \int_0^{x_j} \lambda_{j,k}(t)dt, \tag{2}$$

$$= -c_{mj} \log R_{j,k}(x_j). \tag{3}$$

where $\lambda_{j,k}(t)$, and $R_{j,k}(t)$ stand for the hazard and the reliability functions of the equipment on the kth and during the jth environment. Therefore, the total minimal cost on the first environment is

$$Ct_{m1} = \sum_{k=1}^{n_1} \int_0^{x_1} \lambda_{1,k}(t)dt, \tag{4}$$

$$= -c_{m1} \sum_{k=1}^{n_1} \log R_{1,k}(x_1). \tag{5}$$

We also deduce the total minimal cost on the second environment as follows

$$Ct_{m2} = \sum_{k=n_1+2}^{n_1+n_2+1} \int_0^{x_2} c_{m2}\lambda_{2,k}(t)dt, \tag{6}$$

$$= -c_{m2} \sum_{k=n_1+2}^{n_1+n_2+1} \log R_{2,k}(x_2). \tag{7}$$

In addition, the operation on the $(n_1+1)^{th}$ period also implies a minimal cost. In fact, on this period, the equipment operates on both environments. On the first environment, the equipment operates on y units of time before moving on to the second environment such as $y < x_1$. The minimal cost during this operation is

$$C_{tmy} = c_{m1}\int_0^y \lambda_{1,n_1+1}(t)dt \tag{8}$$

$$= -c_{m1}\log R_{1,n_1+1}(y). \tag{9}$$

After, the equipment moves on to the second environment to operate between $[y; y+x_2]$. In addition, we point out that the second operating environment can be more or less severe than the first. Therefore, to ensure the continuity of reliability function between both operating environments, a transfer function $\phi(t)$ is introduced and defined such as:

$$\begin{aligned} R_{1,n_1+1}(t) &= R_{2,n_1+1}(\phi(t)), \\ \phi(0) &= 0. \end{aligned} \tag{10}$$

That involves a minimal cost on this period

$$C_\phi = c_{m2}\int_{\phi(y)}^{\phi(y)+x_2} \lambda_{2,n_1+1}(t)dt \tag{11}$$

$$= -c_{m2}\big(\log R_{2,n_1+1}(x_2+\phi(y)) - \log R_{2,n_1+1}(\phi(y))\big). \tag{12}$$

To reduce the complexity during computing, we assume that the duration $y = 0$. That involves a total minimal which clearly depends on the Eqs. (5), (7), (12). The total minimal cost on all operating duration is defined by addition

$$C_{Tm} = Ct_{m1} + Ct_{m2} + C_\phi. \tag{13}$$

Indeed, the hypothesis $y = 0$ also impacts the number of preventive actions. In fact, under this latter hypothesis, the number of preventive actions becomes n_1+n_2-1 instead of n_1+n_2 as we indicated in Eq. (1). The total preventive is going to cost

$$C_{TPh} = \begin{cases} n_1C_{p1} + (n_2-1)C_{p2}, \\ (n_1-1)C_{p1} + n_2C_{p2}. \end{cases} \tag{14}$$

Eq. (14) is equivalent to

$$C_{TPh,\gamma} = (n_1 - 1 + \gamma)C_{p1} + (n_2-\gamma)C_{p2}, \tag{15}$$

where $\gamma = 1$ stands for the fact that at the end of n_1^{th} period the equipment is repaired before moving to the second environment, while $\gamma = 0$ corresponds to the reverse.

2.3. Total maintenance cost

From previous Eqs. (13) and (14), we deduce a mathematical formulation of the total maintenance cost according to the set of parameters (n_1, n_2, x_1, x_2) as follows:

$$C(n_1, n_2, x_1, x_2) = \frac{C_{Tm} + C_{TPh,\gamma}}{nx_1 + n_2 x_2}. \tag{16}$$

Based on the equation, the next section is going to analyze the optimality according to the different parameters such as the number and the duration between the preventive repairs.

3. Optimality analysis

Herein, the maintenance cost is rewritten in order to integrated the impacts of preventive maintenance (PM) on the equipment lifetime distribution. We assume that a preventive action allows to reduce the age of equipment to zero and increase the hazard function. **Figures 1** and **2** point out the impact of PM on the equipment hazard and reliability functions. The hazard function is defined after PM as follows

$$\lambda_{j,k}(t) = \beta_j \lambda_{j,k-1}(t), \tag{17}$$

where $j = 1, 2,$ and $\beta_j > 1$. Under these hypotheses, Eq. (5), which represents the total minimal cost on the first environment, is rewritten as

$$C_{tm1} = -c_{m1} \frac{1 - \beta_1^{n_1}}{1 - \beta_1} \log R_1(x_1), \tag{18}$$

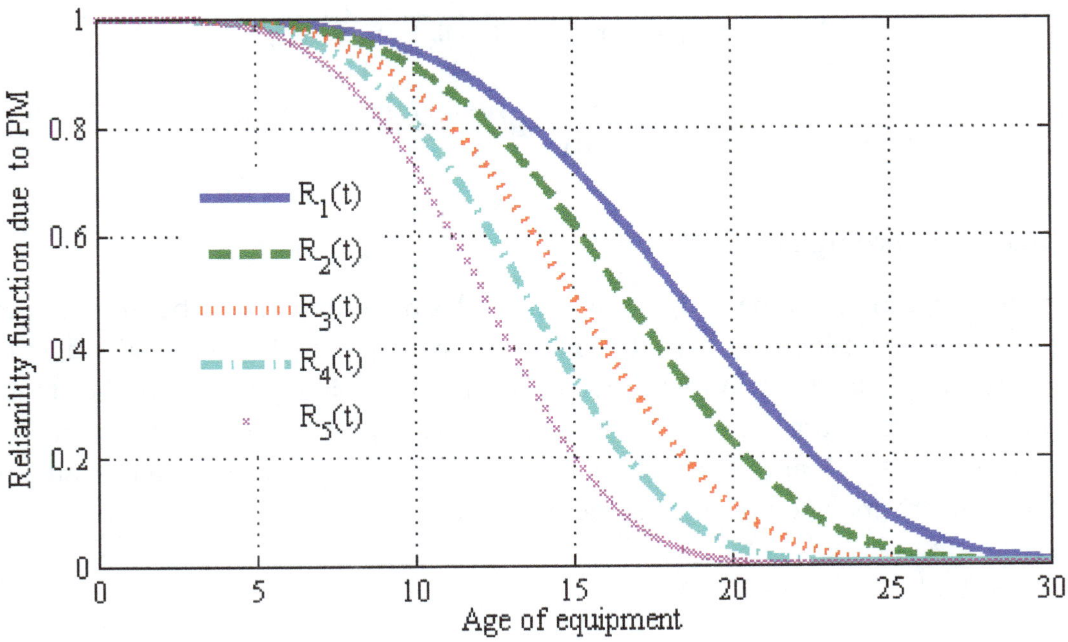

Figure 2. Evolution of reliability function due to preventive maintenance.

with

$$R_1(x_1) = R_{1,1}(x_1). \tag{19}$$

In the second operating environment, the hazard function at $(n+1)^{th}$ is a consequence of $n_1 - 1 + \gamma$ PM in first and $1 - \gamma$ in the second environment.

$$\begin{aligned} \lambda_{2,\,n_1+1}(t) &= \beta_1^{n_1-1+\gamma}\beta_2^{1-\gamma}\lambda_2(t) \\ \lambda_{2,\,n_1+2}(t) &= \beta_1^{n_1-1+\gamma}\beta_2^{2-\gamma}\lambda_2(t) \\ \ldots &= \ldots \\ \lambda_{2,\,n_1+n_2}(t) &= \beta_1^{n_1-1+\gamma}\beta_2^{n_2-\gamma}\lambda_2(t) \end{aligned} \tag{20}$$

with

$$\lambda_2(t) = \lambda_{2,1}(t). \tag{21}$$

The total cost due to the minimal repair in the second environment becomes

$$C_{mt2} = -c_{m2}\beta_1^{n_1-1+\gamma}\beta_2^{1-\gamma}\frac{1-\beta_2^{n_2}}{1-\beta_2}\log R_2(x_2). \tag{22}$$

By considering Eqs. (15), (18), and (19), the total cost per unit of time is rewritten as follows

$$\begin{aligned} C(n_1, n_2, x_1, x_2) &= \frac{1}{n_1x_1 + n_2x_2}\left((n_1 - 1 + \gamma)C_{p1} + (n_2 - \gamma)C_{p2}\right) \\ &- \frac{1}{n_1x_1 + n_2x_2}\left(c_{m1}\frac{1-\beta_1^{n_1}}{1-\beta_1}\log R_1(x_1)\right) \\ &- \frac{1}{n_1x_1 + n_2x_2}\left(c_{m2}\beta_1^{n_1-1+\gamma}\beta_2^{1-\gamma}\frac{1-\beta_2^{n_2}}{1-\beta_2}\log R_2(x_2)\right). \end{aligned} \tag{23}$$

3.1. Optimality according to n_1 and n_2

Let us assume that there is a pair (n_1, n_2) that provides the minimal cost per unit according to the Eq. (20) for given periods (x_1, x_2) between preventive repairs. Then, corresponding cost has to remain the unique lowest bound relative to other pairs of integer. This implies that cost at (n_1, n_2) must be better than the costs from the successive pairs $\{(n_1+1, n_2), (n_1-1, n_2)\}$; $\{(n_1, n_2+1), (n_1, n_2-1)\}$ and $\{(n_1+1, n_2+1), (n_1-1, n_2-1)\}$. The existence and uniqueness of the pairs are analyzed through some propositions.

3.1.1. Local optimality

The local optimality concerns the direct neighbors of the optimal pair such as $\{(n_1+1, n_2), (n_1-1, n_2)\}$ and $\{(n_1, n_2-1), (n_1, n_2+1)\}$. Let pose that

$$L_1(n_1|n_2) = C(n_1, n_2, x_1, x_2) - C(n_1 + 1, n_2, x_1, x_2), \tag{24}$$

$$L_2(n_2|n_1) = C(n_1, n_2, x_1, x_2) - C(n_1, n_2 + 1, x_1, x_2), \tag{25}$$

Proposition 1 *If the lifetime distribution functions are increasing failure rate (IFR) and* $L_1(1|n_2) > 0$, *then there exists a unique optimal number of PM* n_1 *in the first environments in which this* n_1 *ensures the minimal cost per unit time for a fixed pair* (x_1, x_2) *and* n_2.

Proof. As the maintenance cost per unit of time is minimal for (n_1, n_2), then we have

$$\begin{cases} C(n_1, n_2, x_1, x_2) \leq C(n_1 - 1, n_2, x_1, x_2), \\ C(n_1, n_2, x_1, x_2) < C(n_1 + 1, n_2, x_1, x_2). \end{cases} \tag{26}$$

This system is equivalent to

$$\begin{cases} L_1(n_1 - 1|n_2) \geq 0, \\ L_1(n_1|n_2) < 0. \end{cases} \tag{27}$$

with

$$\begin{aligned} L_1(n_1 - 1|n_2) &= -n_2 x_2 C_{p1} + x_1((\gamma - 1)C_{p1} + (n_2 - \gamma)C_{p2}) \\ &-x_1\left(c_{m_1}\frac{1-\beta^{n_1}}{1-\beta_1}\log R_1(x_1) + c_{m2}\beta_1^{n_1-1+\gamma}\beta_2^{1-\gamma}\frac{1-\beta_2^{n_2}}{1-\beta_2}\log R_2(x_2)\right) \\ &+(n_1x_1 + n_2x_2)\left(c_{m1}\beta_1^{n_1-1}\log R_1(x_1) + c_{m2}\beta_1^{n_1-2+\gamma}\beta_2^{1-\gamma}\frac{\beta_1 - 1}{1-\beta_2}(1-\beta_2^{n_2})\log R_2(x_2)\right). \end{aligned}$$

In fact

$$\lim_{n_1 \leftarrow +\infty} L_1(n_1|n_2) = -\infty,$$

and

$$\begin{aligned} L_1(n_1|n_2) - L_1(n_1 - 1|n_2) &= (n_1x_1 + n_2x_2)(c_{m1}\beta_1^{n_1}\log R_1(x_1)) \\ &+(n_1x_1 + n_2x_2)\left(c_{m2}\beta_1^{n_1-1+\gamma}\beta_2^{1-\gamma}\frac{\beta_1 - 1}{1-\beta_2}(1-\beta_2^{n_2})\log R_2(x_2)\right), \end{aligned}$$

The right-hand side of the previous equation shows that $L_1(n_1|n_2) - L_1(n_1 - 1|n_2) < 0$. This implies that $L_1(n_1|n_2)$ decreases with n_1. If $L_1(1|n_2) > 0$, then there exists a unique n_1 which verifies condition (23) and ensures the minimal cost per unit time for given n_2.

Proposition 2 *If the lifetime distribution function of equipment on both environments is IFR and* $L_2(1|n_1) > 0$, *then there exists a unique optimal number of PM* n_2 *in the second environment in which this number ensures the minimal cost per unit time for corresponding fixed pair* (x_1, x_2) *and* n_2.

Proof. As the cost maintenance per unit time is minimal for (n_1, n_2), then we have

$$\begin{cases} C(n_1, n_2, x_1, x_2) \le C(n_1, n_2 - 1, x_1, x_2), \\ C(n_1, n_2, x_1, x_2) < C(n_1, n_2 + 1, x_1, x_2). \end{cases} \tag{28}$$

This is equivalent to

$$\begin{cases} L_2(n_2 - 1|n_1) \ge 0, \\ L_2(n_2|n_1) < 0. \end{cases} \tag{29}$$

with

$$\begin{aligned} L_2(n_2 - 1|n_1) &= -n_1 x_1 C_{p2} + x_2\big((n_1 - 1 + \gamma)C_{p1} - \gamma C_{p2}\big) \\ &-x_2\left(c_{m1}\frac{1-\beta_1^{n_1}}{1-\beta_1}\log R_1(x_1) + c_{m2}\beta_1^{n_1-1+\gamma}\beta_2^{1-\gamma}\frac{1-\beta_2^{n_2}}{1-\beta_2}\log R_2(x_2)\right) \\ &+(n_1x_1 + n_2x_2)\left(c_{m2}\beta_1^{n_1-1+\gamma}\beta_2^{n_2-\gamma}\log R_2(x_2)\right). \end{aligned}$$

This equation implies

$$\lim_{n_2\to\infty} L_2(n_2|n_1) = -\infty,$$

and

$$L_2(n_2|n_1) - L_2(n_2 - 1|n_1) = (n_1x_1 + n_2x_2)\left(c_{m2}\beta_1^{n_1-1+\gamma}\beta_2^{n_2-\gamma+1}\log R_2(x_2)\right) < 0.$$

Therefore, $L_2(n_2|n_1)$ decreases with and for $L_2(1|n_1) > 0$, we have a unique n_2 in which the total per unit of time is minimal for fixed n_1.

3.1.2. Global optimality

The global optimality compares the optimal pair to $\{(n_1 + 1, n_2 + 1), (n_1 - 1, n_2 - 1)\}$. Let us pose that

$$L_3(n_1, n_2) = C(n_1, n_2, x_1, x_2) - C(n_1 + 1, n_2 + 1, x_1, x_2). \tag{30}$$

Proposition 3 *If the lifetime distribution functions are IFR and $L_3(1, 1) > 0$, then there exists a unique optimal number of PM (n_1, n_2) in which this ensures the minimal cost per unit time for a fixed pair (x_1, x_2).*

Proof. As the cost is minimal for (n_1, n_2), then

$$\begin{cases} C(n_1, n_2, x_1, x_2) \le C(n_1 - 1, n_2 - 1, x_1, x_2), \\ C(n_1, n_2, x_1, x_2) < C(n_1 + 1, n_2 + 1, x_1, x_2). \end{cases} \tag{31}$$

This is equivalent to

$$\begin{cases} L_3(n_1-1,n_2-1)\geq 0, \\ L_3(n_1,n_2)<0. \end{cases} \tag{32}$$

with

$$L_3(n_1-1,n_2-1)=(x_1+x_2)\left((\gamma-1)C_{p1}-\gamma C_{p2}\right)$$
$$-(x_1+x_2)\left(c_{m1}\frac{1-\beta_1^{n_1}}{1-\beta_1}\log R_1(x_1)+c_{m2}\beta_1^{n_1-1}\beta_2^{1-\gamma}\frac{1-\beta_2^{n_2}}{1-\beta_2}\right)$$
$$+(n_1x_1+n_2x_2)\left(c_{m1}\beta^{n_1-1}\log R_1(x_1)-c_{m2}\frac{1-\beta_1-\beta_2^{n_2-1}(1-\beta_2)}{1-\beta_2}\beta_1^{n_1-2}\beta_2^{1-\gamma}\right).$$

With

$$L_3(+\infty,+\infty)=-\infty,$$

and

$$L_3(n_1,n_2)-L_3(n_1-1,n_2-1)<0.$$

Therefore, $L_3(n_1,n_2)$ decreases with (n_1,n_2) and for $L_3(1,1)>0$, we have a unique pair (n_1,n_2) in which the total per unit of time is minimal.

3.2. Optimality according to x_1 and x_2

For given number of preventive actions (n_1,n_2), the optimal durations (x_1,x_2) between preventive actions in both environments have to verify

$$\begin{cases} \frac{\partial}{\partial x_1}C(n_1,n_2,x_1,x_2)=0 \\ \frac{\partial}{\partial x_2}C(n_1,n_2,x_1,x_2)=0; \end{cases} \tag{33}$$

This implies

$$\begin{cases} c_{m1}\frac{1-\beta_1^{n_1}}{1-\beta_1}\lambda_1(x_1) = n_1C(n_1,n_2,x_1,x_2), \\ c_{m2}\beta_1^{n_1-1+\gamma}\beta_2^{1-\gamma}\frac{1-\beta_2^{n_2}}{1-\beta_2}\lambda_2(x_2) = n_2C(n_1,n_2,x_1,x_2). \end{cases} \tag{34}$$

By dividing, we obtain

$$\frac{\lambda_1(x_1)}{\lambda_2(x_2)}=\left(\beta_1^{n_1-1+\gamma}\beta_2^{1-\gamma}\right)\frac{n_1}{n_2}\frac{c_{m2}}{c_{m1}}\frac{1-\beta_2^{n_2}}{1-\beta_1^{n_1}}\frac{1-\beta_1}{1-\beta_2} \tag{35}$$

Proposition 4 *If the lifetime functions of the equipment are Weibull-distributed in both environments with the same shape parameter b, then the optimal interval between PM is defined as*

$$\frac{x_1}{x_2} = Cste^* \left(\frac{n_1}{n_2}\right)^{1/(b-1)}. \tag{36}$$

Proof. As lifetime functions are Weibull-distributed with the same parameter b, then the hazard functions are defined as follows

$$\lambda_1(x_1) = \frac{b}{\eta_1}\left(\frac{x_1}{\eta_1}\right)^{b-1}, \tag{37}$$

$$\lambda_2(x_2) = \frac{b}{\eta_2}\left(\frac{x_2}{\eta_2}\right)^{b-1}, \tag{38}$$

and from Eq. (32), we deduce

$$\frac{x_1}{x_2} = \left(\frac{c_{m2}}{c_{m1}}\left(\beta_1^{n_1-1+\gamma}\beta_2^{1-\gamma}\right)\frac{1-\beta_2^{n_2}}{1-\beta_1^{n_1}}\frac{1-\beta_1}{1-\beta_2}\right)^{1/(b-1)}\frac{\eta_1}{\eta_2}\left(\frac{n_1}{n_2}\right)^{1/(b-1)}. \tag{39}$$

The uniqueness is tough to establish due to the number of parameters and the complexity of the proposed cost model here. To make the research of optimal solution easy, we propose a handy heuristic based on the optimal derived conditions in this chapter. The next section describes step by step the proposed heuristic which leads to a suitable solution for our optimization problem.

3.3. Numerical resolution of problem

Herein, an algorithm is drawn in order to find the optimal pairs for (n_1, n_2) and (x_1, x_2). The optimal pairs ensure the minimal cost per unit time defined by Eq. (20). Moreover, the existence of these optimal pairs is discussed in the previous sections. The proposed heuristic makes switching between the research of pairs ((n_1, n_2) and (x_1, x_2)). This algorithm converges surely toward the pair that ensures the minimal cost according to the conditions deduce from the Eq. (20). The next section presents an application of our approach. The algorithm is on the previous propositions and defined as follows.

Algorithm 1 Compute the optimal pairs of number (n_1, n_2) and periods (x_1, x_2) of PM.

Initialize the pair $(n_1, n_2)_0 = (1, 1)$.

Put $(n_1, n_2) = (n_1, n_2)_0$

STEP (A) Research optimal (x_1, x_2) for given (n_1, n_2).

Compute $L_1(1|n_2)$, $L_2(1|n_1)$ and $L_3(1, 1)$.

if $L_1(1|n_2) > 0$ **then**

Research $n_1(a)$ which verifies condition (24) is verified

$n_1(1) = n_1(a)$ and $n_2(1) = n_2$.

$C_1 = C(n_1(1), n_2(1))$

else

$\{L_1(1|n_2) < 0\}$

$C_1 = \infty$

if $L_2(1|n_1) > 0$ **then**

Research $n_2(b)$ which verifies conditions (26).

$n_1(2) = n_1; n_2(2) = n_2(b)$

$C_2 = C(n_1(2), n_2(2))$

else $\{L_2(1|n_1) < 0\}$

$C_2 = \infty$

if $L_3(1,1) > 0$ **then**

Research $n_2(c)$ which verifies $L_3(n_1(c), n_2(c))$ (29).

$n_1(2) = n_1(c); n_2(2) = n_2(c)$

$C_3 = C(n_1(2), n_2(2))$

else

$\{L_3(1,1) < 0\}$

$C_3 = \infty$

$C_{min} = \text{Min}\{C_1, C_2, C_3\}$

$m = m + 1$

$(n_1, n_2)_m = \{(n_1(i), n_2(i)|C_i = C_{min}\}$

if $(n_1, n_2)_m = (n_1, n_2)_{m-1}$ **then**

$(n_1, n_2) = (n_1, n_2)_m$

Keep corresponding (x_1, x_2) **else**

$\{(n_1, n_2)_m \neq (n_1, n_2)_{m-1}\}$

$(n_1, n_2) = (n_1, n_2)_m$

Go to step (A)

end if

End.

4. Numerical application

We consider an equipment whose lifetime distribution function is Weibull with the same shape parameter $b = 2.0$. The equipment has to be used on two environments with different severity. Their severity depends on the scale parameter, such as in first the scale is $\eta_1 = 20$, while $\eta_2 = 10$ stands for the scale parameter in the second environment. This implies that the second environment is twice more severe than first. To reduce the risk of equipment failure of the failure, the equipment undergoes periodic, preventive maintenance. The preventive maintenance costs $C_{p1} = 100$ and $C_{p2} = 150$, respectively, on the first and second. The preventive actions impact the lifetime distribution of equipment. The impact factors due to PM are equal to $\beta_1 = 1.85$ in first and $\beta_2 = 2.5$ in the second environment. In addition, the equipment is minimally repaired at failure. The costs of minimal repair are in both environments $c_{m1} = 80$ and $c_{m2} = 70$. Based on this information, we are going to solve the optimization problem in order to find the number and duration period between PM on each environment which ensure a minimal cost per unit of time. With these parameters, the minimal cost reaches 10.37. This minimal cost involves $n_1 = 1$ and $n_2 = 1$ preventive maintenance (PM) respectively in the first and second environments. The durations between each PM are $x_1 = 26.06$ and $x_2 = 3.03$.

5. Conclusion

This chapter shows how to solve Nakagawa maintenance policy problem for an equipment which operates simultaneously on two environments. Each environment impacts the lifetime distribution function of our equipment. Nakagawa's maintenance problem is modeled under lifetime distribution changing in operation. The proposed model is deeply analyzed in order to derive the conditions under which optimal pairs exist and are reachable. To reach these pairs, algorithm was proposed to find the optimal solution for the periodic preventive maintenance on infinite horizon. The model is handy and suitable for production equipments which have to experience under different operating environments with their own severity degree that impacts the equipment performance such as onshore or offshore.

For future work, we plan to propose a statistical modeling by ignoring the hypothesis on the knowledge of the equipment lifetime distribution and perform an extension of the analysis by considering an finite-time horizon/span.

Author details

Ibrahima dit Bouran Sidibe[1]* and Imene Djelloul[2,3]

*Address all correspondence to: bouransidibe@gmail.com

1 Centre de Formation et de Perfectionnement en Statistique (CFP-STAT) Bamako, Mali

2 Higher School of Applied Sciences of Algiers, Place des Martyres, Alger, Algerie

3 Manufacturing Engineering Laboratory of Tlemcen (MELT), Abou Bekr Belkaid University of Tlemcen, Tlemcen, Algeria

References

[1] Barlow R, Hunter L. Optimum preventive maintenance policies. Operations Research. 1960;**8**(1):90-100

[2] Barlow R, Proschan F. Mathematical Theory of Reliability. New York: John Wiley & Sons; 1965

[3] Nakagawa T, Mizutani S. A summary of maintenance policies for a finite interval. Reliability Engineering & System Safety. 2009;**94**(1):89-96

[4] Nakagawa T. Advanced Reliability Models and Maintenance Policies. Springer series in reliability engineering, London: Springer; 2008

[5] Cho D, Parlar M. A survey of maintenance models for multi-unit systems. European Journal of Operational Research. 1991;**51**(1):1-23

[6] Dekker R. Applications of maintenance optimization models: A review and analysis. Reliability Engineering & Systems Safety. 1996;**51**(3):229-240

[7] Jardine A, Tsang A. Maintenance, Replacement, and Reliability: Theory and Applications. Boca Raton: CRC/Taylor & Francis; 2006

[8] Lugtigheid D, Jardine A, Jiang X. Optimizing the performance of a repairable system under a maintenance and repair contract. Quality and Reliability Engineering International. 2007;**23**(8):943-960

[9] Coolen F, Coolen-Schrijner P, Yan K. Non-parametric predictive inference in reliability. Reliability Engineering & System Safety. 2002;**78**(2):185-193

[10] Coolen-Schrijner P, Coolen F. Non-parametric predictive inference for age replacement with a renewal argument. Quality and Reliability Engineering International. 2004;**20**(3): 203-215

[11] de Jonge B, Klingenberg W, Teunter R, Tinga T. Optimum maintenance strategy under uncertainty in the lifetime distribution. Reliability Engineering & System Safety. 2015;**133**: 59–67.

[12] Nakagawa T. Periodic and sequential preventive maintenance policies. Journal of Applied Probability. 1986;**23**(2):536-542

[13] Thompson WA. On the foundations of reliability. Technometrics. 1981;**23**(1):1-13

13

Obtaining and Using Cumulative Bounds of Network Reliability

Alexey S. Rodionov and Denis A. Migov

Abstract

In this chapter, we study the task of obtaining and using the exact cumulative bounds of various network reliability indices. A network is modeled by a non-directed random graph with reliable nodes and unreliable edges that fail independently. The approach based on cumulative updating of the network reliability bounds was introduced by Won and Karray in 2010. Using this method, we can find out whether the network is reliable enough with respect to a given threshold. The cumulative updating continues until either the lower reliability bound becomes greater than the threshold or the threshold becomes greater than the upper reliability bound. In the first case, we decide that a network is reliable enough; in the second case, we decide that a network is unreliable. We show how to speed up cumulative bounds obtaining by using partial sums and how to update bounds when applying different methods of reduction and decomposition. Various reliability indices are considered: k-terminal probabilistic connectivity, diameter constrained reliability, average pairwise connectivity, and the expected size of a subnetwork that contains a special node. Expected values can be used for unambiguous decision-making about network reliability, development of evolutionary algorithms for network topology optimization, and obtaining approximate reliability values.

Keywords: network reliability, factoring method, network connectivity, random graph, diameter constraint, probabilistic connectivity, pairwise connectivity, network topology optimization, estimation, cumulative updating

1. Introduction

The network reliability analysis is one of the primary tasks in the course of the network topology design and optimization. Usually, random graphs are used when modeling networks with unreliable elements [1–7]; therefore, the network reliability is defined as a connectivity measure in a random graph. There are many reliability indices of networks, for example, probabilistic connectivity [4], average pairwise reliability [5, 6], and diameter constrained reliability [7].

The most widespread network reliability index is the probabilistic connectivity of the corresponding random graph. Depending on the number of terminals (selected nodes that must be connected), there are three types of this measure such as two-terminal, all-terminal (ATR) and k-terminal reliabilities. The average pairwise reliability (APR) describes the network reliability from the point of connection between every two nodes, while a network may be still disconnected. The reliability of a network with a diameter constraint (DCR) is defined as the probability of each pair of nodes connectivity by paths that travel through a limited number of communication channels. This index is more useful, especially for P2P networks [8], wavelength division multiplexing networks, wireless sensor networks, and so on.

Another reliability index we consider in this chapter is the mathematical expectation of the size of a connected subgraph that contains a special node (MENC) [2], which describes the quality of monitoring area coverage.

Despite the fact that problems of exact reliability computations are NP-hard [9, 10], it is possible to conduct the exact calculation for networks of practical interest dimension, taking into account some special features of network topologies. For large-scale networks, various methods of reliability evaluation are widely used [11, 12]. Won and Karray suggested the cumulative updating of reliability bounds for uniqueness of decision on network feasibility [13]. Originally, this approach was presented for ATR. In this chapter, we present our results on cumulative updating of other reliability indices, and some improvements for ATR bounds cumulative updating. In addition, the main ideas of constructing the cumulative bounds for random graph characteristics are discussed along with examples of their usage in obtaining better evaluation of reliability indices and improving bionic optimization algorithms.

2. The basic notations

We simulate a network with perfectly reliable nodes and unreliable links by an undirected random graph $G = (V,E)$ with given presence probabilities $0 \le e_i \le 1$ of the edges. The notations which are used in this chapter are given below. Most of them coincide with the notations from [14].

G—Undirected probabilistic network;

V—Set of n nodes;

E—Set of m edges;

e_i; e_{ij} – i-th edge or edge that connects i-th and j-th nodes, depending on context;

p_j – Operating probability of j-th edge;

w_i – Weight of node i, $WT = (w_1,\ldots,w_n)$;

$W(G)$ – Total weight of nodes of G;

$R(G)$ – All-terminal reliability of G, that is, probability that every two nodes are connected;

K—Set of terminal nodes;

$R_K^d(G)$—Diameter constrained reliability of *G*, that is, the probability that every two nodes from *K* are connected by a path that travel through not more than *d* edges;

LB, UB—Lower and upper bounds of *G* reliability. In case of ATR, we use the original notation form [13]: *RL, RU;*

R_0—Predefined threshold for the network reliability value;

N(*G*) (*M*(*G*))—Mathematical expectation of the number of disconnected (or connected) pairs of nodes in *G*;

CS(*G; s*)—Mathematical expectation of a number of nodes in a connected subgraph that contains a node *s*; if *s* = 1, then CS(*G*);

$\overline{R}$—Average pairwise reliability of *G*;

C—Edge chain composed of edges *e*1,…,*ek*;

G/*C* (*G*/*e*)—Network *G* with a contracted chain *C* (edge *e*);

$G \setminus C$ ($G \setminus e$)—Network *G* without chain *C* (edge *e*); and

G_e^*—Network *G* with an absolutely reliable edge *e;*

3. Factoring method and cumulative updating of network reliability bounds

The most common exact method for calculating various network reliability measures is the factoring method. According to this method, we divide the probability space, which consists of all particular network realizations, into two sets based on the presence or absence of a network element. Further on we refer to such a network element as a factored element, or a pivot element. As a result, for a given network *G* and a factored element *e*, we will obtain the two networks: G_e^*, where *e* is absolutely reliable, and $G \setminus e$, where *e* is absolutely unreliable, so we could remove it. The probability of G_e^* is equal to the pivot element reliability, and the probability of $G \setminus e$ is equal to the pivot element failure probability. For a reliability index *Rel* of the initial network, the following expression holds (the total probability law):

$$Rel(G) = p_e Rel(G_e^*) + (1 - p_e) Rel(G \setminus e) \tag{1}$$

Then the obtained networks are subject to the same factoring procedure. Recursions continue until either an unreliable network is obtained (0 is returned) or an absolutely reliable network is obtained (1 is returned). For ATR (**Figure 1**), expression (1) turns to:

$$R(G) = p_e R(G/e) + (1 - p_e) R(G \setminus e). \tag{2}$$

We can speed up the factoring process by calculating the intermediate networks reliabilities directly, that is, without further factorization. For ATR calculation, the five-vertex graph reliability formula can be used [13]. Another way to accelerate the reliability computing by using

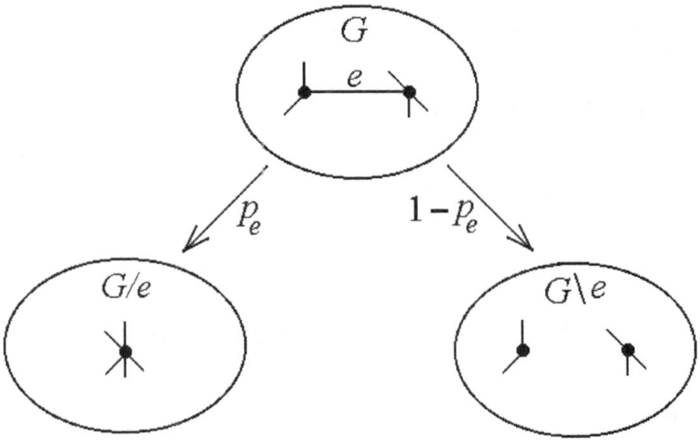

Figure 1. The factoring method.

various reduction [4] and decomposition [17] methods, such as serial-parallel transformation on each recursive call of the factoring procedure, biconnected decomposition, and other methods.

The idea of the cumulative updating method is an incremental updating of exact lower (LB) and upper (UB) reliability bounds and comparing them with a given reliability threshold R_0. If $LB > R_0$, then the network is reliable. If $UB < R_0$, then the network is unreliable. The LB must necessarily be non-decreasing, and the UB must necessarily be non-increasing. Another obligatory condition is the equality of LB, UB at the last step and exact reliability value R. One possible way of bounds updating is the usage of the factoring method. When a final graph or a disconnected graph is obtained, we can update bounds [13].

As a rule, any algorithm of calculation of mathematical expectation for a non-negative function μ of a random graph G practically comes to summarizing non-negative values. According to the definition of mathematical expectation in the discrete case [14]:

$$E[\mu(G)] = \sum_{H\in\Gamma} P(H)\cdot\mu(H), \tag{3}$$

where Γ is a set of all possible final realizations of G that are obtained during factorization.

If some realizations Γ_0 are obtained along with their probabilities, and a function μ is obtained for all these realizations, then the LB is greater or equal to the corresponding partial sum.

Now let us assume that for μ, its possible minimal (μ_m) and maximal (μ_M) values are known. In this case, we easily obtain the following bounds:

$$LB = \sum_{H\in\Gamma_0} P(H)\cdot\mu(H) + \mu_m\ 1 - \sum_{H\in\Gamma_0} P(H)\Big);\quad UB = \sum_{H\in\Gamma_0} P(H)\cdot\mu(H) + \mu_M\ 1 - \sum_{H\in\Gamma_0} P(H)\Big). \tag{4}$$

Finally, both bounds will obviously come to an exact value.

From (4), we easily obtain the equations for improving bounds when the new (i-th) realization H_i is obtained and $\mu(H_i)$ is calculated:

$$LB_i = LB_{i-1} + P(H_i) \cdot [\mu(H_i) - \mu_m]; \quad UB_i = UB_{i-1} - P(H_i) \cdot [\mu_M - \mu(H)]. \qquad (5)$$

For ATR, $\mu_m = 0$ and $\mu_M = 1$, so we arrive to the equations presented by Won and Karray in [13], while for MENC $\mu_m = 1$ and $\mu_M = n$, where n is a total number of nodes in a graph. Initial values of bounds are equal to these values.

4. Improvements for ATR bounds cumulative updating

4.1. The chain branching

Our methods named chain branching (CB) and chain reduction (CR) [15] could be used [16] .as a basis for two kinds of the ATR bound updating (UAB) algorithms named UAB based on CB (UAB_CB), and UAB based on CR (UAB_CR). These methods can be used if a network contains a chain, that is, a sequence of adjacent two-degree nodes. The CB reduces the calculation of ATR for calculating the ATRs of networks obtained from G by removing a pivot chain, or removing it and contracting its terminal nodes, while CR reduces it for calculating the ATR of the network G_R, in which, this chain is substituted by a single edge. Let a chain C consist of k edges, its terminal nodes are s and t, and let an edge e_{st} has a reliability p_{st}, then [16]

$$R(G) = \left[\prod_{i=1}^{k} p_i + p_{st} \sum_{i=1}^{k} (1 - p_i) \prod_{j \neq i} p_j\right] \cdot R(G/C) + (1 - p_{st}) \sum_{i=1}^{k} (1 - p_i) \prod_{j \neq i} p_j \cdot R(G \setminus C). \qquad (6)$$

The best choice is the usage of the longest chain as a pivot. However, it requires finding of all the chains. We may accelerate the calculation by choosing a chain (or an edge) incidental to a node with a minimal degree for obtaining a new chain.

4.2. The usage of Cutnodes

Let a network G have a cutnode u that divides G into two subnetworks G_1 and G_2, therefore, $R(G) = R(G_1)R(G_2)$. $RL^{(1)}$ and $RU^{(1)}$ are lower and upper bounds for $R(G_1)$, respectively. Like wise, $RL^{(2)}$ and $RU^{(2)}$ are the lower and upper bounds for $R(G_2)$, respectively. Then $RL^{(1)} \cdot RL^{(2)} \leq R(G) \leq RU^{(1)} \cdot RU^{(2)}$.

Let us estimate the ATR of subnetworks G_1 and G_2 in turn starting from G_1. While estimating the ATR of G_1, we have no information about the ATR of G_2, thus $RL^{(2)} = 0$ and $RU^{(2)} = 1$, and RL for $R(G)$ is 0, while RU for $R(G)$ is $RU(G_1)$. After some steps of estimating $R(G_1)$, we switch to estimating $RL^{(2)}$ and $RU^{(2)}$, thus improving RL and RU for G, and so on. Therefore, we calculate the bounds for $R(G_1)$ and $R(G_2)$ separately (possibly in parallel). Now let us continue with applying this scheme to calculating the bounds for $R(G_1)$ and $R(G_2)$ by the factoring method.

Similar to [15], we suppose that some networks $H_1, H_1, \ldots, H_I$ are obtained from G_1, and networks $K_1, K_2, \ldots, K_J$ from G_2 by factorization (1). Let us denote $x_i = P(H_i)$, $y_i = R(H_i)$, $1 \leq i \leq I$, and $s_i = P(K_i)$, $t_i = R(K_i)$, $1 \leq i \leq J$. Then [16]:

$$R(G_1) = \sum_{i=1}^{I} x_i y_i = 1 - \sum_{i=1}^{I} x_i(1 - y_i); \quad R(G_2) = \sum_{i=1}^{J} s_i t_i = 1 - \sum_{i=1}^{J} s_i(1 - t_i). \tag{7}$$

Statement 1 *For all a,b such that $1 \le a \le I$, $1 \le b \le J$, the following inequalities hold:*

$$\sum_{i=1}^{a} x_i y_i \sum_{i=1}^{b} s_i t_i \le R(G) \le \left[1 - \sum_{i=1}^{a} x_i(1 - y_i)\right]\left[1 - \sum_{i=1}^{b} s_i(1 - t_i)\right]. \tag{8}$$

Proof of the statement can be found in [16], as well as the algorithm with the use of cutnodes (UAB_C).

4.3. The usage of two-node cuts

Now let us describe the UAB modification that uses two-node cuts. Let the network G has a node cut u,v that divides it into two subnetworks G_1 and G_2 (**Figure 2**). For any network H that contains the nodes u and v we will denote a network that is obtained by contracting these nodes as H'.

For the first time, the approach based on two-node cut was presented by Kevin Wood [17] in 1989. It describes how to perform the reliability-preserving triconnected decomposition of a network for calculating its k-terminal reliability. Later in 2006, we have introduced a general method which allows using an arbitrary node cut for calculating all-terminal reliability [18, 19]. This method makes possible to reduce the reliability calculation to the calculation of reliabilities of a group of networks each one with a smaller dimension. The total number of these networks is 2^*B_W, where B_W is the Bell number and W is the number of nodes in the cut. A particular case of two-node cut along with numerical experiments to demonstrate the efficiency for the ATR calculation was presented in [16, 20], and in [21]—for the k-terminal reliability calculation. Despite the fact that the method from [17] presents a graph transformation for further factoring, and the method from [20] presents an expression for the network reliability, both of them practically lead to similar calculations of the same complexity. Another result was with the ATR calculation using special kind of node cuts [22]: the so-called

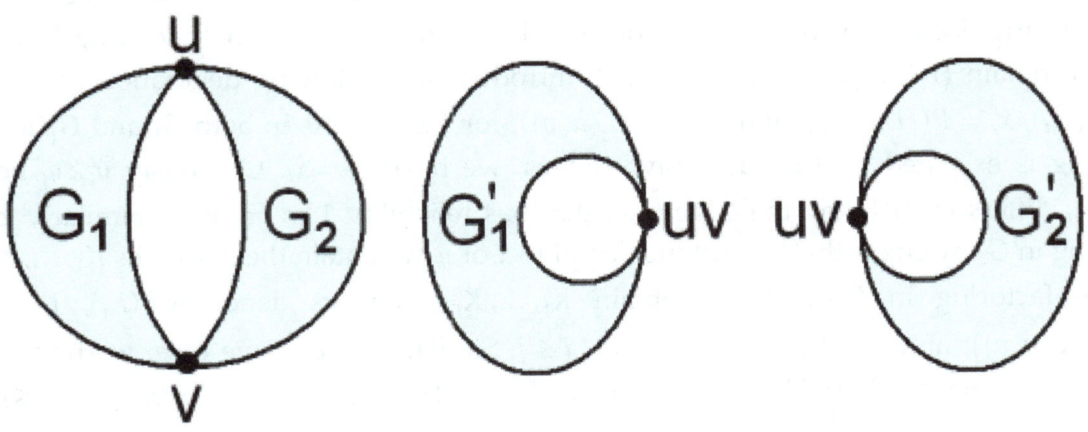

Figure 2. The network G with two-node cut and auxiliary networks.

longitudinal and cycle cuts. All the listed results, which we have obtained, are described in detail in [19]. Following is the corresponding Eq. [20] for the ATR without proof:

Theorem 1 *The following equation holds:*

$$R(G) = R(G_1)\left[R\left(G_2'(u,v)\right) - R(G_2)\right] + R(G_2)\left[R\left(G_1'(u,v)\right) - R(G_1)\right] + R(G_1)R(G_2). \quad (9)$$

The equation for three-node cut is sufficiently more complicated [18, 19]:

$$\begin{aligned} R(G) = \frac{1}{2}\Big[&R\left(G_1^{1|23}\right)\left(R\left(G_2^{12|3}\right) + R\left(G_2^{13|2}\right) - R\left(G_2^{1|23}\right)\right) \\ &+R\left(G_1^{12|3}\right)\left(R\left(G_2^{1|23}\right) + R\left(G_2^{13|2}\right) - R\left(G_2^{12|3}\right)\right) + R\left(G_1^{13|2}\right)\left(R\left(G_2^{12|3}\right) + R\left(G_2^{1|23}\right) - R\left(G_2^{13|2}\right)\right) \\ &-R(G_1)\left(R\left(G_2^{12|3}\right) + R\left(G_2^{13|2}\right) + R\left(G_2^{1|23}\right)\right) - R(G_2)\left(R\left(G_1^{12|3}\right) + R\left(G_1^{13|2}\right) + R\left(G_1^{1|23}\right)\right) \\ &+R(G_1)R(G_2)\Big] + R(G_1)R\left(G_2^{123}\right) + R\left(G_1^{123}\right)R(G_2), \end{aligned} \quad (10)$$

where nodes 1, 2, and 3 composes the threee-node cut, $G_i^{x|yz}$ is a graph that is obtained by merging y and z nodes in G_i, and G_i^{xyz} is obtained by merging x, y, and z nodes. The expression for four-node cut has also been obtained [19], but we do not present it due to its huge size.

However, our results on using an arbitrary node cut for all-terminal reliability calculation were published only in Russian [18, 19], and only the two-node cut case was presented in English [16, 20]. Maybe, this is the reason that these results are not so widely known, and therefore, the same ideas were proposed again by Juan Manuel Burgos and Franco Robledo Amoza [23, 24] in 2016. As we have done it in [16–20], in [23, 24], the authors show how to calculate the all-terminal reliability of a network with a node cut via reliabilities of a group of smaller networks, but in a somewhat different way. As a result, Eqs. (9–10) were presented.

Let us describe how we can use expression (9) and update the ATR bounds separately for both subgraphs G_1 and G_2 (**Figure 2**) for obtaining the ATR bounds of the whole graph G.

Suppose that some networks $H_1, \ldots, H_I$ are obtained from G_1 during the factoring process. If this factoring does not involve edges incidental to u or v, then parallel factoring in $G_1'(u,v)$ leads to obtain $H_1'(u,v), \ldots, H_I'(u,v)$. Let us introduce the following denotations: $x_i = P(H_i)$, $y_i = R(H_i)$, $x_i' = P\left(H_i'(u,v)\right)$, and $y_i' = R\left(H_i'(u,v)\right)$, for $1 \le i \le I$. As in both G_1 and $G_1'(u,v)$, the factoring is executed by the same pivot edges, we have $x_i' = x_i$. Obviously $y_i' \ge y_i$ because $H_1'(u,v)$ differs from H_i by the edge (u,v) that has reliability 1 in $H_1'(u,v)$. Similarly, by the factoring in G_2 by edges that are not incidental to x or y, we obtain the networks $K_1, \ldots, K_J$, and by the factoring in $G_2'(u,v)$, we obtain $K_1', \ldots, K_J'$. Let us denote $s_i P(K_i)$, $t_i = R(K_i)$, $s_i' = P\left(K_i'(u,v)\right)$, and $t_i' = R\left(K_i'(u,v)\right)$ for $1 \le i \le J$. Similar to the above case, we have $s_i' = s_i$, $t_i' \ge t_i$. Let $L = \max\{I, J\}$, and let $x_i = y_i = y_i' = 0$ for $I+1 \le i \le L$, $s_i = t_i = t_i' = 0$ for $J+1 \le i \le L$.

Let us obtain $RL_{1,i}, RU_{1,i}, RL_{1,i}', RU_{1,i}', RL_{2,i}, RU_{2,i}, RL_{2,i}', RU_{2,i}'$ for $1 \le i \le L$:

$$RL_i^1 = \sum_{j=1}^{i} x_j y_j,\ RU_i^1 = 1 - \sum_{j=1}^{i} x_j\left(1 - y_j\right),\ RL'^1_i = \sum_{j=1}^{i} x_j y_{j'},\ RU'^1_i = 1 - \sum_{j=1}^{i} x_j\left(1 - y_{j'}\right),$$
$$RL_i^2 = \sum_{j=1}^{i} s_j t_j,\ RU_i^2 = 1 - \sum_{j=1}^{i} s_j(1 - t_j),\ RL'_i{}^2 = \sum_{j=1}^{i} s_j t_{j'},\ RU'^2_i = 1 - \sum_{j=1}^{i} s_j\left(1 - t_{j'}\right).$$

As is shown in [13], these values are the upper and lower bounds for reliabilities of the corresponding networks; these bounds are exact for $i = L$. In the case $i = 0$, the lower bounds are 0 and the upper bounds are 1.

Note that $RL'_{1,i} \geq RL_{1,i}$, $RU'_{1,i} \geq RU_{1,i}$, as $y'_i \geq y_i$; $RL'_{2,i} \geq RL_{2,i}$, $RU'_{2,i} \geq RU_{2,i}$, as $t'_i \geq t_i$.

Now we obtain RL_j, and RU_j for odd and even j separately. For $j = 2i$, $0 \leq i \leq L$:

$$\begin{aligned} RU_{2i} &= RU_i^1 RU'^2_i + RU'^1_i RU_i^2 - RU_i^1 RU_i^2, \\ RL_{2i} &= RL_i^1\left(RL'^2_i - RL_i^2\right) + RL_i^2\left(RL'^1_i - RL_i^1\right) + RL_i^1 RL_i^2. \end{aligned} \tag{11}$$

For $j = 2i + 1, 0 \leq i \leq L - 1$:

$$\begin{aligned} RU_{2i+1} &= RU_{i+1}^1 RU'^2_i + RU'^1_{i+1} RU_i^2 - RU_{i+1}^1 RU_i^2, RL_{2i+1} \\ &= L_{i+1}^1\left(RL'^2_i - RL_i^2\right) + RL_i^2\left(RL'^1_{i+1} - RL_{i+1}^1\right) + RL_{i+1}^1 RL_i^2. \end{aligned} \tag{12}$$

Statement 2 *For any* $1 \leq j \leq 2\,L$, *we have*

$$RU_{j-1} \geq RU_j \geq R(G) \geq RL_j \geq RL_{j-1}. \tag{13}$$

If $j = 2\,L$, then the second and the third inequalities become equalities. The proof of the statement can be found in [16], as well as the algorithm (UAB_2NC) with two-node cuts.

Note that both CB and CR can be used for speeding up the 2NC algorithm under the condition that the nodes u and v do not belong to a pivot chain. In the case of CR $RU_j^1 := RU_j^1 - x_j(1 - p)$, $RU'^1_j := RU'^1_j - x_j(1 - p)$, where p is a multiplier by which the reliability of a reduced network must be multiplied in CR [4].

4.4. Case studies

We have compared the algorithms presented with node cuts UAB_C and UAB_2NC to the algorithm CR from [13] that does not take possible cutnodes or two-node cuts into account. The results [16] of the comparison are presented in **Tables 1–3**. The CR was used while factoring for better performance, and the five-node networks were considered as simple ones. The PC with Intel Core Duo 2.93 GHz was used for testing.

We have chosen the two networks G_1 and G_2 for testing UAB_C, and the two networks G_3 and G_4 for testing UAB_2NC.

Network	R_0	Algorithm	Time	Recursions
G_1	0.9	UAB_CR	39 s	5,245,491
G_1	0.9	UAB_C	<1 ms	2504
G_1	0.94	UAB_CR	34 s	639,562
G_1	0.94	UAB_C	<1 ms	2643
G_2	0.5	UAB_CR	2 s	259,879
G_2	0.5	UAB_C	60 ms	4322
G_2	0.901	UAB_CR	3 min 22 s	24,765,569
G_2	0.901	UAB_C	45 ms	3233

Table 1. Comparison of UAB_CR and UAB_C.

R_0	*UAB_CR*		*UAB_2NC*	
	Time (s)	Recursions	Time (s)	Recursions
0.5	2.2	257,084	0.08	8696
0.6	43.2	5,055,553	0.18	19,399
$R(G_3)$	—	—	0.28	27,180
0.92	13	1,532,293	0.08	8725
0.922	2.5	281,565	0.08	8696

Table 2. Results of the numerical experiments for the network G_3.

R_0	*UAB_CR*		*UAB_2NC*	
	Time (s)	Recursions	Time (s)	Recursions
0.2	0.12	16,814	0.1	10,963
0.3	27	3,630,494	0.87	80,756
$R(G_4)$	—	—	8,8	805,358
0.996	113	14,878,088	0.02	1747
0.998	0.36	45,374	< 1 ms	35

Table 3. Results of the numerical experiments for the network G_4.

The network G_1 consists of two 9-vertex complete graphs which are connected by a cutnode. Thus, it contains 17 nodes and 72 edges.

The network G_2 consists of two 5×5 lattices which are connected by a cutnode. As a cutnode, we choose the corner. The resulting network contains 49 nodes and 80 edges. **Table 1** shows the results of numerical experiments.

The network G_3 (48 nodes and 79 edges) is composed of two 5×5 lattices which are connected by two nodes (**Figure 3**).

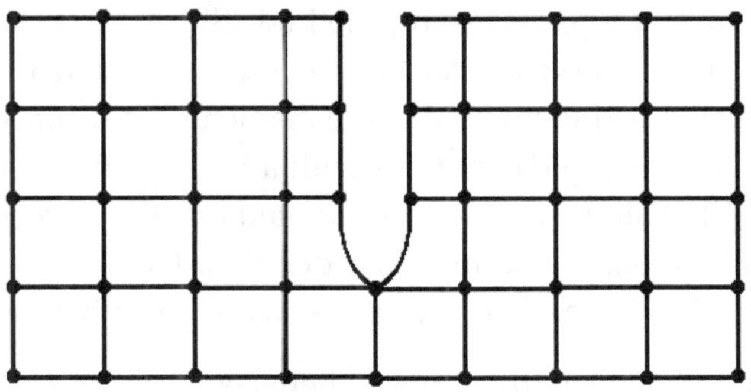

Figure 3. The union of two 5 × 5 lattices by two nodes.

The network G_4 (22 nodes and 108 edges) is composed of two 11-vertex complete graphs which are connected by two nodes without edge connecting them. A similar network that is composed of two 5-vertex complete graphs is presented in **Figure 4**.

The edges G_1 and G_4 are equally reliable with $p = 0.5$, while the edges G_2 and G_3 are equally reliable with $p = 0.9$. $R(G_1) = 0.9307194$, $R(G_2) = 0.883248$, $R(G_3) = 0.903168801959$, and $R(G_4) = 0.982472649148$.

Tables 2 and **3** present the results for G_3 and G_4, respectively. The third row contains the results when the threshold value coincides with the exact reliability value. The results for the factoring algorithm with chain reduction are shown in the second column. The results for the UAB_2NC are shown in the third column.

As can be seen, the approach proposed has a great advantage. The efficiency of UAB as it is depends on the closeness of the threshold value R_0 to the reliability value $R(G)$.

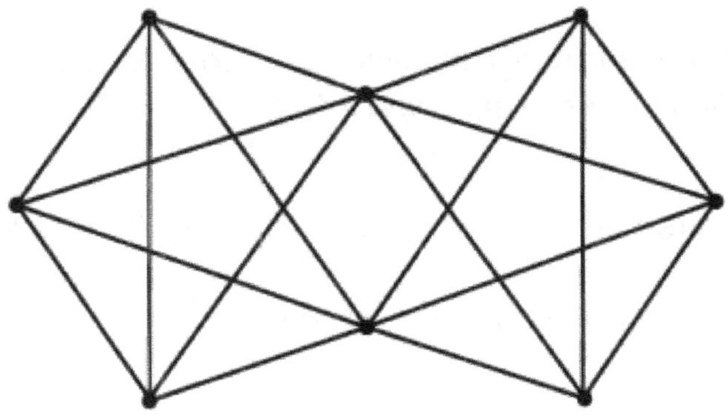

Figure 4. The union of two complete five-node networks by two nodes without connecting edge.

5. Cumulative updating of diameter constrained network reliability

For the DCR, expression (1) takes the following form:

$$R_K^d(G) = p_e R_K^d(G_e^*) + (1 - p_e) R_K^d(G \backslash e). \tag{14}$$

We denote this method as a simple factoring method (SFM). Cancela and Petingi have proposed a modified factoring method for calculating DCR [7], which is much faster than the already described classical factoring method in the diameter constrained case (14). The Cancela and Petingi factoring method (CPFM) operates with a list of paths instead of graphs. For any terminals *s*, *t*, the list $P_{st}(d)$ of all the paths with a limited length between *s*, *t* is generated. By P_d the union of $P_{st}(d)$ for all the pairs of terminals is denoted. *P*(*e*) is composed of the paths of P_d which include the edge *e*. The list of CPFM arguments are given below [7]:

- np_{st}: the number of paths of length at most *d* between *s* and *t* in *G*;
- $links_p$: the number of non-perfect edges (edges *e* such that $p(e) < 1$) in the path *p*, for every $p \in P_d$.
- $feasible_p$: this is a flag, which has the value False when the path is no longer feasible, that is, it includes an edge which failed; and True otherwise;
- $connected_{st}$: this is a flag, which has the value True when *s* and *t* are connected by a perfect path of length at most *d* and False otherwise;
- *connectedPairs*: the number of pairs of terminals which are connected by a perfect path of length at most d.

Pseudocode of the method proposed for the DCR bounds cumulative updating [27], which is based on CPFM, is presented below:

Input: *G* = (*V*;*E*), *d*, P_d, *P*(*e*), *np*(*s*; *t*), *links*(*p*), *feasible*(*p*), *connected*(*s*,*t*), *connectedPairs*, *RL* = 0, *RU* = 1, *Pl* = 1.
Function FACTO (*np*(*s*,*t*), *links*(*p*), *feasible*(*p*), *connected*(*s*,*t*),*connectedPairs*, *Pr*)
if $RL > R_0$ *or* $RU < R_0$ **then**
  e is an arbitrary edge: $0 < p_e < 1$
  contractEdge (*np*(*s*; *t*), *links*(*p*), *feasible*(*p*), *connected*(*s*,*t*),*connectedPairs*, *Pr*)
  deleteEdge (*np*(*s*; *t*), *links*(*p*), *feasible*(*p*), *connected*(*s*,*t*),*connectedPairs*, *Pr*)
end FACTO

Function contractEdge(*np*(*s*; *t*), *links*(*p*), *feasible*(*p*), *connected*(*s*,*t*),*connectedPairs*, *Pr*)
Pr = Pr * p_e
for each *p* = (*s*,..,*t*) *in P*(*e*) *such that feasible*(*p*) = *true* **do**
  links(*p*) = *links*(*p*) - 1
  if *connected*(*s*,*t*) = *false* **and** *links*(*p*) = 0 **then**
    connected(*s*,*t*) *true*
    connectedPairs = *connectedPairs* + 1
    if *connectedPairs* = |K|*(|K|-1)/2 **then** *RL* = *RL* + *Pr*
FACTO (*np*(*s*,*t*), *links*(*p*), *feasible*(*p*), *connected*(*s*,*t*), *connectedPairs*, *Pr*)
end contractEdge

Function deleteEdge(*np*(*s*,*t*), *links*(*p*), *feasible*(*p*), *connected*(*s*,*t*), *connectedPairs*, *Pr*)
Pr = *Pr* *(1-p_e)
For each *p* = (*s*,…,*t*) *in P*(*e*) *such that feasible*(*p*) = *true* **do**

feasible(*p*) = *false*
np(*s*,*t*) = *np*(*s*,*t*) - 1
if *np*(*s*,*t*) = 0 **then** *RU* = *RU* - *Pr*
FACTO (*np*(*s*,*t*), *links*(*p*), *feasible*(*p*), *connected*(*s*,*t*), *connectedPairs*, *Pr*)
end deleteEdge

For the speeding up of DCR calculation, it is possible to apply methods of reduction and decomposition. In our previous studies [25–27], we have obtained such methods which can calculate the DCR faster. These methods are the analogue of the well-known series-parallel transformation for CPFM, and the pivot edge selection strategy. Also, we have obtained decomposition methods for calculating DCR in the case with two terminals. The methods obtained allow us to significantly reduce the number of recursive calls in CPFM and the complexity of DCR computation. In the above algorithm, we assume that the series-parallel reduction has been already performed and the pivot edge selection strategy is used.

Below we show how *RL* and *RU* are changing during the procedure proposed for the Intellinet network topology (**Figure 5**). The edge reliability is equal to 0.7 for each edge and the diameter value is equal to 12. The threshold value was equal to the exact DCR value, which was previously

Figure 5. The Intellinet network.

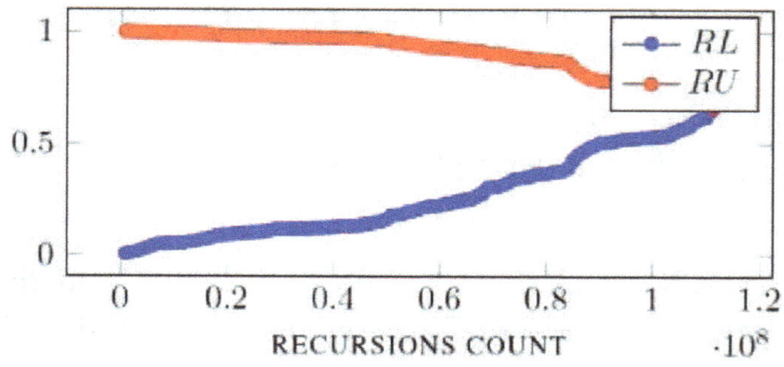

Figure 6. The behavior of DCR bounds for the Intellinet network.

calculated. The calculation time was about 12 min. Experiments were performed on Intel Xeon E31240 3.3 GHz, 8 cores (**Figure 6**).

6. Special problems of updating reliability bounds

When special methods of graph reduction or decomposition are applied for the speeding up of calculations, the corresponding equations for updating the bounds must be used. Derivation of such equations for different functions μ(G) is based on taking into account all "inner" changes of bounds that are concealed inside intermediate steps thus leading to final results of such a reduction or a decomposition. For example, when "branching by chain" [15] is applied, consequent factoring by edges of a chain is used for derivation, and each factoring may require a change of bounds.

Now let us turn to an example of a simple reduction, which is removing a dangling node. Let the graph G have some attached trees in its structure is shown in **Figure 7**.

In the case of the ATR, we have the trivial equality $R(G) = p_e R(G \setminus e)$, where e is the edge that connects the dangling node to the graph. As removing e with probability $1 - p_e$ leads to graph

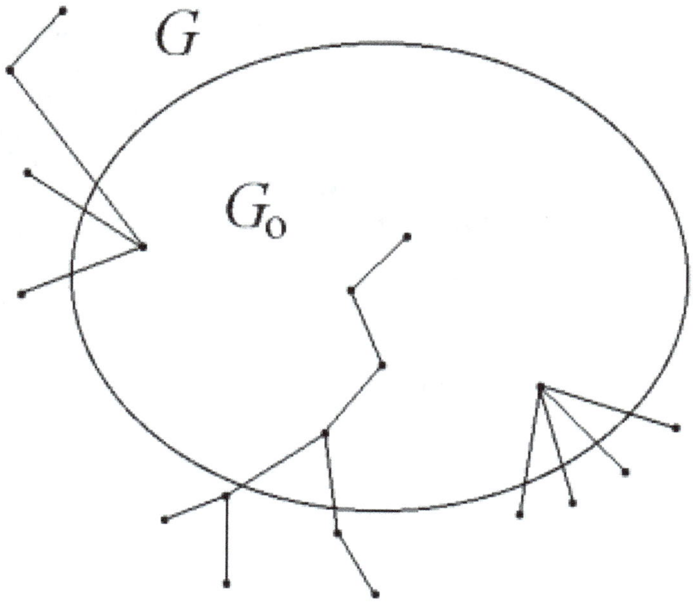

Figure 7. The graph with attached trees.

disconnection, the upper bound must be decreased by $1-p_e$. The lower bound rests unchanged, but probabilities of all further realizations of $G\backslash e$ must be multiplied by p_e. If there are some attached trees as shown in **Figure 7**, then the ATR of the graph G is equal to the ATR of this graph without all attached trees (G_0) multiplied by the product of reliabilities of all the edges in these trees (let us denote it as *Pr*): $R(G) = \Pr \cdot R(G_0)$. Thus, we continue with the graph G_0, *LB* does not change, *Pr* must be included as a multiplier in probabilities of further realizations, and *UB* is reduced by probability of at least one edge of attached trees, that is, 1-*Pr*. If the reduction takes place at the initial step, then the upper bound starts from *Pr*.

The case of APR is not so simple. It is known that the task of obtaining this reliability index is equivalent to the one of obtaining mathematical expectation of the number of disconnected pairs of nodes in a random graph (EDP, see [28, 29]). Indeed, the following equations are valid:

$$\overline{R}(G) = \frac{C_n^2 - N(G)}{C_n^2}; N(G) = C_n^2(1-\overline{R}(G)). \tag{15}$$

From these equations, bounds are easily obtained:

$$LB_{APR} = \frac{C_n^2 - UB_{EDP}}{C_n^2}, UB_{APR} = \frac{C_n^2 - LB_{EDP}}{C_n^2}. \tag{16}$$

In [14], the following equation was derived with deleting the dangling node t that is incidental to the node s:

$$N(G) = N(G_0) + (1-p_{st})w_t W(G\backslash e_{st}). \tag{17}$$

Here w_t is the weight of the node t, initially 1. This weight shows an expected number of nodes that are merged in this node during the graph transformations: when using the factoring method, for example, one should remember the number of nodes that are merged if we assume that an edge between them is reliable. This weight is used for calculating the number of disconnected pairs of nodes when the graph is divided. Let $W(G)$ be the total weight of the all nodes (initially, it is equal to the number of nodes) in G. G_0 equals G with the only difference: the weight of the node s becomes equal to $w_s + p_{st}w_t$.

From this, we obtain the following changes in the bounds: LB_{EDP} is increased by $(1-p_{st})$ $w_t[W(G) - w_t]$ and the UB_{EDP} is decreased by $p_{st}w_t[W(G) - w_t]$. Contrary to case of the ATR, the result of removing attached trees highly depends on their structures, so dangling nodes must be removed one at a time. The only known exception with the derived equation is the case of the attached chain [6].

7. Using cumulative bounds for network reliability approximation

A simple approximation of the function μ(G) through its bounds is their average: μ(G) = (*LB* + *UB*)/2. However, the bounds tend to the exact solution with different rates. It seems to be reasonable assuming that cross of lines, obtained by the linear approximation of curves for *LB* and *UB*, may be a better approximation of μ(*G*). Let us have the bounds LB_i and uB_i at a step $i > 0$ and let $\mu_i(G)$ be the corresponding approximation. The following proportion takes place:

$$(\mu_M - UB_i) : (LB_i - \mu_m) = (UB_i - \widehat{\mu}_i(G)) : (\widehat{\mu}_i(G) - LB_i). \tag{18}$$

Hence, we have

$$\widehat{\mu}_i(G) = \frac{\mu_M \cdot LB_i - \mu_m \cdot UB_i}{\mu_M - UB_i - \mu_m + LB_i}. \tag{19}$$

Let us name this approximation as the approximation based on trends (ABT).

In the case of ATR, $\mu_m = 0$ and $LB = UB = \mu(G)$, so

$$\widehat{R}_i(G) = \frac{LB_i}{1 - UB_i + LB_i}, \tag{20}$$

while in the case of MENC, $\mu_m = 1$ and $\mu_M = C_n^2$, so

$$\widehat{CS}_i(G) = \frac{C_n^2 \cdot LB_i - UB_i}{C_n^2 - UB_i - 1 + LB_i}. \tag{21}$$

At the last step, $LB = UB = \mu(G)$, and from (19) we have that the proposed approximation also equals this value.

In **Figures 8–10**, the behavior of bounds for the EDP of 4x4 lattice ($p = 0.7$), for the MENC of the same lattice with *c*-node 1, and for the probability of the flow transmission (PFT) between two

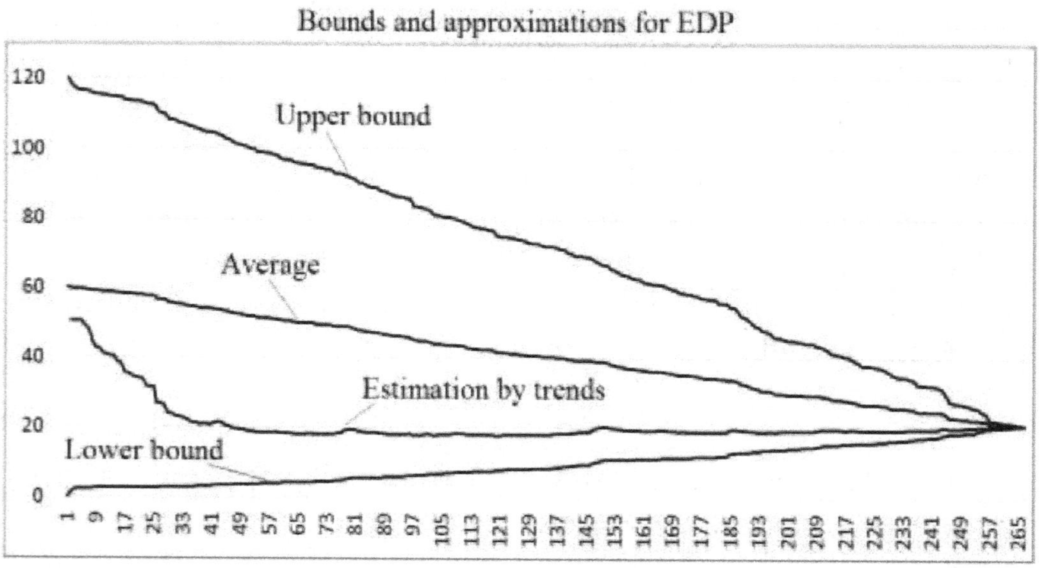

Figure 8. The behavior of bounds and approximations of EDP, the exact value is 20.915633.

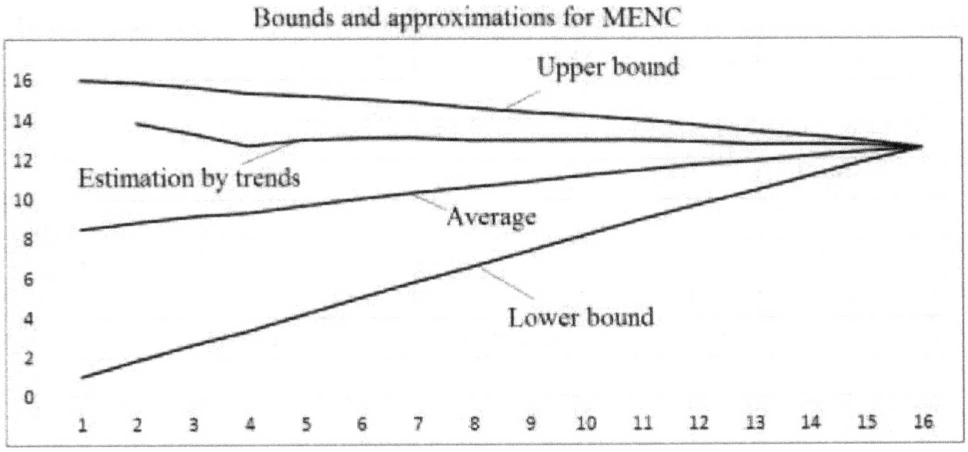

Figure 9. The behavior of bounds and approximations of MENC, the exact value is 12.562672.

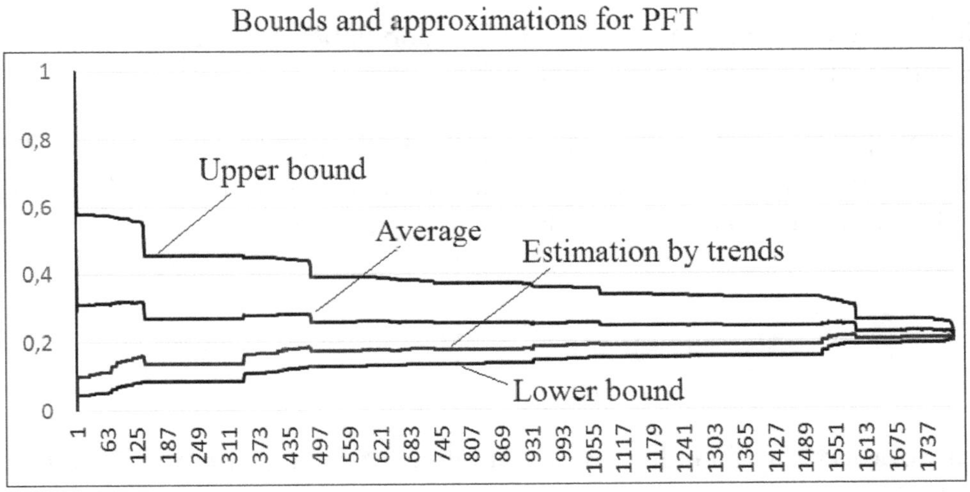

Figure 10. The behavior of bounds and approximations of PFT, the exact value is 0.196600.

diagonal corners of this lattice with throughput of all edges 1 and the edges reliability uniformly distributed between 0.5 and 1, are presented, respectively.

As we can see the ABT has become better than the average of *LB* and *UB* from the first step or very fast.

We consider the practical usage of cumulative bounds and ABT in the following section.

8. Using cumulative bounds in network topology optimization

When optimizing the network topology by a certain criteria, a fitness function (FF) must be calculated for each alternative. In our case (k-terminal reliability, MENC, and APR) this means using NP-hard algorithms for its obtaining. Note that using approximate algorithms may lead to wrong decisions when structures with close values of FF are compared.

We propose using the cumulative bounds for decision-making about whether a new solution obtained by crossover or mutation deserves including into population.

The ideas are simple (without the loss of generality, we assume that a task is maximizing some function of an unreliable network):

1. A wittingly inappropriate solution may be rejected before carrying out the exhaustive calculation of the corresponding FF. If the upper bound of the FF becomes smaller than the FF value of the worst solution in the current population, then this potential solution is rejected.

2. At initial stages, a new solution may be included into the population based on the LB value. If it exceeds the FF value of the worst solution in the current population, then this new solution substitutes the worst one with assigning ABT as an approximate value of the FF.

Note that inevitable narrowing of bounds of FF in population leads to seldom updating of the population and to a better approximation of FF.

The experiments [30, 31] with network topology optimization by the criteria of the ATR and the DCR show that using the cumulative bounds allow the speeding up of calculations up to two times without loss of precision or with negligible loss (about 10^{-7}).

9. Conclusion

The general approach of obtaining the cumulative bounds of random graph functions is presented. We have shown the possibility of computing the cumulative bounds by partial sums and updating the cumulative bounds in case of applying different methods of reduction and decomposition. For example, various indices of the network reliability were considered: the all-terminal reliability, the diameter constrained reliability, the average pairwise connectivity, and the expected size of a subnetwork that contains a special node. Also, we have described how the cumulative bounds approach can be used for network reliability evaluation and development of evolutionary algorithms for the network topology optimization. Future works can include methods of computing the cumulative bounds for other reliability indices along with methods of reduction and decomposition, further improvement of evolutionary algorithms for the network topology optimization, and parallel algorithms for the network reliability analysis.

Acknowledgements

This research was supported by the Russian Foundation for Basic Research under grants 17-07-00775, 17-47-540977.

Author details

Alexey S. Rodionov and Denis A. Migov*

*Address all correspondence to: mdinka@rav.sscc.ru

Institute of Computational Mathematics and Mathematical Geophysics of SB RAS, Novosibirsk, Russia

References

[1] Colbourn CJ. The Combinatorics of Network Reliability. New York, NY, USA: Oxford University Press; 1987. 160 p

[2] Jereb L. Network reliability: Models, measure and analysis. In: 6th IFIP Workshop on Performance Modeling and Evaluation of ATM Networks; 1998. p. T02/1–T02/10

[3] Moore EF, Shannon CE. Reliable circuits using less reliable relays. Journal of The Franklin Institute. 1956;**262**(4b):191-208

[4] Shooman AM. Algorithms for network reliability and connection availability analysis. In: Electro/International 1995, Boston, MA, USA. 1995, pp. 309-333

[5] Sun F, Shayman MA. On pairwise connectivity of wireless multihop networks. International Journal of Security and Networks. 2007;**2**(1/2):37-49

[6] Rodionov AS, Rodionova OK. Exact bounds for average pairwise network reliability. In: The 7th International Conference on Ubiquitous Information Management and Communication; Malaysia. ACM; 2013. Article No. 45

[7] Cancela H, Petingi L. Diameter constrained network reliability: Exact evaluation by factorization and bounds. In: Int. Conf. on Industrial Logistics; Okinawa, Japan. 2001. p. 359-356

[8] Pandurangan G, Raghavan P, Upfal E. Building low-diameter peer-to-peer networks. IEEE Journal on Selected Areas in Communications. 2003;**21**(6):995-1002

[9] Valiant LG. The complexity of enumeration and reliability problems. SIAM Journal on Computing. 1979;**8**(3):410-421

[10] Bodlaender HL, Wolle T. A note on the complexity of network reliability problems. IEEE Transactions on Information Theory. 2004;**47**:1971-1988

[11] Deuermeyer LA. New approach for network reliability analysis. IEEE Transactions on Reliability. 1982;**31**(4):350-354

[12] Goyal NK, Misra RB, Chaturvedi SK. Snem: A new approach to evaluate terminal pair reliability of communication networks. Journal of Quality in Maintenance Engineering. 2005;**11**(3):239-253

[13] Won JM, Karray F. Cumulative update of all-terminal reliability for faster feasibility decision. IEEE Transactions on Reliability. 2010;**59**(3):551-562

[14] Rodionov AS. Cumulative estimated values of structural Network's reliability indices and their usage. In: Dynamics of Systems, Mechanisms and Machines; Omsk, Russia. IEEE Press; 2016. p. 1-4

[15] Rodionova OK, Rodionov AS, Choo H. Network probabilistic connectivity: Exact calculation with use of chains. In: International Conference on Computational Science and Its Application. Lecture Notes in Computer Science, vol. 3036; Heidelberg: Springer; 2004. p. 565-568

[16] Rodionov AS, Migov DA, Rodionova OK. Improvements in the efficiency of cumulative updating of all-terminal network reliability. IEEE Transactions on Reliability. 2012;**61**(2): 460-465

[17] Wood RK. Triconnected decomposition for computing K–terminal network reliability. Networks. 1989;**19**:203-220

[18] Migov DA. Using node cuts for exact calculation of network probabilistic connectivity. In: Aryn EM, editor. Proc. Int. Conf. on Computational Technologies in Science, Engineering and Education, vol. 2; 20–22–09-2006; Pavlodar, Kazakhstan. 2006. p. 51-58. (in Russian)

[19] Migov DA Random Graph Probabilistic Connectivity Calculation Using Node Cuts [Dissertation] ICM&MG SB RAS, Novosibirsk, Russia: 2008. 97 p. (in Russian) Abstract available from: http://www.sscc.ru/Diss_sov/Migov.pdf

[20] Migov DA, Rodionov AS, Rodionova OK, Choo H. Network probabilistic connectivity: Using node cuts. In: EUC Workshop. LNCS, vol. 4097; Springer; 2006. p. 702-709

[21] Migov DA. Calculation of K-terminal network probabilistic connectivity using 2-node cuts. Bull. Inst. of Comp. Math. Geoph. Ser. Computer Science. Proc. Int. Conf. on Problems of Operation of Information Networks; 7–11-08-2006; Berdsk, Russia. 2006;**6**:134-138. (in Russian)

[22] Migov DA Network probabilistic connectivity computation with use of special kind vertex cuts. In: Proc. Russian Conf. of Young Scientists on Mathematical Modeling and Information Technologies; 1–03–11-2004; Novosibirsk, Russia. ICT SB RAS:2004. (in Russian) Available from: http://www.ict.nsc.ru/ws/YM2004/8566/Paper.htm

[23] Burgos JM, Amoza FR. Factorization of network reliability with perfect nodes I: Introduction and statements. Discrete Applied Mathematics. 2016;**198**:82-90

[24] Burgos JM. Factorization of network reliability with perfect nodes II: Connectivity matrix. Discrete Applied Mathematics. 2016;**198**:91-100

[25] Migov DA. Computing diameter constrained reliability of a network with junction points. Automation and Remote Control. 2011;**72**(7):1415-1419

[26] Migov DA, Rodionov AS. Decomposing graph with 2-node cuts for diameter constrained network reliability calculation. In: 7th Int. Conference on Ubiquitous Information Management and Communication; Malaysia. New York: ACM; 2013. Article No. 39

[27] Migov DA, Nesterov SN. Methods of speeding up of diameter constrained network reliability calculation. In: International Conference on Computational Science and its Application. LNCS, Vol. 9156(2); Springer. 2015. pp. 121-133

[28] Rodionov AS, Rodionova OK. Network probabilistic connectivity: Expectation of a number of disconnected pairs of nodes. In: High Performance Computing and Communications. Lecture Notes in Computer Science, vol. 4208; Berlin: Springer; 2006. p. 101-109

[29] Rodionov AS, Rodionova OK, Choo H. On the expected value of a number of disconnected pairs of nodes in unreliable network. In: Computational Science and its Applications. Lecture Notes in Computer Science, vol. 4707; Berlin: Springer; 2007. p. 534-543

[30] Nechunaeva K, Migov D. Speeding up of genetic algorithm for network topology optimization with use of cumulative updating of network reliability. In: Conference on Ubiquitous Information Management and Communication; Bali, Indonesia. New York: ACM; 2015. p. Article No. 42

[31] Migov DA, Nechunaeva KA, Nesterov SN, Rodionov AS. Cumulative updating of network reliability with diameter constraint and network topology optimization. In: Computational Science and its Applications. Lecture Notes in Computer Science, vol. 9786; Berlin. Springer; 2016. p. 141-152

Permissions

All chapters in this book were first published in SR, by InTech Open; hereby published with permission under the Creative Commons Attribution License or equivalent. Every chapter published in this book has been scrutinized by our experts. Their significance has been extensively debated. The topics covered herein carry significant findings which will fuel the growth of the discipline. They may even be implemented as practical applications or may be referred to as a beginning point for another development.

The contributors of this book come from diverse backgrounds, making this book a truly international effort. This book will bring forth new frontiers with its revolutionizing research information and detailed analysis of the nascent developments around the world.

We would like to thank all the contributing authors for lending their expertise to make the book truly unique. They have played a crucial role in the development of this book. Without their invaluable contributions this book wouldn't have been possible. They have made vital efforts to compile up to date information on the varied aspects of this subject to make this book a valuable addition to the collection of many professionals and students.

This book was conceptualized with the vision of imparting up-to-date information and advanced data in this field. To ensure the same, a matchless editorial board was set up. Every individual on the board went through rigorous rounds of assessment to prove their worth. After which they invested a large part of their time researching and compiling the most relevant data for our readers.

The editorial board has been involved in producing this book since its inception. They have spent rigorous hours researching and exploring the diverse topics which have resulted in the successful publishing of this book. They have passed on their knowledge of decades through this book. To expedite this challenging task, the publisher supported the team at every step. A small team of assistant editors was also appointed to further simplify the editing procedure and attain best results for the readers.

Apart from the editorial board, the designing team has also invested a significant amount of their time in understanding the subject and creating the most relevant covers. They scrutinized every image to scout for the most suitable representation of the subject and create an appropriate cover for the book.

The publishing team has been an ardent support to the editorial, designing and production team. Their endless efforts to recruit the best for this project, has resulted in the accomplishment of this book. They are a veteran in the field of academics and their pool of knowledge is as vast as their experience in printing. Their expertise and guidance has proved useful at every step. Their uncompromising quality standards have made this book an exceptional effort. Their encouragement from time to time has been an inspiration for everyone.

The publisher and the editorial board hope that this book will prove to be a valuable piece of knowledge for researchers, students, practitioners and scholars across the globe.

List of Contributors

Bernard Tonderayi Mangara
Central University of Technology, Free State (CUT), Bloemfontein, South Africa

Joseph B. Bernstein
Ariel University, Ariel, Israel

Marco Antônio Ferreira Boaski, Caio dos Santos, Mauricio Sperandio and Daniel Pinheiro Bernardon
Federal University of Santa Maria, Santa Maria, Brazil

Maicon Jaderson Ramos and Daniel Sperb Porto
RGE Sul Power Utility, Brazil

Rabah Medjoudj
Lamos Research Unit, Faculty of Technology, Electrical Engineering Department, University of Bejaia, Bejaia, Algeria

Hassiba Bediaf
SOPERIE, Electrical Engineering Society, Bejaia, Algeria

Djamil Aissani
Lamos Research Unit, Faculty of Exact Sciences, Operational Research Department, University of Bejaia, Bejaia, Algeria

Carlos Henrique Barriquello, Vinícius Jacques Garcia, Magdiel Schmitz and Daniel Pinheiro Bernardon
Federal University of Santa Maria, Santa Maria, Brazil

Júlio Schenato Fonini
RGE Sul, Brazil

Jon T. Selvik
University of Stavanger, Stavanger, Norway

Eric P. Ford
International Research Institute of Stavanger (IRIS), Stavanger, Norway

Atanda K. Raji
Centre for Distributed Power and Electronic Systems, Cape Peninsula University of Technology, Bellville Campus, Cape Town, South Africa

Yi Xiao-Jian
Department of Overall Technology, China North Vehicle Research Institute, China
School of Mechatronical Engineering, Beijing Institute of Technology, China

Shi Jian
Academy of Mathematics and Systems Science, Chinese Academy of Sciences, China
School of Mathematical Sciences, University of Chinese Academy of Sciences, China

Hou Peng
School of Mechatronical Engineering, Beijing Institute of Technology, China

Michał Szewczyk
Silesian University of Technology, Gliwice, Poland

Seong-woo Woo
Reliability Association of Korea, Seoul, Korea

Magdiel Schmitz, Maria Clara Ferreira Almeida da Silva, Vinícius Jacques Garcia, Daniel Bernardon, Lynceo Favigna Braghirolli
Federal University of Santa Maria, Rio Grande do Sul, Brazil

Júlio Fonini
RGE Sul Power Utility, Rio Grande do Sul, Brazil

Ibrahima dit Bouran Sidibe
Centre de Formation et de Perfectionnement en Statistique (CFP-STAT) Bamako, Mali

Imene Djelloul
Higher School of Applied Sciences of Algiers, Place des Martyres, Alger, Algerie
Manufacturing Engineering Laboratory of Tlemcen (MELT), Abou Bekr Belkaid University of Tlemcen, Tlemcen, Algeria

Alexey S. Rodionov and Denis A. Migov
Institute of Computational Mathematics and Mathematical Geophysics of SB RAS, Novosibirsk, Russia

Index

www.ingramcontent.com/pod-product-compliance
Lightning Source LLC
LaVergne TN
LVHW060513210926
848485LV00005B/1075